$$\rho_d A_0 V_0^2 = \rho_{m0} A_m v_m^2. \tag{2.65}$$

The combination of Eqs. (2.63), (2.64) and (2.65) gives us the following equation for the velocity of the mixture (cf. Eq. (2.51)):

$$v_m \equiv \frac{\mathrm{d}z}{\mathrm{d}t}_{|m} = \frac{2V_0}{1+\sqrt{a+bz}}, \tag{2.66}$$

where

$$a = 1 + 4(1-\alpha_d)\tilde{\rho}_a; \qquad b = \frac{64(1-\alpha_d)D_t\tilde{\rho}_a}{V_0 D_0^2}. \qquad \tilde{\rho}_a = \rho_g/\rho_d,$$

Integration of Eq. (2.66) gives:

$$3bz + 2(a+bz)^{3/2} = 6V_0 bt. \tag{2.67}$$

For sufficiently large z we can assume that $bz \gg a$ and simplify Eq. (2.67) to

$$z = \left(\frac{9V_0^2}{b}\right)^{1/3} t^{2/3} = \frac{V_0}{4}\left(\frac{9D_0^2}{(1-\alpha_d)\tilde{\rho}_a D_t}\right)^{1/3} t^{2/3}. \tag{2.68}$$

The dependence of z on t can be compared with the one predicted by Eq. (2.59). In the limit $\alpha_d \to 0$ Eqs. (2.59) and (2.68) predict the same penetration if:

$$D_t = \frac{9}{64} \frac{V_0^{3/2}(\tan\theta)^{3/2}\sqrt{d_0 t}}{\tilde{\rho}_a^{1/4}}. \tag{2.69}$$

Considering the values of parameters for experimental results discussed in [136] (case 1): $V_0 = 318.3\,\mathrm{m/s}, d_0 = 0.2\,\mathrm{mm}, \theta = 13°, t = 1\,\mathrm{ms}, \tilde{\rho}_a = 19.7/760 = 0.025$, we obtain $D_t = 0.1\ \mathrm{m}^2/s$. Tentatively, the turbulent diffusivity coefficient can be roughly estimated from the Tchen formula $D_t = \sigma_u^2 T_L^*$ (strictly valid only in homogeneous turbulence for a long time limit). Then the values of $\sigma_u^2 = 2k/3$ and $T_L = \mathcal{O}(k/\varepsilon)$, if experimentally available, can be used for a cross-check of the above number.

There is much uncertainty regarding the experimental observations of spray penetration as discussed in [135]. The general conclusion inferred in [135] is that in the case of low pressure injection sprays the shape of the spray is close to conical and spray penetration is approximately proportional to $\sqrt{t}$. For the high pressure injection, the shape of the spray is close to parabolic, described by Eq. (2.61). In this case we would expect that spray penetration is fairly well described by Eq. (2.68) and proportional to $t^{2/3}$. We have checked this conclusion using data from a high-speed video recording of a Diesel spray injected at 100 MPa into air with density of 49 kg/m^3. The comparison between the results of experimental measurements of spray penetration and the best fits of experimental data by the curves $\propto t^{1/2}$ and $\propto t^{2/3}$ showed that the $t^{2/3}$ plot provides a noticeably better fit than the $\propto t^{1/2}$ plot. The maximum and mean deviations for the plot $\propto t^{2/3}$ (37.8 and 9.6 %) were clearly less than the maximum and mean deviations for the plot $\propto t^{1/2}$ (65.0 and 21.1 %) [116]. This supports our approach to linking spray penetration with turbulent dispersion of droplets, although more studies in this direction are needed.

The effects of turbulence on the initial stage of spray penetration were also studied in [116], based on the analysis of turbulent diffusion of the liquid phase (Eulerian approach, see [17]) or turbulent perturbation of individual droplet trajectories (Lagrangian approach, see [114, 115]). In both approaches the analysis was based on the assumption that spray at the initial stage can be approximated as an array of non-interacting droplets. As mentioned in Sect. 2.2.1, this assumption is highly questionable.

Note that spray penetration is closely linked with its momentum flux. Results of detailed numerical and experimental investigation of these fluxes in high pressure Diesel sprays are presented in [113].

Effects of cavitation and nozzle geometry on spray penetration were studied in [149, 150].

The analysis of spray dynamics presented so far has been based mainly on the Eulerian-Lagrangian approach (Eulerian for the carrier phase and Lagrangian for the dispersed phase). An alternative technique for the analysis of this process could make use of vortex-based algorithms (e.g. [80, 160]). In the latter paper such an algorithm was combined with the fully Lagrangian approach to modelling the dispersed phase. This approach is sometimes known as the Osiptsov method [53]. In the next section the vortex-based approach is used for the analysis of vortex ring-like structures in sprays.

2.3 Vortex Ring-like Structures in Sprays

Our analysis so far has been focused on the basic processes leading to spray formation and penetration, ignoring a number of important details. These details include the oscillations of sprays and the formation/dynamics of vortex ring-like structures near the spray leading edges. The former process was considered in a number of papers, including [123]. The focus of this section will be on the latter phenomenon.

The vortex ring-like structures are not observed for all sprays. For example, they are not observed for sprays in Diesel engine conditions, where liquid fuel is injected into a high pressure gas, except at the very initial stage of the process [16]. At the same time, these structures are typical for gasoline engine sprays, where liquid fuel is injected into gas at pressures close to atmospheric pressure. A typical spray image in gasoline engine-like conditions (direct injection (G-DI) injector [5]) is shown in Fig. 2.2.

As can be seen in Fig. 2.2, the shape of the spray is rather chaotic, but vortex ring-like structures can be clearly recognised. There are some recognisable

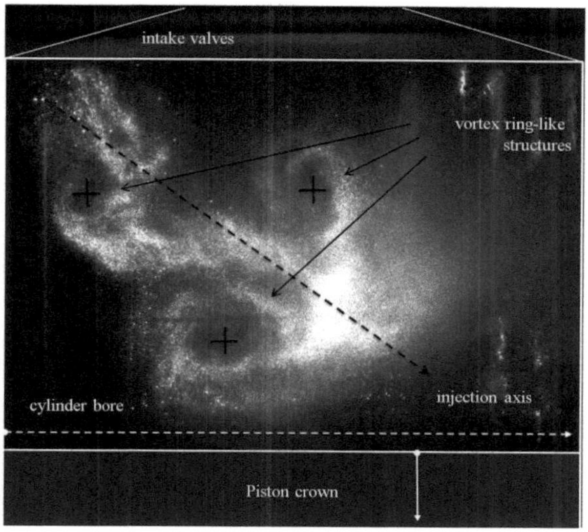

Fig. 2.2 A typical image acquired using a G-DI injector. Positions of the points where radial and axial components of the velocity are equal to zero in the image of the vortex ring are shown as crosses. Reprinted from Ref. [70], Copyright Elsevier (2010)

similarities between these structures and the conventional well organised vortex rings formed, for example, during the injection of water into water with the help of a round piston (e.g. [87, 144]). However, the early attempts to apply the theory of conventional vortex rings to the analysis of the above-mentioned vortex ring-like structures were not successful [137]. In our recent papers [69, 70] the conventional vortex ring theory was generalised to take into account the effect of turbulence. The new model developed in these papers turned out to be successful in predicting some features of the vortex ring-like structures shown in Fig. 2.2, including their translational velocities.

In what follows in this section we will give a brief overview of historic developments of the conventional vortex ring theory. Then recent developments presented in [69, 70] are summarised. Finally, some predictions of the theory developed [69, 70] are compared with experimental data referring to vortex ring-like structures similar to those shown in Fig. 2.2.

2.3.1 Conventional Vortex Rings

A schematic sketch of the vortex ring is presented in Fig. 2.3. R_0 in this figure is the radius of the vortex ring (distance from the vortex ring axis to the area of zero vorticity); ℓ is the characteristic vortex ring thickness. Only the case of a one-phase fluid is considered in this section and the effects of turbulence are ignored. In the

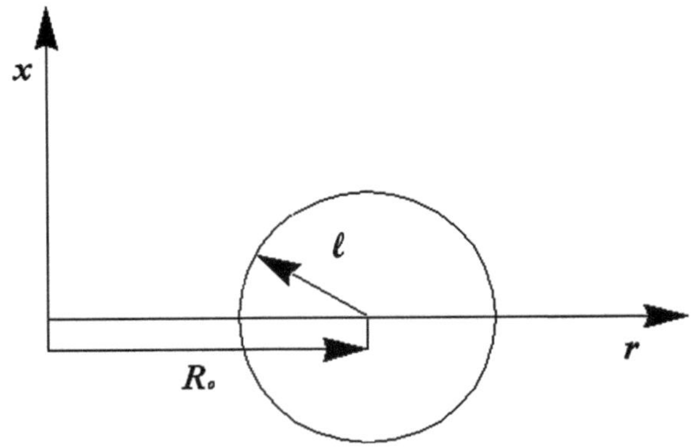

Fig. 2.3 A schematic view of a vortex ring. This figure is reproduced from [141] with permission from JUMV—Society of automotive engineers in Serbia

rest of this section, an overview of historic developments of the conventional vortex ring theory is presented mainly following [70].

Two approaches were used in theoretical studies of vortex ring translational velocities and energies. In the first approach, the relation between velocity and vorticity was used to obtain the formulae for thin cored rings: $\varepsilon = \ell/R_0 \ll 1$ [31, 34]. A more general approach valid for arbitrary ε, developed in [54] (see also [77]), is based on the Helmholtz-Lamb formula for the ring's translational velocity U in the form:

$$U = \frac{\pi}{2M} \int_0^{\infty} \int_{-\infty}^{\infty} \left(\Psi - 6x \frac{\partial \Psi}{\partial r} \right) \zeta \, \mathrm{dx} \, \mathrm{dr}, \tag{2.70}$$

where ζ and Ψ are the vorticity and streamfunction, respectively, and $M = I/\rho$ is the momentum of vorticity per unit density.

Using Eq. (2.70), Saffman [131] (see also [132]) derived an explicit expression for the translational velocity of a thin-cored viscous vortex ring in the form:

$$U_s = \frac{\Gamma_0}{4\pi R_0} \left[\ln \left(\frac{4R_0}{\sqrt{\nu t}} \right) - 0.558 + O\left(\frac{\nu t}{{R_0}^2} \ln \left(\frac{\nu t}{{R_0}^2} \right) \right) \right], \tag{2.71}$$

where Γ_0 is the initial circulation of the ring, t is time and ν is the fluid kinematic viscosity. The vorticity distribution inside this ring corresponds to the Lamb-Oseen vortex filament [77]. This asymptotic formula is valid for the description of the initial stage of viscous vortex ring development when $\nu t << {R_0}^2$. The final stage of viscous vortex ring decay ($\nu t >> {R_0}^2$) can be described based on the Phillips self-similar solution for the vorticity (ζ_f) and streamfunction (Ψ_f) distributions [111]:

$$\zeta_f = \frac{Mr}{16\pi^{3/2}(\nu t)^{5/2}} \exp\left(-\frac{s_*^2}{2}\right), \tag{2.72}$$

$$\Psi_f = \frac{M}{4\pi}\left(\operatorname{erf}\left(\frac{s_*}{\sqrt{2}}\right) - s_*\sqrt{\frac{2}{\pi}}\exp\left(\frac{-s_*{}^2}{2}\right)\right)\frac{r^2}{\left(r^2+x^2\right)^{3/2}}, \tag{2.73}$$

where

$$s_* = \sqrt{\frac{r^2+x^2}{2\nu t}},$$

x, r are cylindrical coordinates for the axisymmetric vortex ring. The derivation of the translational velocity in this case is not straightforward. Since Formula (2.70) was derived based on the full Navier-Stokes equation, the substitution of Expressions (2.72) and (2.73) into (2.70) leads to inconsistency.

Attempts to account for the second-order effects of the non-linear convective terms of the vorticity equation were made by Kambe and Oshima [65]. However their results are not uniformly valid. Rott and Cantwell [128, 129] studied this case taking into account the flow dynamics in the potential flow region surrounding the vortical region. They showed that the asymptotic translational velocity of the ring can be predicted by the following formula:

$$U_f = \frac{7M}{15\,(8\pi\nu t)^{3/2}} = 0.0037038\frac{I/\rho}{(\nu t)^{3/2}}. \tag{2.74}$$

Another approach to this problem was developed in [7, 64, 67, 68]. These authors obtained a first-order solution to the Navier-Stokes equation with the origin in the centre of the vortex centroid, valid in the limit of small Reynolds numbers Re defined as:

$$\mathrm{Re} = \zeta_0 L^2/\nu,$$

where $\zeta_0 = At^\lambda$ is the vorticity scale; constant A is to be specified from the conservation of M.

The translational velocity of the viscous vortex ring was derived in the form [67]:

$$U = \frac{M\theta\sqrt{\pi}}{4\pi^2 R_0^3}\left\{3\exp\left(-\frac{\theta^2}{2}\right)\mathrm{I}_1\left(\frac{\theta^2}{2}\right) + \frac{\theta^2}{12}{}_2\mathrm{F}_2\left(\left\{\frac{3}{2},\frac{3}{2}\right\},\left\{\frac{5}{2},3\right\},-\theta^2\right)\right.$$
$$\left. - \frac{3\theta^2}{5}{}_2\mathrm{F}_2\left(\left\{\frac{3}{2},\frac{5}{2}\right\},\left\{2,\frac{7}{2}\right\},-\theta^2\right)\right\}, \tag{2.75}$$

where $\theta = R_0/\ell = \varepsilon^{-1}$, I_1 is the first-order Bessel function and ${}_2\mathrm{F}_2$ is the generalised hypergeometric function [98].

Similarly, the kinetic energy and circulation were derived in the form [68]:

$$E = \frac{M^2\theta\sqrt{\pi}}{2\pi^2 R_0^3}\left\{\frac{1}{12}{}_2\mathrm{F}_2\left(\left\{\frac{3}{2},\frac{3}{2}\right\},\left\{\frac{5}{2},3\right\},-\theta^2\right)\right\}, \tag{2.76}$$

$$\Gamma = \frac{M}{\pi R_0^2}\left\{1-\exp\left(-\frac{\theta^2}{2}\right)\right\}. \tag{2.77}$$

Note that apart from the definition of Re given above, at least two other definitions of this number, have been used in the literature:

$$\mathrm{Re}_u = U_p D/\nu, \tag{2.78}$$

based on the ejection velocity U_p and orifice diameter D, and

$$\mathrm{Re}_{\Gamma_0} = \Gamma_0/\nu, \tag{2.79}$$

where Γ_0 is the initial circulation carried by the ring.

The closed-form representations (2.75)–(2.77) enable us to analyse the asymptotic behaviour of these parameters. In the limit of small θ, these equations reduce to:

$$U_f = \frac{M\theta^3}{4\pi^2 R_0^3}\sqrt{\pi}\left(\frac{7}{30}-\frac{11\theta^2}{140}\right)+O\left(\theta^4\right), \tag{2.80}$$

$$E_f = \frac{M^2\theta^3}{2\pi^2 R_0^3}\sqrt{\pi}\left(\frac{1}{12}-\frac{\theta^2}{40}\right)+O\left(\theta^4\right), \tag{2.81}$$

$$\Gamma_f = \frac{M\theta^2}{2\pi R_0^2}. \tag{2.82}$$

In the limit of large θ, they are reduced to:

$$U_s = \frac{M\sqrt{\pi}}{4\pi^2 R_0^3}\left(\frac{2\log(\theta)+3-\gamma-2\varphi(3/2)}{2}\right)+O\left(\frac{1}{\theta^4}\right), \tag{2.83}$$

$$E_s = \frac{M^2\sqrt{\pi}}{2\pi^2 R_0^3}\left(\log(\theta)-\gamma/2-\varphi(3/2)\right)+O\left(\frac{1}{\theta^4}\right), \tag{2.84}$$

$$\Gamma_s = \frac{M}{\pi R_0^2}, \tag{2.85}$$

where $\gamma \approx 0.57721566$ is the Euler constant and φ is the di-gamma function defined as

$$\varphi = \frac{d \log \Gamma(x)}{\mathrm{dx}},$$

and $\Gamma(x)$ is the Gamma function.

Stanaway et al. [153] performed direct numerical simulation of the Navier-Stokes equation for an axisymmetric vortex ring at small and moderate Reynolds numbers. They showed that Formula (2.75) compares fairly well with their result at a small Reynolds number [35]. The large-Reynolds-number asymptotics was discussed in [36].

An alternative approach to estimate the temporal evolution of the vortex ring translational velocity was suggested by Saffman [131], using simple dimensional analysis. He derived the following equation:

$$U = \frac{M}{k}\left(R_0{}^2 + k'\nu t\right)^{-3/2}, \tag{2.86}$$

where k and k' are adjustable constants.

To obtain these constants, Weingand and Gharib [163] compared their experimental results for $830 < Re_{\Gamma_0} < 1650$ with those predicted by Eq. (2.86). This comparison led them to the following values: $k = 14.4$ and $k' = 7.8$. Later $k = 10.15$ and $k' = 8.909$ were obtained theoretically by Fukumoto and Kaplanski [35].

2.3.2 *Turbulent Vortex Rings*

In contrast to the aforementioned laminar vortex ring models, the theory of turbulent vortex rings is far less developed. To the best of the author's knowledge, the first attempt to investigate turbulent vortex ring flow structures was made by Lugovtsov [92, 93] who based his analysis on the introduction of the time-dependent, turbulent (eddy) viscosity (cf. [76, 78]):

$$\nu_* \propto \ell\ell', \tag{2.87}$$

where $\ell' = \mathrm{d}\ell/\mathrm{d}t$ and $\ell \propto t^{1/4}$. Equation (2.87) follows from a simple dimensional analysis [9], remembering that ℓ has the dimension of length, while ℓ' has the dimension of velocity. Using Eq. (2.87), Lugovtsov [92, 93] developed a turbulent vortex ring model with turbulent viscosity ν_*.

Equations (2.75)–(2.77) were originally derived for $\ell = \sqrt{2\nu t}$ (laminar vortex ring). Later, in [69] it was shown that they remain valid in a more general case when $\ell = at^b$, where a and b are constants ($1/4 \le b \le 1/2$). The model based on this presentation of ℓ was called the generalised vortex ring model. This model incorporates both the laminar model for $b = 1/2$ and the fully turbulent model for $b = 1/4$. For $a = \sqrt{2\nu}$, $b = 1/2$ and for large times (small θ), the leading order term of (2.75) is identical with the one predicted by Eq. (2.74). For small times, $\nu t << R_0^2$, the vorticity is concentrated on a circle of radius R_0 and tends to a

Gaussian form [67]. The leading term of Expression (2.75) coincides with Saffman's formula (2.71) [35].

For $b = 1/4$ in the generalised vortex ring model, the leading term in (2.80) corresponds to

$$U_f = 0.0105 M t^{-3/4}/a^3. \tag{2.88}$$

The same asymptotic behaviour of the translational velocity for the turbulent vortex ring was found theoretically and tested experimentally in [2, 45]. Some preliminary results of the comparison of the predictions of the above results, referring to the generalised vortex ring model, with the published experimental data were reported in [69].

One of the important limitations of the model described above is that it is based upon the assumption that $R_0 = \text{const}$. Possible alternative approach in which this assumption is relaxed for conventional laminar vortex rings is based on Eq. (2.86). The latter equation was generalised by [69] to take into account the effects of turbulence. These authors started with the dimensionally correct equation:

$$U = \frac{M}{k R^3}, \tag{2.89}$$

where k is a proportionality constant. The decay of circulation can be described by the second dimensionally correct equation:

$$\frac{\mathrm{d}(U R)}{\mathrm{d}t} = -k' \frac{\nu_* U}{R}, \tag{2.90}$$

where k' is another proportionality constant. The viscosity ν_* is defined by Eq. (2.87). In contrast to the case considered by [131] and [163], ν_* depends on time. Having substituted Eq. (2.89) into Eq. (2.90) and integrating the latter equation from $t = t_0 = 0$ to t, one obtains:

$$R^2 - R_0^2 = \frac{k' a^2}{2} t^{2b}. \tag{2.91}$$

For $b = 1/2$ and $a = \sqrt{2\nu}$, Eq. (2.91) reduces to the one derived by [131] and [163]. Substituting Eq. (2.91) into Eq. (2.89) gives:

$$U = \frac{M}{k\,(R_0^2 + \frac{k' a^2}{2} t^{2b})^{3/2}}. \tag{2.92}$$

For $b = 1/2$ and $a = \sqrt{2\,\nu}$, Eq. (2.92) reduces to Eq. (2.86). Hence, Eq. (2.92) can be considered as the generalisation of Eq. (2.86) to the case of arbitrary a and b, including the case of turbulent vortex rings. For further analysis of Eq. (2.92) see the original paper [69].

As follows from the further analysis of the generalised vortex ring model (Eq. (2.75) for $\ell = at^b$ performed in [70], the vorticity distributions, predicted numerically (for realistic $\mathrm{Re} = \zeta_0 L^2/\nu$) and analytically (for $\mathrm{Re} \to 0$) look rather different. In the numerical results, the contours expand and shift along the axis of symmetry with time, while forms predicted by the analytical solution only expand during the time interval under consideration. Results of calculations of the time evolution of the vortex ring velocities and energies, based on vorticity distributions, predicted by the analytical formulae and numerical solution, however, show that both variables are not sensitive to Re. Hence, analytical formulae are expected to be good approximations for realistic values of vortex ring velocities and energies even for high Reynolds numbers.

The underlying physics behind these effects was clarified by the newly found analytical solution for the normalised vorticity distribution in the limit of long times in the form $\omega_0 + \mathrm{Re}\omega_1$, where ω_0 is the value of vorticity predicted by the classical Phillips solution in the limit $\mathrm{Re} \to 0$ [111]. Results of further investigations of the effects of Re on vortex rings, leading to the formation of elliptical vortex rings, are described in [66].

As pointed out in [5, 70]. The main advantage of the generalised vortex ring model is that it incorporates an additional parameter b, which can potentially make it applicable to the analysis of not only classical vortex rings, but also complex vortex-ring-like structures similar to those shown in Fig. 2.2. Some results of this comparison for translational velocities of these structures are presented and discussed in the next section.

2.3.3 *Translational Velocities of the Vortex Rings-like Structures*

The analyses in this section is focused on typical non-evaporating sprays in gasoline engine-like conditions, similar to those shown in Fig. 2.2. The experimental conditions are described in detail in [5]. A summary is presented here, following [70].

Preliminary investigations were performed on two production gasoline injectors; a low pressure port fuel injector (PFI) and a high-pressure, direct injection (G-DI) injector [5]. In both cases, iso-octane was injected into air under atmospheric conditions. High-speed photography was used to observe the formation and translation of vortex ring-like structures. Within both sprays, vortex ring-like structures were observed. However, the most stable and consistent results were obtained in the experiments with the high-pressure injector. In the G-DI spray (see Fig. 2.2), four vortex-ring-like structures could be identified within the experimental data. The two most persistent (and clearly defined) structures were chosen in the comparison with the predictions of the model described in the previous section.

A Dantec Dynamics phase Doppler anemometer (PDA) was used to study the dynamics of fuel droplets in the vortex structures. Measurements of the fuel droplet velocities and diameters were performed in a fine measurement grid that bisected the spray axis and the region of the ring-like structures. The optimisation of the

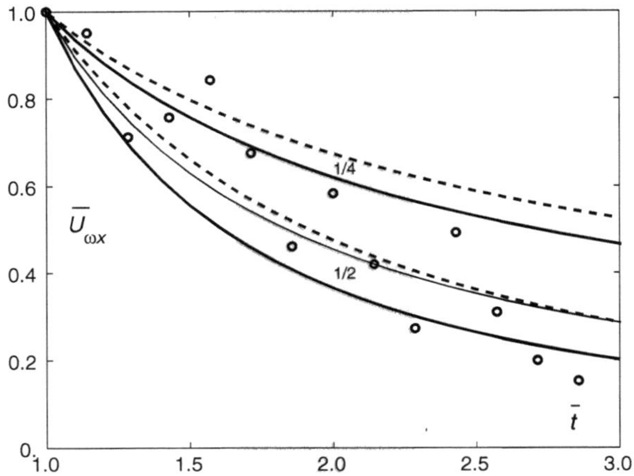

Fig. 2.4 The values of normalised velocities $\bar{U}_{\omega x}(\bar{t}) = U_{\omega x}(t)/U_{\omega x}(t_{\text{init}})$ versus normalised time $\bar{t} = t/t_{\text{init}}$ as obtained from the experimental data (*circles*), their approximation ($\bar{U}_{\omega x}(t) = \bar{t}^{-1.14}$) (*thin solid curve*) and calculated from Eq. (2.93) for $\tilde{t}_{\text{init}} = t_{\text{init}}/t_0 = 0.5$ (dashed curves) and 5 (*thick solid curves*), and $b = 1/2$ and $1/4$ (numbers near the *curves*). Reprinted from Ref. [70], Copyright Elsevier (2010)

PDA operating parameters and the statistical accuracy of the measurements formed an important part of the study. These PDA data were used to study the integral characteristics of the vortex rings including their translational axial velocities.

In Fig. 2.4, the values of the observed axial velocities, normalised by the value of the velocity at the time when the vortex ring was initially observed $\bar{U}_{\omega x}(\bar{t}) = U_{\omega x}(t_{\text{init}})$ versus normalised time $\bar{t} = t/t_{\text{init}}$ are shown by circles. Following [5], the experimental results were approximated by the function $\bar{U}_{\omega x}(t) = \bar{t}^{B_n}$, where $B_n = -1.14$ was found using the best fit technique. The plot $\bar{U}_{\omega x}(t) = \bar{t}^{-1.14}$ is shown in the same Fig. 2.4.

In [5] these results were compared with the predictions of the generalised vortex ring model in the limit of long times. In what follows, the same comparison is made for arbitrary time, following [70] and using Eq. (2.75). Note that this equation cannot be used directly for comparison with experimental data, as it refers to vortex ring velocity, while the measured variable is the fluid velocity in the region of maximal vorticity $U_{\omega x}(t)$. These two variables are related by the equation [69]:

$$U_{\omega x} \equiv V_{\omega x}/v_n = U_x + 2\pi\theta^2 \int_0^\infty \mu \operatorname{erfc}\left(\frac{\mu}{\sqrt{2}}\right) J_1(\theta\mu)\, J_0(\sigma_{\max}\mu) \mathrm{d}\mu. \quad (2.93)$$

The direct comparison between the experimental results and Eq. (2.93) is still complicated by the fact that the relation of the value of t_{init} and the instant of time the vortex ring is expected to be initiated (t_0) cannot be inferred from experimental data. Following [5], two well separated values of $\tilde{t}_{\text{init}} = t_{\text{init}}/t_0$: $\tilde{t}_{\text{init}} = 0.5$ and $\tilde{t}_{\text{init}} = 5$

were considered, and the corresponding values of $\bar{U}_{\omega x}(\bar{t}) = U_{\omega x}(t)/U_{\omega x}(t_{\text{init}})$ versus $\bar{t}$ for $b = 1/2$ and $b = 1/4$, where $\bar{t} = \tilde{t}/\tilde{t}_{\text{init}} = t/t_{\text{init}}$, were calculated. The plots are shown in the same Fig. 2.4. Note that

$$\bar{t}_{\text{init}} = \frac{t_{\text{init}}}{t_0} = \theta_{\text{init}}^b. \tag{2.94}$$

For $b = 1/2$, the range of $\bar{t}_{\text{init}}$: (0.5, 5) corresponds to the range of θ_{init}: (0.7, 2.2). For $b = 1/4$, the same range of $\bar{t}_{\text{init}}$ corresponds to the range of θ_{init}: (0.8, 1, 5).

As follows from Fig. 2.4, most experimental points (8 out of 13) lie between the theoretical curves corresponding to $\tilde{t}_{\text{init}} = 5$ and $b = 1/4$ and $1/2$. The points outside this range still look reasonably close to this region remembering the scatter of experimental data. At the same time only 4 out of 13 points lie between the theoretical curves corresponding to $\tilde{t}_{\text{init}} = 0.5$ and $b = 1/4$ and $1/2$. Hence, one can conclude that the observed vortex ring-like structures refer to the late stage of vortex ring development with the values of b lying between $b = 1/2$, corresponding to the laminar case, and $b = 1/4$, corresponding to a fully developed turbulent case. The most probable value of $b = 1.14/3 = 0.38$ also lies between $1/4$ and $1/2$.

References

1. Abdelghaffar, W. A., Elwardany, A. E., & Sazhin, S. S. (2010). Modeling of the processes in diesel engine-like conditions: Effects of fuel heating and evaporation. *Atomization Sprays*, *20*, 737–737.
2. Afanasyev, Y. D., & Korabel, V. N. (2004). Starting vortex dipoles in a viscous fluid: Asymptotic theory, numerical simulation, and laboratory experiments. *Physics of Fluids*, *16*(11), 3850–3858.
3. Allocca, L., Belardini, P., Bertoli, C., Corcione, F. E., & de Angelis, F. (1992). Experimental and numerical analysis of a diesel spray. SAE report 920576.
4. Arienti, M., & Sussman, M. (2014). An embedded level set method for sharp-interface multiphase simulations of diesel injectors. *International Journal of Multiphase Flow*, *59*, 1–14.
5. Begg, S., Kaplanski, F., Sazhin, S. S., Hindle, M., & Heikal, M. (2009). Vortex ring structures in gasoline engines under cold-start conditions. *International Journal of Engine Research*, *10*, 195–214.
6. Bellman, R., & Pennington, R. H. (1954). Effects of surface tension and viscosity on Taylor instability. *Quarterly of Applied Mathematics*, *12*, 151–162.
7. Berezovski, A., & Kaplanski, F. (1995). Vorticity distributions for thick and thin vortex pairs and rings. *Archives of Mechanics*, *47*(6), 1015–1026 (in Russian).
8. Borman, G. L., & Ragland, K. W. (1998). *Combustion engineering*. New York: McGraw-Hill.
9. Cantwell, B. (2002). *Introduction to symmetry analysis*. Cambridge: Cambridge University Press.
10. Cebeci, T., & Smith, A. M. O. (1974). *Analysis of turbulent boundary layers. Applied mathematics and mechanics* (Vol. 15). NY: Academic Press.
11. Chehroudi, B., & Bracco, F. V. (1988). Structure of a transient hollow cone spray. SAE report 880522.
12. Chesnel, J., Menard, T., Reveillon, J., & Demoulin, F. -X. (2011). Subgrid analysis of liquid jet atomization. *Atomization Sprays*, *21*, 41–67.

13. Chigier, N., & Reitz, R. D. (1998). Regimes of jet breakup and breakup mechanisms (physical aspects). In Kuo, K. K. (Ed.), *Recent advances in spray combustion: Spray atomization and drop burning phenomena* (pp. 109–135). Reston: American Institute of Aeronautics and Astronautics Inc.
14. Chryssakis, C., & Assamis, D. N. (2008). A unified fuel spray breakup model for internal combustion engine applications. *Atomization Sprays*, *18*, 375–426.
15. Cossali, G. E. (2001). An integral model for gas entrainment into full cone sprays. *Journal of Fluid Mechanics*, *439*, 353–366.
16. Crua, C., Shoba, T., Heikal, M., Gold, M., & Higham, C. (2010). High-speed microscopic imaging of the initial stage of Diesel spray formation and primary breakup. SAE Technical, Report 2010–01-2247.
17. Csanady, G. T. (1973). *Turbulent diffusion in the environment*. Dordrecht-Holland: D. Reidel Publishing Comp.
18. Dai, Z., & Faith, G. M. (2001). Temporal properties of secondary drop breakup in the multi-mode breakup regime. *International Journal of Multiphase Flow*, *27*, 217–236.
19. Dent, J. C. (1971). A basic for the comparison of various experimental methods for studying spray penetration. SAE report 710571.
20. Desantes, J. M., Payri, R., Salvador, F. J., & Gil, A. (2006). Development and validation of a theoretical model for diesel spray penetration. *Fuel*, *85*, 910–917.
21. Desantes, J. M., Payri, R., Garcia, J. M., & Salvador, F. J. (2007). A contribution to the understanding of isothermal diesel spray dynamics. *Fuel*, *86*, 1093–1101.
22. Desantes, J. M., Payri, R., Salvador, F. J., & de la Morena, J. (2010). Influence of cavitation phenomenon on primary break-up and spray behavior at stationary conditions. *Fuel*, *89*, 3033–3041.
23. Devassy, B.M., Habchi, C., & Daniel, E. (2013). A new atomization model for high speed liquid jets using a turbulent, compressible, two-phase flow model and a surface density approach. *Proceedings of ILASS—Europe 2013, 25th European Conference on Liquid Atomization and Spray Systems*, Chania, Greece, 1–4 September 2013, paper 32.
24. Dombrowski, N., & Johns, W. R. (1963). The aerodynamic instability and disintegration of viscous liquid sheets. *Chemical Engineering Science*, *18*, 203–214.
25. Douglas, J. F., Gasiorek, J. M., Swaffield, J. A., & Jack, L. B. (2005). *Fluid mechanics* (5th ed.). Singapore: Pearson.
26. Drazin, P. G., & Reid, W. H. (2004). *Hydrodynamic stability* (2nd ed.), Cambridge: Cambridge University Press.
27. Dumouchel, C., & Grout, S. (2009). Application of the scale entropy diffusion to describe a liquid atomization process. *International Journal of Multiphase Flow*, *35*, 952–962.
28. Dumouchel, C., & Grout, S. (2011). On the scale diffusivity of a 2-D liquid atomization process analysis. *Physica A*, *390*, 1811–1825.
29. Dumouchel, C., & Blaisot, J. -B. (2013) Multi-scale analysis of liquid atomization processes and sprays. *Proceedings of ILASS—Europe 2013, 25th European Conference on Liquid Atomization and Spray Systems*, Chania, Greece, 1–4 September 2013, paper 25.
30. Duret, B., Menard, T., Reveillon, J., & Demoulin, F. X. (2013) Improving primary atomization modelling through DNS of two-phase flows. *Proceedings of ILASS—Europe 2013, 25th European Conference on Liquid Atomization and Spray Systems*, Chania, Greece, 1–4 September 2013, paper 110.
31. Dyson, F. W. (1893). The potential of an anchor ring-part ii. *Philosophical transactions of the Royal Society of London*, *A184*, 1041–1106.
32. Eggers, J., & Villermaux, E. (2008). Physics of liquid jets. *Reports on Progress in Physics* 616, 79 p.
33. Faeth, G. M., Hsiang, L. -P., & Wu, P. -K. (1995). Structure and breakup properties of sprays. *International Journal of Multiphase Flow*, *21*, Suppl., 99–127.
34. Fraenkel, L. E. (1972). Examples of steady vortex rings of small cross-section in an ideal fluid. *Journal of Fluid Mechanics*, *51*, 119–135.

35. Fukumoto, Y., & Kaplanski, F. (2008). Global time evolution of an axisymmetric vortex ring at low Reynolds numbers. *Physics of Fluids, 20*, 053103.
36. Fukumoto, Y., & Moffatt, H. K. (2008). Kinematic variational principle for motion of vortex rings. *Physica D, 237*, 2210–2217.
37. Fuster, D., Bagué, A., Boeck, T., Le Moyne, L., Leboissetier, A., Popinet, S., et al. (2009). Simulation of primary atomization with an octree adaptive mesh refinement and VOF method. *International Journal of Multiphase Flow, 35*, 550–565.
38. Ghasemi, A., Barron, R. M., & Balachandar, R. (2014). Spray-induced air motion in single and twin ultra-high injection diesel sprays. *Fuel, 121*, 284–297.
39. Ghosh, S., & Hunt, J. C. R. (1994). Induced air velocity within droplet driven sprays. *Proceedings of the Royal Society of London, A444*, 105–127.
40. Ghosh, S., & Hunt, J. C. R. (1998). Spray jets in a cross-flow. *Journal of Fluid Mechanics, 365*, 109–136.
41. Giannadakis, E., Gavaises, M., & Arcoumanis, C. (2006). Modelling of cavitation in diesel injector nozzles. *Journal of Fluid Mechanics, 616*, 153–193.
42. Girin, A. G. (2012). On the mechanism of inviscid drop breakup at relatively small Weber numbers. *Atomization Sprays, 22*, 921–934.
43. Girin, A. G. (2013). Deformation and acceleration of drop. *Proceedings of ILASS—Europe 2013, 25th European Conference on Liquid Atomization and Spray Systems*, Chania, Greece, 1–4 September 2013, paper 43.
44. Girin, A. G., & Ivanchenko, Y. A. (2012). Model of liquid film disintegration at 'bag' mode of drop breakup. *Atomization Sprays, 22*, 935–949.
45. Glezer, A., & Coles, D. (1990). An experimental study of a turbulent vortex ring. *Journal of Fluid Mechanics, 211*, 243–283.
46. Gorokhovski, M. A., & Saveliev, V. L. (2003). Analysis of Kolmogorov's model of breakup and its application into Lagrangian computation of liquid sprays under air-blast atomisation. *Physics of Fluids, 15*, 184–192.
47. Gorokhovski, M., & Herrmann, M. (2008). Modeling primary atomization. *Annual Review of Fluid Mechanics, 40*(1), 343–366.
48. Gorokhovski, M. A., & Saveliev, V. L. (2008). Statistical universalities in fragmentation under scaling symmetry with a constant frequency of fragmentation. *Journal of Physics D: Applied Physics, 41*, 085405.
49. Grout, S., Dumouchel, C., Cousin, J., & Nuglisch, H. (2007). Fractal analysis of atomizing liquid flows. *International Journal of Multiphase Flow, 33*, 1023–1044.
50. Günther, A., & Wirth, K. -E. (2013). Evaporation phenomena in superheated atomization and its impact on the generated spray. *International Journal of Heat and Mass Transfer, 64*, 952–965.
51. Habchi, C. (2011). The energy spectrum analogy breakup (SAB) model for the numerical simulation of sprays. *Atomization Sprays, 21*, 1033–1057.
52. Habchi, C. (2013). A Gibbs free energy relaxation model for cavitation simulation in Diesel injectors. *Proceedings of ILASS—Europe 2013, 25th European Conference on Liquid Atomization and Spray Systems*, Chania, Greece, 1–4 September 2013, paper 93.
53. Healy, D. P., & Young, J. B. (2005). Full Lagrangian methods for calculating the particle concentration and velocity fields in dilute gas-particle flows. *Proceedings of the Royal Society of London. Series A, 461*, 2197–2225.
54. Helmholtz, H. (1858) On integrals of the hydrodynamical equations which express vortex-motion. P. G. Tait, Trans. with a letter by Lord Kelvin (W. Thompson) in *London Edinburgh and Dublin Philosophy Magazine and Journal of Science, 33*, 485–512 (Fourth series).
55. Herrmann, M. (2011). On simulating primary atomization using the refined level set grid method. *Atomization Sprays, 21*, 283–301.
56. Herrmann, M. (2013). On simulating primary atomization. *Atomization and Sprays, 23* (11–12), v–ix.
57. Heywood, J. B. (1988). *Internal combustion engines fundamentals*. New York: McGraw-Hill Book Company.

58. Hiroyasu, H. & Kadota, T. (1974). Fuel droplet size distribution in a diesel combustion chamber. SAE Paper 740715.
59. Hong, J. G., Ku, K. W., & Lee, C. -W. (2011). Numerical simulation of the cavitating flow in an elliptical nozzle. *Atomization Sprays, 21*, 237–248.
60. Hsiang, L. -P., & Faeth, G. M. (1993). Drop properties after secondary breakup. *International Journal of Multiphase Flow, 19*, 721–735.
61. Huh, K. Y., Lee, E., & Koo, J. -Y. (1998). Diesel spray atomization model considering nozzle exit turbulence conditions. *Atomization Sprays, 8*, 453–459.
62. Jiang, X., Siamas, G. A., Jagus, K., & Karayiannis, T. G. (2010). Physical modelling and advanced simulations of gas-liquid two-phase jet flows in atomization and sprays. *Progress in Energy Combustion Science, 36*, 131–167.
63. Juniper, M. P. (2008). The effect of confinement on the stability of non-swirling round jet/wake flows. *Journal of Fluid Mechanics, 605*, 227–252.
64. Kaltaev, A. (1982). *Investigation of dynamic characteristics of a vortex ring of viscous fluid. Continuum dynamics* (pp. 63–70). Alma-Ata: Kazah State University (in Russian).
65. Kambe, T., & Oshima, Y. (1975). Generation and decay of viscous vortex rings. *Journal of the Physical Society of Japan, 38*, 271–280.
66. Kaplanski, F., Fukumoto, Y., & Rudi, U. (2012). Reynolds-number effects on vortex ring evolution in a viscous fluid. *Physics Fluids, 24*, 033101.
67. Kaplanski, F., & Rudi, U. (1999). Dynamics of a viscous vortex ring. *International Journal of Fluid Mechanics Research, 26*, 618–630.
68. Kaplanski, F., & Rudi, Y. (2005). A model for the formation of 'optimal' vortex rings taking into account viscosity. *Physics of Fluids, 17*, 087101.
69. Kaplanski, F., Sazhin, S. S., Fukumoto, Y., Begg, S., & Heikal, M. (2009). A generalised vortex ring model. *Journal of Fluid Mechanics, 622*, 233–258.
70. Kaplanski, F., Sazhin, S. S., Begg, S., Fukumoto, Y., & Heikal, M. (2010). Dynamics of vortex rings and spray induced vortex ring-like structures. *European Journal of Mechanics - B/Fluids, 29*(3), 208–216.
71. Karimi, K. (2007). Characterisation of multiple-injection diesel sprays at elevated pressures and temperatures. Ph.D. Thesis, University of Brighton, Brighton, United Kingdom.
72. Kolakaluri, R., Li, Y., & Kong, S. -C. (2010). A unified spray model for engine spray simulation using dynamic mesh refinement. *International Journal of Multiphase Flow, 36*, 858–869.
73. Kolmogorov, A. N. (1941). On the log-normal distribution of particle sizes during the breakup process. *Doklady Akademii Nauk SSSR, 31*, 99.
74. Kostas, J., Honnery, D., & Soria, J. (2009). Time resolved measurements of the initial stages of fuel spray penetration. *Fuel, 88*, 2225–2237.
75. Kasyap, T. V., Sivakumar, D., & Raghunandan, B. N. (2009). Flow and breakup characteristics of elliptical liquid jets. *International Journal of Multiphase Flow, 35*, 8–19.
76. Kovasznay, L. S. G., Fujita, H., & Lee, R. L. (1974). Unsteady turbulent puffs. *Advances in Geophysics, 18B*, 253–263.
77. Lamb, H. (1932). *Hydrodynamics*. New York: Dover Publishers.
78. Lavrentiev, M. A., & Shabat, B. V. (1973). *Problems of hydrodynamics and mathematical models*. Moscow: Nauka Publishing House. (in Russian).
79. Lebas, R., Menard, T., Beau, P. A., Berlemont, A., & Demoulin, F. X. (2009). Numerical simulation of primary break-up and atomization: DNS and modelling study. *International Journal of Multiphase Flow, 35*, 247–260.
80. Lebedeva, N. A., Osiptsov, A. N., & Sazhin, S. S. (2013). A combined fully Lagrangian approach to mesh-free modelling of transient two phase flows. *Atomization Sprays, 23*, 47–69.
81. Lee, C. H., & Reitz, R. D. (2013). CFD simulations of diesel spray tip penetration with multiple injections and with engine compression ratios up to 100:1. *Fuel, 111*, 289–297.
82. Lee, C. S., & Park, S. W. (2002). An experimental and numerical study on fuel atomization characteristics of high-pressure diesel injection sprays. *Fuel, 81*, 2417–2423.
83. Lefebvre, A. H. (1989). *Atomization and sprays*. Bristol, PA: Taylor & Francis.
84. Li, X. (1995). Mechanism of atomisation of a liquid jet. *Atomization Sprays, 5*, 89–105.

85. Li, Y., & Umemura, A. (2014). Two-dimensional numerical investigation on the dynamics of ligament formation by Faraday instability. *International Journal of Multiphase Flow, 60*, 64–75.
86. Lightfoot, M. (2009). Fundamental classification of atomization processes. *Atomization Sprays, 19*, 1065–1104.
87. Lim, T., & Nickels, T. (1995). Vortex rings. In S. I. Green (Ed.), *Fluid vortices* (pp. 95–153). Dordrecht: Kluwer.
88. Lin, S. P., & Rietz, R. D. (1998). Droplet and spray formation from a liquid jet. *Annual Review of Fluid Mechanics, 30*, 85–105.
89. Liu, F. -S., Zhou, L., Sun, B. -G., Li, Z. -J., & Schock, H. J. (2008). Validation and modification of wave spray model for diesel combustion simulation. *Fuel, 87*, 3420–3427.
90. Liu, Z., & Liu, Z. (2006). Linear analysis of three-dimensional instability of non-newtonian liquid jets. *Journal of Fluid Mechanics, 559*, 451–459.
91. Lozano, A., Barreras, F., Hauke, G., & Dopazo, C. (2001). Longitudinal instabilities in an air-blasted liquid sheet. *Journal of Fluid Mechanics, 437*, 143–173.
92. Lugovtsov, B. A. (1970). On the motion of a turbulent vortex ring and its role in the transport of passive contaminant. *In Some problems of mathematics and mechanics* (dedicated to the 70th anniversary of M.A. Lavrentiev) (pp. 182–187), Leningrad: Nauka Publishing House (in Russian).
93. Lugovtsov, B. A. (1976). On the motion of a turbulent vortex ring. *Archives of Mechanics, 28*, 759–766.
94. Malaguti, S., Fontanesi, S., Cantore, G., Montanaro, A., & Allocca, L. (2013). Modelling of primary breakup process of a gasoline direct engine multi-hole spray. *Atomization and Sprays, 23*, 861–888.
95. Marmottant, P., & Villermauz, E. (2004). On spray formation. *Journal of Fluid Mechanics, 498*, 73–111.
96. Martinez, L., Benkenida, A., & Cuenot, B. (2010). A model for the injection boundary conditions in the context of 3D simulation of diesel spray: methodology and validation. *Fuel, 89*, 219–228.
97. Martynov, S. B., Mason, D. J., & Heikal, M. R. (2006). Numerical simulation of cavitation flows based on their hydrodynamic similarity. *International Journal of Engine Research, 7*, 1–14.
98. MATHEMATICA (2007). Book version 6.0.0, Wolfram Research Inc. Available at http://functions.wolfram.com. Retrieved July 25, 2008.
99. Ménard, T., Tanguy, S., & Berlemont. A. (2007). Coupling level set/VOF/ghost fluid methods: Validation and application to 3D simulation of the primary break-up of a liquid jet. *International Journal of Multiphase Flow, 33*, 510–524.
100. Narasimha, R., & Sreenivasan, K. R. (1979). Relaminarization of fluid flows. *Advances in Applied Mechanics, 19*, 221–301.
101. Nichols, J. (1972). Stream and droplet breakup by shock waves. In D.T. Harrje & F.H. Reardon (Eds.), *Liquid propellant rocket combustion instability* (pp. 126–128), NASA SP-194, Washington, DC.
102. Nordin, N. (2001). Complex chemistry modeling of Diesel spray combustion, Ph.D. Thesis, Chalmers University of Technology, Gothenburg, Sweden.
103. O'Rourke, P. J. (1981). Collective drop effects on vaporizing liquid sprays. Ph.D. thesis, Princeton University.
104. O'Rourke, P. J., & Amsden, A. A. (1987). The TAB method for numerical calculation of spray droplet breakup. SAE report 872089.
105. Panton, R. L. (1996). *Incompressible flow*. New York, Chichester: John Wiley & Sons Inc.
106. Park, K. S., & Heister, S. D. (2010). Nonlinear modeling of drop size distributions produced by pressure-swirl atomizers. *International Journal of Multiphase Flow, 36*, 1–12.
107. Pastor, J. V., López, J. J., García, J. M., & Pastor, J. M. (2008). A 1D model for the description of mixing-controlled inert diesel sprays. *Fuel, 87*, 2871–2885.

108. Pastor, J. V., Garca-Oliver, J. M., Nerva, J.-G., & Giménez, B. (2011). Fuel effect on the liquid-phase penetration of an evaporating spray under transient diesel-like conditions. *Fuel*, *90*, 3369–3381.
109. Patterson, M. A., & Reitz, R. D. (1998). Modelling of the effects of fuel spray characteristics on Diesel engine combustion and emission. SAE report 980131.
110. Payri, R., Salvador, F. J., Gimeno, J., & Zapata, L. D. (2008). Diesel nozzle geometry influence on spray liquid-phase fuel penetration in evaporative conditions. *Fuel*, *87*, 1165–1176.
111. Phillips, O. M. (1956). The final period of decay of non-homogeneous turbulence. *Proceedings of the Cambridge Philosophical Society*, *252*, 135–151.
112. Poole, D. R., Barenghi, C. F., Sergeev, Y. A., & Vinen, W. F. (2005). The motion of tracer particles in helium II. *Physical Review B*, *71*, 064514-1-16.
113. Postrioti, L., Mariani, F., & Battistoni, M. (2012). Experimental and numerical momentum flux evaluation of high pressure diesel spray. *Fuel*, *98*, 149–163.
114. Pozorski, J., & Minier, J.-P. (1998). On the Lagrangian turbulent dispersion models based on the Langevin equation. *International Journal of Multiphase Flow*, *24*, 913–945.
115. Pozorski, J., & Minier, J.-P. (1999). PDF modeling of dispersed two-phase turbulent flows. *Physical Review E*, *59*, 855–863.
116. Pozorski, J., Sazhin, S. S., Wacławczyk, M., Crua, C., Kennaird, D., & Heikal, M. R. (2002). Spray penetration in a turbulent flow. *Flow, Turbulence and Combustion*, *68*(2), 153–165.
117. Ranger, A. A., & Nicholls, J. A. (1969). The aerodynamic shattering of liquid drops. *AIAA Journal*, *3*, 285–290.
118. Reitz, R. D. (1987). Modelling atomization processes in high-pressure vaporizing sprays. *Atomisation and Spray Technology*, *3*, 309–337.
119. Reitz, R. D., & Bracco, F. V. (1982). Mechanism of atomization of a liquid jet. *Physics of Fluids*, *25*, 1730–1742.
120. Reitz, R. D., & Bracco, F. V. (2009). Mechanisms of breakup of round jets. In N. Cheremisnoff (Ed.), *Encyclopedia of fluid mechanics* (Vol. 3, pp. 012101). Houston: Gulf Publishing (Chapter 10).
121. Reitz, R. D., & Diwakar, R. (1986). Effect of drop breakup on fuel sprays. SAE report 860469.
122. Reitz, R. D., & Diwakar, R. (1987). Structure of high-pressure fuel sprays. SAE report 870598.
123. Rewse-Davis, Z., Nouri, J., Gavaises, M., & Arcoumanis, C. (2013). Near-nozzle instabilities in gasoline direct injection sprays. *Proceedings of ILASS—Europe 2013, 25th European Conference on Liquid Atomization and Spray Systems*, Chania, Greece, 1–4 September 2013, paper 141.
124. Rimbert, N. (2010). Simple model for turbulence intermittencies based on self-avoiding random vortex stretching. *Physical Review E*, *81*, 056315.
125. Rimbert, N., Séro-Guillaume, O. (2004). Log-stable laws as asymptotic solutions to a fragmentation equation: Application to the distribution of droplets in a high Weber-number spray. *Physical Review E*, *69*, 056316.
126. Roisman, I. V., Araneo, L., & Tropea, C. (2007). Effect of ambient pressure on penetration of a diesel spray. *International Journal of Multiphase Flow*, *33*, 904–920.
127. Rotondi, R., Bella, G., Grimaldi, C., & Postrioti, L. (2001). Atomization of high-pressure Diesel spray: experimental validation of a new breakup model. SAE report 2001–01-1070.
128. Rott, N., & Cantwell, B. (1993). Vortex drift. i: Dynamic interpretation. *Physics of Fluids*, *A5*, 1443–1450.
129. Rott, N., & Cantwell, B. (1993). Vortex drift. ii: The flow potential surrounding a drifting vortical region. *Physics of Fluids*, *A5*, 1451–1455.
130. Ruo, A. C., Chen, F., & Chang, M. H. (1992). Linear instability of compound jets with nonaxisymmetric disturbances. *Physics of Fluids*, *21*, 681–689.
131. Saffman, P. G. (1970). The velocity of viscous vortex rings. *Studies in Applied Mathematics*, *49*, 371–380.
132. Saffman, P. G. (1992). *Vortex dynamics*. Cambridge: Cambridge University Press.
133. Sarimeseli, A., & Kelbaliev, G. (2004). Modelling of the break-up of deformable particles in developed turbulent flow. *Chemical Engineering Science*, *59*, 1233–1240.

134. Sakaguchi, D., Yamamoto, S., Ueki, H., & Ishdia, M. (2010). Study of heterogeneous structure in diesel fuel spray by using micro-probe L2F. *Journal of Fluid Science and Technology, 5*, 75–85.
135. Savich, S. (2001). Spray dynamics and in-cylinder air motion. Ph.D. Thesis, The University of Brighton.
136. Sazhin, S. S., Feng, G., & Heikal, M. R. (2001). A model for fuel spray penetration. *Fuel, 80*(15), 2171–2180.
137. Sazhin, S. S., Kaplanski, F., Feng, G., Heikal, M. R., & Bowen, P. J. (2001). A fuel spray induced vortex ring. *Fuel, 80*(13), 1871–1883.
138. Sazhin, S. S., Crua, C., Kennaird, D., & Heikal, M. R. (2003). The initial stage of fuel spray penetration. *Fuel, 82*(8), 875–885.
139. Sazhin, S. S., Crua, C., Hwang, J. -S., No, S. -Y., & Heikal, M. (2005). Models of fuel spray penetration. *Proceedings of the Estonian Academy of Sciences: Engineering* 11(2), 154–160.
140. Sazhin, S. S., Martynov, S. B., Kristyadi, T., Crua, C., & Heikal, M. R. (2008). Diesel fuel spray penetration, heating, evaporation and ignition: Modelling versus experimentation. *International Journal of Engineering Systems Modelling and Simulation, 1*, 1–19.
141. Sazhin, S. S., Kaplanski, F., Begg, S., & Heikal, M. (2009). Vortex ring-like structures in gasoline fuel sprays, *Proceedings of the JUMV International Automotive Conference and Exhibition (XXII Science and Motor Vehicles 2009)*, Belgrade 14–16 April 2009, paper 31 (CD). Published by the Society of Automotive Engineers of Serbia.
142. Schugger, C., Meingast, U., & Renz, U. (2000). Time-resolved velocity measurements in the primary breakup zone of a high pressure diesel injection nozzle. *Proceedings of ILASS-Europe*, Darmstadt, Germany (2000).
143. Senecal, P. K., Schmidt, D. P., Nouar, I., Rutland, C. J., Reitz, R. D., & Corradini, M. L. (1999). Modeling high-speed viscous liquid sheet atomization. *International Journal of Multiphase Flow, 25*, 1073–1097.
144. Shariff, K., & Leonard, A. (1992). Vortex rings. *Annual Review of Fluid Mechanics, 24*, 235–279.
145. Sher, I., & Sher, E. (2011). Analytical criterion for droplet breakup. *Atomization and Sprays, 21*, 1059–1063.
146. Shinjo, J., & Umemura, A. (2004). Simulation of liquid jet primary breakup: Dynamics of ligament and droplet formation. *International Journal of Multiphase Flow, 36*, 513–532.
147. Snegirev, A., Talalov, V., Sheinman, I., & Sazhin, S. S. (2011). An enhanced spray model for flame suppression simulations. *Proceedings of the Fifth European Combustion Meeting ECM*, Cardiff, 29th June to 1st July 2011 paper 072–1 (CD).
148. Som, S., & Aggarwal, S. K. (2009). Assessment of atomization models for diesel engine simulations. *Atomization and Sprays, 19*, 885–903.
149. Som, S., & Aggarwal, S. K. (2010). Effects of primary breakup modeling on spray and combustion characteristics of compression ignition engines. *Combustion and Flame, 157*, 1179–1193.
150. Som, S., Ramirez, A. I., Longman, D. E., & Aggarwal, S. K. (2011). Effect of nozzle orifice geometry on spray, combustion, and emission characteristics under diesel engine conditions. *Fuel, 90*, 1267–1276.
151. Squire, H. B. (1933). On the stability for three-dimensional disturbances of viscous fluid flow between parallel walls. *Proceedings of the Royal Society of London, 142*, 621–628.
152. Srinivasan, V., Salazar, A. J., & Saito, K. (2008). Numerical investigation on the disintegration of round turbulent liquid jets using LES/VOF techniques. *Atomization and Sprays, 18*, 571–617.
153. Stanaway, S., Cantwell, B. J., & Spalart, P. R. (1988). A numerical study of viscous vortex rings using a spectral method, NASA Technical Memorandum 101041.
154. Stapper, B. E., Sowa, W. A., & Samuelson, G. S. (1992). An experimental study of the effects of liquid properties on the breakup of a two-dimensional liquid sheet. *ASME Journal of Engineering for Gas Turbines and Power, 114*, 39–45.

155. Su, T. F., Patterson, M. A., Reitz, R. D., & Farrell, P. V. (1996). Experimental and numerical studies of high pressure multiple injection sprays. SAE report 960861.
156. Tanner, F. X. (2004). Development and validation of a cascade atomization and drop breakup model for high-velocity dense sprays. *Atomization and Sprays*, *14*, 211–242.
157. Turner, M. R., Healey, J. J., Sazhin, S. S., & Piazzesi, R. (2011). Stability analysis and break-up length calculations for steady planar liquid jets. *Journal of Fluid Mechanics*, *668*, 384–411.
158. Turner, M. R., Healey, J. J., Sazhin, S. S., & Piazzesi, R. (2012). Wave packet analysis and break-up length calculations for accelerating planar liquid jets. *Fluid Dynamics Research*, *44*(1), article number: 015503, DOI: 10.1088/0169-5983/44/1/015503.
159. Turner, M. R., Sazhin, S. S., Healey, J. J., Crua, C., & Martynov, S. B. (2012). A breakup model for transient diesel fuel sprays. *Fuel*, *97*, 288–305.
160. Uchiyama, M. (2009). Numericall simulation of non-evaporating spray jet by the vortex method. *Atomization and Sprays*, *19*, 917–928.
161. Versteeg, H. K., & Malalasekera, W. (2007). *An introduction to computational fluid dynamics. The finite volume method* (2nd ed.). Harlow: Pearson Prentice Hall.
162. Vogel, T., Rieβ, S., Lutz, M., Wenzing, M., & Leipertz, A. (2011). Comparison of current spray models under high pressure and high temperature engine relevant conditions. *Proceedings of the 24th European Conference on Liquid Atomization and Spray Systems (ILASS)—Europe 2011*, Estoril, Portugal, 5–7 September 2011.
163. Weigand, A., & Gharib, M. (1997). On the evolution of laminar vortex rings. *Experiments in Fluids*, *22*, 447–457.
164. Yang, H. Q. (1992). Asymmetric instability of a liquid jet. *Physics of Fluids*, *A4*, 681–689.
165. Yi, Y., & Reitz, R. D. (2004). Modelling the primary breakup of high-speed jets. *Atomization and Sprays*, *14*, 53–80.
166. Zhao, H., Liu, H. -F., Cao, X, -K., Li, W. -F., & Xu, J. -L. (2011). Breakup characteristics of liquid drops in bag regime by a continuous and uniform air jet flow. *International Journal of Multiphase Flow*, *37*, 530–534.
167. Zhang, K., Wang, Z., Wang, J., & Wang, Z. (2012). Spray model based on step response theory. *Fuel*, *95*, 499–503.

Chapter 3
Heating of Non-evaporating Droplets

Abstract The analysis of heating of non-evaporating droplets is based on the assumptions that they retain their spherical form and that the temperature over their whole surface is the same (although it can vary with time). The models for convective and radiative heating of droplets are considered separately. The following cases of convective heating are considered: coupled analytical solutions for stationary droplets, separate solutions for gas and liquid phases in some limiting cases, and heating of moving droplets (steady-state and transient cases). The radiative heating of droplets is considered, using the following approaches: the most general Mie theory, analysis of integral absorption of radiation in droplets, and geometric optics analysis.

Anyone wishing to model the heating of non-evaporating droplets must take into account a number of important processes. These include the deformation of droplets in the air stream and the inhomogeneity of the temperature distribution at the droplet surface. A rigorous solution of this general problem, however, would not only be difficult, but also would have rather limited practical applications. Indeed, in realistic engineering applications, simultaneous modelling of the heating of a large number of droplets would be required. Moreover, this modelling would have to be performed alongside gas dynamics, turbulence, and chemical modelling. This leads to a situation in which the parameters of gas around droplets may be estimated with substantial errors, which are often difficult to control. This is why the focus has to be on finding a reasonable compromise between accuracy and computer efficiency of the models, rather than on the accuracy of the models alone.

The most commonly used assumption is that the droplets retain their spherical form even in the process of their movement. This assumption is made in this chapter as well. The generalisation of the models to droplets of arbitrary shapes have been discussed in a number of books and articles (see [58, 103, 142]). These generalisations could be applied to simple models of droplet heating, but it is not obvious how they can be applied to the more sophisticated models discussed below.

Another simplification widely used in droplet heating models is the assumption that the temperature over the whole droplet surface is the same (although it can vary

S. Sazhin, *Droplets and Sprays*, DOI: 10.1007/978-1-4471-6386-2_3,

with time). This assumption effectively allows the separation of the analysis of heat transfer in gaseous and liquid phases. It is expected to be a good approximation in the case of a stationary or very rapidly moving droplet, when the isotherms almost coincide with streamlines [6]. The errors introduced by this assumption in intermediate conditions are generally assumed to be acceptable.

In what follows, the models for convective and radiative heating of droplets are considered separately (Sects. 3.1 and 3.2).

3.1 Convective Heating

3.1.1 Stagnant Droplets

3.1.1.1 Coupled Analytical Solutions

In the case of stagnant droplets, there is no bulk motion of gas relative to the droplets, and the problem of their heating by the ambient gas reduces to a conduction problem. The heat conduction equation can be solved separately in the droplet and the gas, and the solutions are matched at the droplet surface. Assuming the spherical symmetry of the problem, its mathematical formulation is based on the solution of the equation [125, 126]

$$\frac{\partial T}{\partial t} = \kappa \left(\frac{\partial^2 T}{\partial R^2} + \frac{2}{R} \frac{\partial T}{\partial R} \right), \tag{3.1}$$

where

$$\kappa = \begin{cases} \kappa_l = k_l/(c_l \rho_l) & \text{when } R \leq R_d \\ \kappa_g = k_g/(c_{pg} \rho_g) & \text{when } R_d < R \leq \infty, \end{cases} \tag{3.2}$$

$\kappa_{l(g)}, k_{l(g)}, c_{l(pg)}$, and $\rho_{l(g)}$ are the liquid (gas) thermal diffusivity, thermal conductivity, specific heat capacity, and density respectively, R is the distance from the centre of the droplet, t is time, and the subscripts $_l$ and $_g$ refer to liquid and gas respectively.

Equation (3.1) is based on the Fourier law, assumed to be valid in most engineering applications. The model based on this equation is known as the diffusion model [83]. Other heat conduction models and their generalisations are discussed in [29, 61, 81, 83, 97, 110, 146, 155].

This equation should be solved subject to the following initial and boundary conditions:

$$T|_{t=0} = \begin{cases} T_{d0}(R) & \text{when } R \leq R_d \\ T_{g0}(R) & \text{when } R_d < R \leq \infty, \end{cases} \tag{3.3}$$

$$T|_{R=R_d-0} = T|_{R=R_d+0}; \quad k_l \left.\frac{\partial T}{\partial R}\right|_{R=R_d-0} = k_g \left.\frac{\partial T}{\partial R}\right|_{R=R_d+0}; \quad T|_{R=\infty} = T_{g\infty}. \tag{3.4}$$

Numerical solutions of (3.1) have been widely discussed in the literature (e.g., [109]). The analysis presented below is focused on various analytical solutions and their implementations into numerical codes.

Assuming that $T_{d0}(R) = T_{d0} = \text{const}_1$, and $T_{g0}(R) = T_{g\infty} = \text{const}_2$, Cooper [32] solved this problem analytically, using the Laplace transform method. His solution can be represented as follows:

$$T(R \leq R_d) = T_{g\infty} + \frac{2k_l}{\pi k_g}\sqrt{\frac{\kappa_g}{\kappa_l}}\left(T_{g\infty} - T_{d0}\right)\frac{R_d}{R} \times \int_0^\infty \mathrm{d}u \Phi(u) \exp\left(-u^2 \mathrm{Fo}\frac{\kappa_l}{\kappa_g}\right) \sin\left(u\frac{R}{R_d}\right), \tag{3.5}$$

$$T(R \geq R_d) = T_{g\infty} + \frac{2k_l}{\pi k_g}\sqrt{\frac{\kappa_g}{\kappa_l}}\left(T_{g\infty} - T_{d0}\right)\frac{R_d}{R}\int_0^\infty \frac{\mathrm{d}u}{u}\Phi(u) \exp\left(-u^2 \mathrm{Fo}\frac{\kappa_l}{\kappa_g}\right) \times \left\{\cos(\gamma_c u)\sin u + \sqrt{\frac{\kappa_g}{\kappa_l}}\sin\gamma_c\left[\frac{k_l}{k_g}(u\cos u - \sin u) + \sin u\right]\right\}, \tag{3.6}$$

where

$$\Phi(u) = \frac{(u\cos u - \sin u)}{u^2\sin^2 u + \frac{\kappa_g}{\kappa_l}\left[\frac{k_l}{k_g}(u\cos u - \sin u) + \sin u\right]^2},$$

$\mathrm{Fo} = t\kappa_g/R_d^2$ (Fouriernumber),

$$\gamma_c = \sqrt{\frac{\kappa_l}{\kappa_g}}\left[\frac{R - R_d}{R}\right]u.$$

As expected, this solution predicts diffusion of heat from gas to droplets if $T_{g\infty} > T_{d0}$. As a result, droplet and gas temperatures approach each other in every finite domain. Initially, the heat flux from gas to droplets predicted by the equation

$$\dot{q} = -k_g \left.\frac{\partial T}{\partial R}\right|_{R=R_d+0} \tag{3.7}$$

is infinitely large. It approaches zero at $t \to \infty$.

This solution was originally applied to the problem of heating a stationary liquid sodium sphere in UO_2 atmosphere [32]. However, the applicability of this solution to more general problems, involving the time variations of gas temperature due to external factors and evaporation effects, is questionable. The only practical approach to solving this general problem is currently based on the application of computational fluid dynamics (CFD) codes. In this case, one would need to take into account the distribution of the temperature inside the droplets at the beginning of each time step and the finite size of the computational cells. These effects were taken into account in the solution suggested in [125, 126]. In the models described in these papers, the condition $R \leq \infty$ in (3.2) and (3.3) were replaced with the condition $R \leq R_g < \infty$, where R_g is the external radius of the gaseous sphere, and $T(R \geq R_g)$ is assumed to be constant. The value of $R_g - R_d$ was interpreted as the thickness of the thermal boundary layer (see [126] for details).

The analysis of [125] was based on the assumption that the initial distribution of temperature inside droplets is a function of the distance from the droplet centre, while the initial gas temperature was assumed to be constant. In the more general model, described in [126], both initial droplet and gas temperatures were assumed to be functions of the distance from the droplet centre when $R \leq R_g$. In what follows, the main findings of [126] are briefly summarised.

Having replaced the condition $R \leq \infty$ in (3.2) and (3.3) with the condition $R \leq R_g < \infty$, we can represent the solution to (3.1) as [126] (see the solution given in Appendix B, in which the effects of thermal radiation are ignored)

$$T(R,t) = T_{g0}(R_g) + \frac{1}{R}\sum_{n=1}^{\infty}\left[\exp\left(-\lambda_n^2 t\right)\frac{1}{||v_n||^2}\left(\int_0^{R_d}(-(T_{g0}(R_g) - T_{d0}(R))Rv_n(R)c_l\rho_l \mathrm{d}R + \int_{R_d}^{R_g}(-(T_{g0}(R_g) - T_{g0}(R))Rv_n(R)c_{pg}\rho_g \mathrm{d}R\right)\right]v_n(R), \tag{3.8}$$

where

$$v_n(R) = \begin{cases} \dfrac{\sin(\lambda_n a_d R)}{\sin(\lambda_n a_d R_d)} & \text{when } R < R_d \\ \dfrac{\sin(\lambda_n a_g (R - R_g))}{\sin(\lambda_n a_g (R_d - R_g))} & \text{when } R_d \leq R \leq R_g, \end{cases} \tag{3.9}$$

$$||v_n||^2 = \frac{c_l\rho_l R_d}{2\sin^2(\lambda_n a_d R_d)} + \frac{c_{pg}\rho_g(R_g - R_d)}{2\sin^2(\lambda_n a_g(R_d - R_g))} - \frac{k_l - k_g}{2R_d\lambda_n^2},$$

A countable set of positive eigenvalues λ_n is found from the solution to the equation

$$\sqrt{k_b c_l \rho_l}\cot(\lambda a_d R_d) - \sqrt{k_g c_{pg}\rho_g}\cot(\lambda a_g(R_d - R_g)) = \frac{k_l - k_g}{R_d\lambda}. \tag{3.10}$$

These are arranged in ascending order $0 < \lambda_1 < \lambda_2 < \cdots$, $a_d = \sqrt{\frac{c_l \rho_l}{k_l}}$, $a_g = \sqrt{\frac{c_{pg} \rho_g}{k_g}}$.

Having introduced new dimensionless variables

$$\tilde{T} = \frac{T(R,t)}{T_{g0}(R_g)}, \quad \tilde{T}_d = \frac{T_{d0}(R)}{T_{g0}(R_g)}, \quad \tilde{T}_g = \frac{T_{g0}(R)}{T_{g0}(R_g)}, \quad r = \frac{R}{R_d}, \quad r_g = \frac{R_g}{R_d},$$

we can simplify (3.8) to

$$\tilde{T} = 1 + \frac{R_d}{r} \sum_{n=1}^{\infty} \left[\exp\left(-\lambda_n^2 t\right) \frac{1}{||v_n||^2} \left(\int_0^1 (-(1 - \tilde{T}_d) r v_n(R_d r) c_l \rho_l \mathrm{d}r \right.\right.$$
$$\left.\left. + \int_1^{r_g} (-(1 - \tilde{T}_g) r v_n(R_d r) c_{pg} \rho_g \mathrm{d}r \right) \right] v_n(R_d r). \tag{3.11}$$

If $T_{g0}(R) = T_{g0}(R_g) = \text{const}$ and T_{d0} does not depend on R, then (3.8) can be simplified to

$$T(R,t) = T_{g0} + \frac{1}{R} \sum_{n=1}^{\infty} \left[\exp\left(-\lambda_n^2 t\right) \frac{(T_{g0} - T_{d0})\sqrt{k_l c_l \rho_l}}{\lambda_n ||v_n||^2} \right.$$
$$\left. \times \left[R_d \cot(\lambda_n a_d R_d) - \frac{1}{\lambda_n a_d} \right] \right] v_n(R). \tag{3.12}$$

This solution was studied in detail in [125].

Although the above solutions are strictly valid for stationary droplets only, they can be applied to moving droplets as well. The droplet movement primarily affects the values of $R_g - R_d$, as discussed in the Sect. 3.1.1.2.

3.1.1.2 Thermal Film Thickness

The physical meaning of the value of $R_g - R_d$ can be interpreted in terms of the so-called film theory [6]. The key concept of this theory is thermal film thickness δ_T, the expression for which is derived from the requirement that the rate of a purely molecular transport by thermal conduction through the film must be equal to the actual intensity of the convective heat transfer between the body surface and the external flow. For the case of heat conduction at the surface of a sphere, this requirement can be written as [69]

$$q_s{''} = \frac{k_g \Delta T}{R_d - \frac{R_d^2}{R_d + \delta_{T0}}} = h \Delta T, \tag{3.13}$$

where $q_s'' = |\dot{q}_s|/(4\pi R_d^2)$ is the value of the heat flux at the surface of the droplet, $\Delta T = T_g - T_s$, the index $_0$ here indicates that the effects of the Stefan flow are not taken into account (no evaporation), h is the convective heat transfer coefficient. From (3.13), it follows that

$$\delta_{T0} = \frac{2R_d}{\mathrm{Nu}_0 - 2}, \tag{3.14}$$

where Nu_0 is the Nusselt number of the nonevaporating droplet ($\mathrm{Nu}_0 = 2R_d h/k_g$).

The value of $R_g - R_d$ in our model is identified with δ_{T0}. Following [6], Nu_0 is estimated as

$$\mathrm{Nu}_0 = 1 + (1 + \mathrm{RePr})^{1/3} \max\left[1,\ \mathrm{Re}^{0.077}\right], \tag{3.15}$$

where Re and Pr are the Reynolds and Prandtl numbers, respectively. A number of other approximations for Nu_0 are presented and discussed in [38].

If we impose an additional requirement that the initial heat rate inside the 'film' does not depend on R, we get the following equation for $T_{g0}(R)$:

$$\frac{4\pi k_g \left(T_{g0}(R_g) - T_{d0}\right)}{\frac{1}{R_d} - \frac{1}{R_g}} = \frac{4\pi k_g \left(T_{g0}(R) - T_{d0}\right)}{\frac{1}{R_d} - \frac{1}{R}}. \tag{3.16}$$

The introduction of non-zero Reynolds number affects our earlier assumption about the spherical symmetry of the problem and $h = k_g/R_d$. This can be overcome if we replace k_g by

$$k_{g,\mathrm{eff}} = k_g \mathrm{Nu}_0/2$$

to satisfy (3.13). Also, k_l would need to be replaced by the effective liquid thermal conductivity, following the effective thermal conductivity model. These effects are considered later.

Note that the 'film theory' based on (3.13)–(3.16) was developed under the assumption that droplet heating is quasisteady. This obviously contradicts the unsteady formulation of the problem (3.1)–(3.4). This contradiction, however, seems to be unavoidable, since the value of $R_g - R_d$ in our model needs to be imposed 'externally' as an input parameter. In [125], a range of values of $R_g - R_d$ was considered without any attempt to link them with the underlying physics of the phenomenon.

3.1.1.3 Analysis of the Coupled Solutions

Let us consider typical values of parameters for the case in which Diesel fuel droplets with an initial temperature of 300 K are injected into a gas at temperature 900 K and pressure 30 atm (situation typical for Diesel engines [118]):

$$\rho_l = 600\,\mathrm{kg/m^3} \quad k_l = 0.145\,\mathrm{W/(mK)} \quad c_l = 2830\,\mathrm{J/(kgK)}$$

$$\rho_g = 23.8\,\mathrm{kg/m^3} \quad k_g = 0.061\,\mathrm{W/(mK)} \quad c_{pg} = 1120\,\mathrm{J/(kgK)}.$$

This leads us to the following estimates of thermal diffusivities of the liquid and gas as defined by (3.2):

$$\kappa_l = 8.54 \times 10^{-8}\,\mathrm{m}^2/\mathrm{s}; \quad \kappa_g = 2.29 \times 10^{-6}\,\mathrm{m}^2/\mathrm{s}.$$

Note that we took gas temperature slightly higher than that used in [125], where it was assumed that $T_{g0}(R_g) = 800$ K. The values of transport coefficients for gas were taken to be the same as in [118, 125]. The differences in the values of these coefficients for these two temperatures were ignored as in [125].

We assume that the droplets can be treated as a body whose temperature is initially homogeneous, while $T_{g0}(R_g) = 900$ K and $R_d = 10$ μm. Here Pr is assumed to be equal to 0.7, and two values of Re are considered: 1 and 5.

Recalling (3.14), this leads to the following values of R_g:

$$R_{g1} = 3.301\, R_b \quad \text{and} \quad R_{g2} = 11.337\, R_b.$$

Two cases of the initial distribution of gas temperature in the range $R_d < R \leq R_g$ are considered. Firstly, we assume that $T_{g0}(R)$ satisfies (3.16), which leads to the following expression:

$$T_{g0}(R) = T_{d0} + \left[T_{g0}(R_g) - T_{d0}\right] \frac{\frac{1}{R_d} - \frac{1}{R}}{\frac{1}{R_d} - \frac{1}{R_g}}. \tag{3.17}$$

Secondly, we assume that

$$T_{g0}(R) = T_{g0}(R_g). \tag{3.18}$$

The latter case is identical to the one considered in [125]. An analysis of the effects of thermal radiation is not considered for the time being. This will be discussed in the Sect. 3.1.1.4.

The analysis is focused on the dimensionless time (Fourier number), distance, and temperature defined as

$$\mathrm{Fo} = t\kappa_g/R_d^2, \quad r = R/R_d, \quad \hat{T}_{(s)} = (T_{g0}(R_g) - T_{(s)}(R,t))/(T_{g0}(R_g) - T_{d0}).$$

The calculations were performed using Mathematica 6.0 on a 3.0-GHz Kernel. One hundred terms of the series were taken.

Plots of $\hat{T}$ versus r for $R_g = 3.301\, R_d$ and four values of Fo are shown in Fig. 3.1. The plots are shown for both initial distributions of gas temperature in $R_d < R \leq R_g$, defined by expressions (3.17) and (3.18). As follows from this figure, for Fo $= 0.1$, most of the interior of the droplet is not affected by high gas temperature for both initial distributions of $T_{g0}(R)$, but the droplet temperatures near the surface are affected more strongly by gas for distribution (3.18) than for distribution (3.17). The difference in gas temperatures ($r > 1$) for these initial distributions of $T_{g0}(R)$ is clearly visible, as expected. For Fo $= 1$ and Fo $= 10$, a more rapid heating of

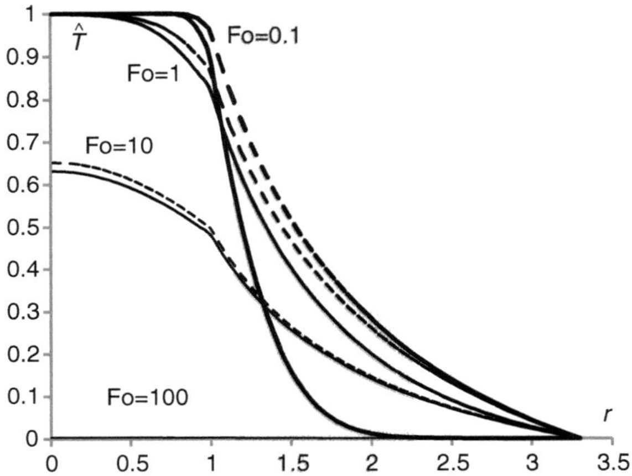

Fig. 3.1 Plots of $\hat{T} \equiv (T_{g0}(R_g) - T(R,t))/(T_{g0}(R_g) - T_{d0})$ versus $r = R/R_b$ for $r_g \equiv R_g/R_d =$ 3.301 and four values of Fo (indicated near the *curves*). *Solid curves* refer to the initial distribution (3.18), while *dashed curves* refer to the initial distribution (3.17). The thickness of the curves is inversely proportional to Fo. Reprinted from Ref. [126]. Copyright Elsevier (2011)

the droplet for distribution (3.18), compared with distribution (3.17), is seen much more clearly, compared with the case Fo $= 0.1$. Gas temperatures predicted by both distributions in these cases are much closer than in the case Fo $= 0.1$. For Fo $= 100$, for both initial temperature distributions, both droplet and gas temperatures become very close to $T_{g0}(R_g)$.

The plots, similar to those shown in Fig. 3.1 but for $R_g = 11.337\ R_b$, are presented in Fig. 3.2. Comparing Figs. 3.1 and 3.2, one can see that the trends of the curves in both figures are essentially the same, although the difference in the droplet heat-up for Fo $= 1$ and Fo $= 10$, predicted for distributions (3.17) and (3.18), is more clearly visible in Fig. 3.2 than in Fig. 3.1.

The plots of $\hat{T}_s$ versus Fo for $R_g = 3.301\ R_d$, $R_g = 11.337\ R_b$ and both initial distributions of $T_{g0}(R)$ are shown in Fig. 3.3. As follows from this figure, the droplet surface is always heated more quickly for distribution (3.18) compared with distribution (3.17) as expected. Also, the droplet is heated more quickly for $R_g = 3.301\ R_d$ than for $R_g = 11.337\ R_d$. All these results are consistent with those shown in Figs. 3.1 and 3.2.

Ignoring the effects of the droplet movement, we can estimate the heat flux arriving at its surface as

$$q'' = k_g \left.\frac{\partial T}{\partial R}\right|_{R=R_d+0} = k_l \left.\frac{\partial T}{\partial R}\right|_{R=R_d-0}. \tag{3.19}$$

On the other hand, from Newton's law of cooling, it follows that

$$q''_N = h\left(T_{g0}(R_g) - T_s\right), \tag{3.20}$$

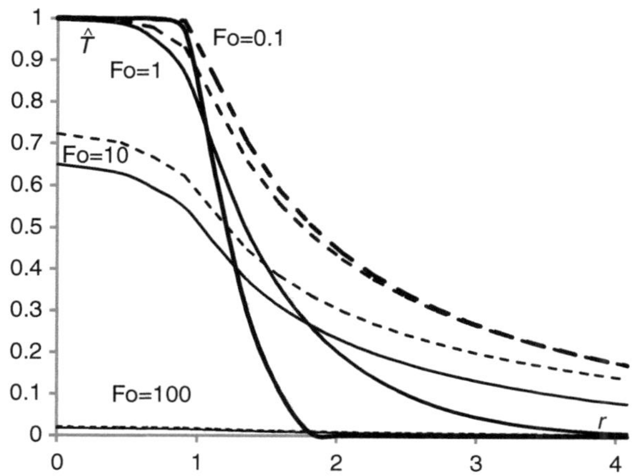

Fig. 3.2 The same as Fig. 3.1, but for $r_g = 11.337$. Reprinted from Ref. [126]. Copyright Elsevier (2011)

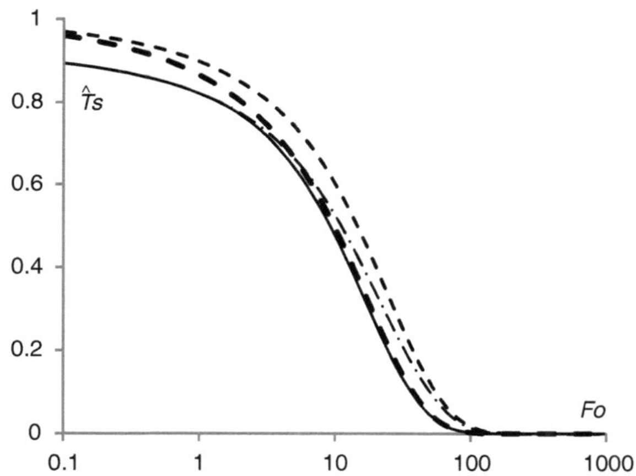

Fig. 3.3 Plots of $\hat{T}_s \equiv (T_{g0}(R_g) - T_s(R,t))/(T_{g0}(R_g) - T_{d0})$ versus Fo for $r_g = 3.301$ and distribution (3.18) (*solid*), $r_g = 11.337$, and distribution (3.18) (*dashed-dotted*), $r_g = 3.301$, and distribution (3.17) (*thick dashed*), $r_g = 11.337$, and distribution (3.17) (*thin dashed*). Reprinted from Ref. [126]. Copyright Elsevier (2011)

where $h = \frac{k_g}{R_d}$ (note that this law is sometimes called the Newton–Richmann law of cooling [35]). In the steady-state limit, $q'' = q''_N$. However, in the general transient case, they are linked by the equation

$$q'' = \chi q''_N, \tag{3.21}$$

where

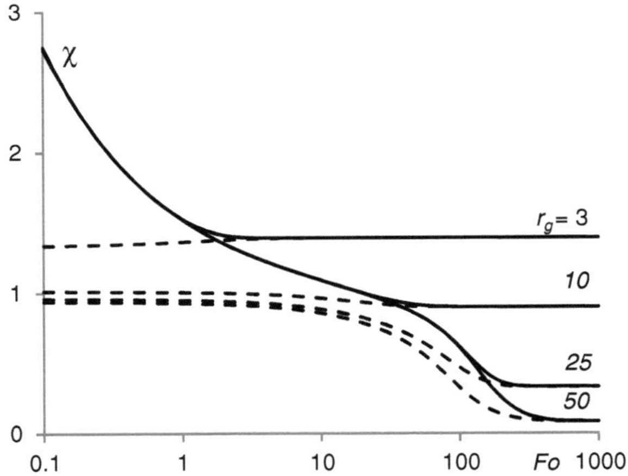

Fig. 3.4 Plots of χ versus Fo for four values of r_g (indicated near the *curves*). *Solid curves* refer to the initial distribution (3.18), while *dashed curves* refer to the initial distribution (3.17). Reprinted from Ref. [126]. Copyright Elsevier (2011)

$$\chi = \frac{k_g \left.\frac{\partial T}{\partial R}\right|_{R=R_d+0}}{\frac{k_g}{R_d}\left(T_{g0}(R_g) - T_s\right)} = \frac{R_d \left.\frac{\partial T}{\partial R}\right|_{R=R_d+0}}{T_{g0}(R_g) - T_s}. \tag{3.22}$$

An explicit expression for χ also follows from Cooper's solution [32] in the limit when $R_g \to \infty$:

$$\chi \equiv \chi_t = \frac{-\frac{k_l}{k_g}\int_0^\infty (u\cos u - \sin u)\Phi(u)\exp(-u^2 \text{Fo}\frac{\kappa_l}{\kappa_g})\text{d}u}{\int_0^\infty \sin u\Phi(u)\exp(-u^2 \text{Fo}\frac{\kappa_l}{\kappa_g})\text{d}u}. \tag{3.23}$$

Here $\Phi(u)$ is the same as in (3.5) and (3.6). If Newton's law is valid, then $\chi = 1$.

The plots of χ versus Fo for various $r_g \equiv R_g/R_d$ and both initial distributions of $T_{g0}(R)$ are shown in Fig. 3.4. The solid plots referring to distribution (3.18) are identical to those presented in [125]. The solid plot referring to $R_g = 50R_d$ is practically indistinguishable from the one that follows from the analysis by Cooper [32], obtained in the limit $R_g = \infty$ using an approach totally different from the one considered in [125, 126] (see (3.23)). This coincidence confirms the validity of both approaches.

The dashed curves, obtained for the initial distribution (3.18), coincide with the solid curves referring to distribution (3.17) in the limit of large Fo. For small Fo, the deviations between the curves corresponding to distributions (3.17) and (3.18) are clearly seen. For distribution (3.18), χ rapidly increases with decreasing Fo ($\chi \to \infty$ when Fo $\to 0$). For distribution (3.17), χ approaches finite values when Fo $\to 0$, being always less than predicted for distribution (3.18). The values of χ in the limit Fo $\to 0$ can be estimated analytically from (3.18) and (3.21) as

$$\chi = \frac{\tilde{R}_g}{\tilde{R}_g - 1}. \tag{3.24}$$

Note that the values of χ for Fo $= 0.1$ can differ by up to about 8% from those predicted by (3.24) (although the values of temperature were calculated with errors less than about 0.5 %). This is related to the very slow convergence of the corresponding series in (3.8) for the derivative of the temperature in the vicinity of the droplet surface (up to 3000 terms in this series were taken).

Although the solutions discussed in this chapter so far clearly show the importance of coupling between the liquid and gas phases, it remains unclear how they can be used in most practical engineering applications. All the available CFD codes are based on separate solutions for gas and liquid phases, followed by their coupling [64, 88, 111]. Hence, some kind of separation of the solutions for gas and liquid phases would be essential to make them compatible with these codes.

The required separation between the solutions could be achieved based on a comparison between the thermal diffusivities of gas and liquid. As previously, let us consider typical values of parameters for Diesel fuel spray droplets and assume that these droplets have initial temperature 300 K and are injected into a gas at temperature 800 K and pressure 30 atm [118]. For these values of parameters we obtain $\kappa_l = 8.53 \times 10^{-8}$ m^2/s and $\kappa_g = 2.28 \times 10^{-6}$ m^2/s. This allows us to assume

$$\kappa_l \ll \kappa_g. \tag{3.25}$$

This condition tells us that gas responds much more quickly than liquid to changes in the thermal environment. As a zeroth approximation, we can ignore the changes in liquid temperature altogether, and assume that the droplet surface temperature remains constant over time. This immediately allows us to decouple the solution of (3.1) for the gas phase from the trivial solution to this equation for liquid phase ($T(R \le R_d) =$ const). The former solution will be presented and discussed in Sect. 3.1.1.4.

3.1.1.4 Analytical Solution for the Gas Phase

Assuming that $T(R \le R_d) =$ const, the solution for the gas phase can be represented as

$$T(R > R_d) = T_{g\infty} + \frac{R_d}{R}(T_{d0} - T_{g\infty})\left[1 - \mathrm{erf}\left(\frac{R - R_d}{2\sqrt{\kappa_g t}}\right)\right], \tag{3.26}$$

where

$$\mathrm{erf}(x) = \frac{2}{\sqrt{\pi}} \int_0^x \exp(-t^2)\mathrm{d}t.$$

In the limit $R = R_d$, Eq. (3.26) gives $T = T_{d0}$. In the limit $t \to 0$ but $R \neq R_d$, this equation gives $T = T_{g\infty}$.

Having substituted (3.26) into (3.7), we obtain the following equation for the heat flux from gas to droplets [136]:

$$\dot{q} = \frac{k_g(T_{d0} - T_{g\infty})}{R_d}\left(1 + \frac{R_d}{\sqrt{\pi \kappa_g t}}\right). \tag{3.27}$$

The same expression follows from the analysis reported later in [54, 119], whose authors were apparently not aware of the original paper [136]. Moreover, this expression might have been derived even earlier, since in 1971 it was referred to in [30] as the 'well known conduction solution' without giving any references.

For $t \gg t_d \equiv R_d^2/(\pi \kappa_g)$. Equation (3.27) can be further simplified and rewritten as

$$|\dot{q}| = h(T_{g\infty} - T_{d0}), \tag{3.28}$$

where h is the convection heat transfer coefficient defined as

$$h = \frac{k_g}{R_d}. \tag{3.29}$$

From the fact that the convection heat transfer is commonly described by the Nusselt number $\mathrm{Nu} = 2R_d h/k_g$, Eq. (3.29) is seen to be equivalent to the statement that $\mathrm{Nu} = 2$. The solution (3.28) can be obtained directly from (3.1) if the time derivative of temperature is ignored (steady-state solution). It gives us the well-known Newton's law for heating of stationary droplets so long as the boundary layer around the droplets has had enough time to develop. Note that for the values of the parameters mentioned above, $t_d = 3.5$ μs. That means that except at the very start of droplet heating, this process can be based on (3.28) and (3.29). These equations are widely used in CFD codes.

Comparing (3.27), (3.28), and (3.29), it can be seen that Newton's law (3.28), (3.29) can be used to describe the transient process discussed above if the gas thermal conductivity k_g is replaced by the 'time-dependent' gas thermal conductivity $k_{g(\mathrm{t})}$, defined as [119, 136]

$$k_{g(\mathrm{t})} = k_g\left(1 + \zeta_t\right), \tag{3.30}$$

where

$$\zeta_t = R_d\sqrt{\frac{c_{pg}\rho_g}{\pi k_g t}}. \tag{3.31}$$

This is applicable only at the very start of droplet heating (at the start of calculations when a droplet is injected into the gas). Unless abrupt changes in gas temperature occur, one may assume that the boundary layer around the droplet has had enough time to adjust to varying gas temperature. This would justify the application of Newton's law in its original formulation (3.28) and (3.29).

Note that in the limit $t \to \infty$, the solution (3.26) simplifies

$$\Delta T \equiv T - T_{g\infty} = \frac{R_d}{R}(T_{d0} - T_{g\infty}), \tag{3.32}$$

where ΔT indicates the local changes in gas temperature after the boundary layer around the droplet has been formed. The change of gas enthalpy, due to the presence of the droplet, in this case can be obtained as

$$\Delta H_e = \int_{R_d}^{\infty} \rho_g c_{pg} \Delta T 4\pi R^2 \mathrm{d}R = 4\pi R_d (T_{d0} - T_{g\infty}) \rho_g c_{pg} \int_{R_d}^{\infty} R \mathrm{d}R = \infty. \tag{3.33}$$

Thus the establishment of the required boundary layer of a single droplet leads to an infinitely large change in the enthalpy of the gas. It seems that Todes [136] was the first to draw attention to this fact. One can see, however, that if the value of ΔT had been calculated for any $t \neq \infty$ from (3.26), then ΔH_e would have remained finite. This appears to be due to the fact that at any $t < \infty$, the second term in the square brackets in (3.26) cannot be ignored at $R - R_d > 2\sqrt{\kappa_g t}$. This means that (3.32) is not valid at these radii, and ΔH_e cannot be calculated using (3.33). In practice, ΔH_e has never been calculated from (3.33), to the best of the author's knowledge. Instead, the amount of heat gained or lost by the droplet during a certain period of time Δt has been calculated based on the values of Nu. This gives reasonable results, provided that the assumption that the droplet surface temperature remains constant in time is valid.

To summarise this section, we may say that the simple Eqs. (3.28) and (3.29) seem to be the most useful for practical applications in CFD codes. At the initial stage of heating, the corrections described by (3.30) can be introduced if required. The range of applicability of these equations for transient heating, however, has not been rigorously justified (cf. (3.33)). The next stage of the analysis is focused on the processes in the liquid phase.

3.1.1.5 Analytical Solutions for the Liquid Phase

Assuming that $k_l \to \infty$, which implies that $T_d(R \leq R_d) = \mathrm{const}$, the droplet temperature can be found from the energy balance equation

$$\frac{4}{3}\pi R_d^3 \rho_l c_l \frac{\mathrm{d}T_d}{\mathrm{d}t} = 4\pi R_d^2 h (T_{g\infty} - T_d). \tag{3.34}$$

This equation merely indicates that all the heat supplied from gas to droplet is spent on raising the temperature of the droplet. It has a straightforward solution:

$$T_d = T_{g\infty} + (T_{d0} - T_{g\infty}) \exp\left(-\frac{3ht}{c_l \rho_l R_d}\right), \tag{3.35}$$

where $T_d(t = 0) = T_{d0}$.

Equation (3.34) and its solution (3.35) are widely used in various applications. In conjunction with the mass transfer equation (described in Chap. 4), Eq. (3.34) was used to determine experimentally the heat transferred by convection to droplets [25]. Other recent applications of this equation and its solution are discussed in [1, 16, 20, 36, 37, 52, 68, 71–73, 78, 107, 115, 132, 135, 148, 154]. Solution (3.35) is widely used in most CFD codes. Sometimes, this is justified by the fact that liquid thermal conductivity is much higher than that of gas. However, the main parameter which controls droplet transient heating is not its conductivity, but its diffusivity. As shown earlier, in the case of Diesel engine sprays, the diffusivity for liquid is more than an order of magnitude less than that for gas. This raises the question whether the assumption $k_l \to \infty$ is applicable to modelling fuel droplets in these engines. The only reasonable way to answer this question is to consider the model, which takes into account the effect of finite liquid thermal conductivity.

The model, taking into account the effect of finite k_l, is based on the solution of (3.1) inside the droplet with the following boundary conditions:

$$h(T_g - T_s) = k_l \left. \frac{\partial T}{\partial R} \right|_{R=R_d-0}, \tag{3.36}$$

$\left.\frac{\partial T}{\partial R}\right|_{R=0} = 0$, and the initial condition $T(t = 0) = T_{d0}(R)$, where $T_s = T_s(t)$ is the droplet's surface temperature, $T_g = T_g(t)$ is the ambient gas temperature, and the subscript $_\infty$ has been omitted.

Note that strictly speaking, the condition $\left.\frac{\partial T}{\partial R}\right|_{R=0} = 0$ can be replaced by a more general requirement that $T(R, t)$ be a continuously twice differentiable function at $R \le R_d$. Both approaches would lead to the same solution.

Assuming that h = const, this solution can be represented as [65, 122] (see the solution given in Appendix A, in which the effects of thermal radiation are ignored)

$$\begin{aligned} T(R,t) = \frac{R_d}{R} \sum_{n=1}^{\infty} \Bigg\{ & q_n \exp\left[-\kappa_R \lambda_n^2 t\right] - \frac{\sin \lambda_n}{\| v_n \|^2 \lambda_n^2} \mu_0(0) \exp\left[-\kappa_R \lambda_n^2 t\right] \\ & - \frac{\sin \lambda_n}{\| v_n \|^2 \lambda_n^2} \int_0^t \frac{\mathrm{d}\mu_0(\tau)}{\mathrm{d}\tau} \exp\left[-\kappa_R \lambda_n^2 (t-\tau)\right] \mathrm{d}\tau \Bigg\} \\ & \times \sin\left[\lambda_n \left(\frac{R}{R_d}\right)\right] + T_g(t), \end{aligned} \tag{3.37}$$

where the λ_n are solutions to the equation

$$\lambda \cos \lambda + h_0 \sin \lambda = 0, \tag{3.38}$$

$$\| v_n \|^2 = \frac{1}{2}\left(1 - \frac{\sin 2\lambda_n}{2\lambda_n}\right) = \frac{1}{2}\left(1 + \frac{h_0}{h_0^2 + \lambda_n^2}\right),$$

$$q_n = \frac{1}{R_d \parallel v_n \parallel^2} \int_0^{R_d} \tilde{T}_0(R) \sin\left[\lambda_n \left(\frac{R}{R_d}\right)\right] \mathrm{d}R,$$

$$\kappa_R = \frac{k_l}{c_l \rho_l R_d^2}, \qquad \mu_0(t) = \frac{h T_g(t) R_d}{k_l},$$

$h_0 = (hR_d/k_l) - 1$, $\tilde{T}_0(R) = RT_{d0}(R)/R_d$. The solution to (3.38) gives a set of positive eigenvalues λ_n numbered in ascending order ($n = 1, 2, \ldots$). A proof of the convergence of the series in (3.37) is given in [122]. In the limit $k_l \to \infty$, solution (3.37) reduces to solution (3.35), as expected [116] (see Appendix A for the details).

Solution (3.37) was generalised to the case of almost constant convection heat transfer coefficient [122]. In the case of the general time-dependent convection heat transfer coefficient, the solution of the differential equation can be reduced to the solution of the Volterra integral equation of the second kind [122]. However, the practical applicability of these solutions to CFD codes turned out to be limited [123].

The benefits of taking into account the finite thermal conductivity of fuel droplets in the CFD modelling of combustion processes in Diesel engines was first demonstrated in [15], whose authors performed the calculations using KIVA II CFD code with the conventional model for droplet heating based on (3.35) (called the Spalding model), and with the models taking into account the finite thermal conductivity of droplets. Note that the latter effect in this paper was taken into account based on the direct numerical solution of (3.1), rather than on the analytical solution to this equation (see solution (3.37)). Their results clearly demonstrated that taking into account the effects of finite thermal conductivity of droplets leads to better agreement with experimental data, compared with the conventional model.

The first direct experimental evidence of the need to take into account the effects of finite k_l seems to have been provided in [90]. In that paper, results of experimental studies and the modelling of heating and evaporation of monodisperse ethanol and acetone droplets in two regimes were presented. Firstly, pure heating and evaporation of droplets in a flow of air of prescribed temperature were considered. Secondly, droplet heating and evaporation in a flame produced by previously injected combusting droplets were studied. The phase Doppler anemometry technique was used for droplet velocity and size measurements. Two-colour laser-induced fluorescence thermometry was used to estimate droplet temperatures. The experiments were performed for various distances between droplets and various initial droplet radii and velocities. The experimental data were compared with the results of modelling based on given gas temperatures measured by coherent anti-Stokes Raman spectroscopy, and Nusselt and Sherwood numbers calculated using measured values of droplet relative velocities. In estimating the latter numbers, the finite distance between droplets was taken into account. The model was based on the analytical solution (3.37). It was pointed out that for relatively small droplets (initial radii about 65 μm), the experimentally measured droplet temperatures are close to the predicted average droplet temperatures, while for larger droplets (initial radii about 120 μm), the experimentally measured droplet temperatures are close to the temperatures predicted at the centre of the droplets.

Although nobody seems to contest the benefits of taking into account the finite thermal conductivity of fuel droplets in CFD modelling, the developers of CFD codes do not embrace this modelling opportunity too eagerly. The main reason for this is the additional cost of CPU time involved. This has stimulated efforts to develop a model taking into account the effect of finite thermal conductivity of droplets, but with minimal extra demands on CPU time. The idea of this new model may have been prompted by looking at the character of the plots $T(R)$ predicted by (3.37). Assuming that initially, $T(R)$ is constant, one can see that except at the very start of heating, the shape of the curve $T(R)$ looks very close to a parabola. This allows us to approximate $T(R)$ as [44]

$$T(R,t) = T_c(t) + [T_s(t) - T_c(t)]\left(\frac{R}{R_d}\right)^2, \qquad (3.39)$$

where T_c and T_s are the temperatures at the centre ($R = 0$) and on the surface ($R = R_d$) of the droplet respectively. Approximation (3.39) is obviously not valid at the very beginning of the heating process when $T = T_c$ in most of the droplet.

Substitution of (3.39) into the boundary condition at $R = R_d$ (Eq. (3.36)) gives [44]

$$T_s - T_c = \frac{\zeta}{2}(T_g - T_s), \qquad (3.40)$$

where $\zeta = \frac{\mathrm{Nu}}{2}\frac{k_g}{k_l}$, Nu is the Nusselt number introduced earlier. Equation (3.39) does not satisfy (3.1), but it should satisfy the equation of thermal balance, which can be obtained from integrating (3.1) along the radius R:

$$\rho_l c_l \frac{R_d}{3}\frac{\mathrm{d}\overline{T}}{\mathrm{d}t} = h\left(T_g - T_s\right), \qquad (3.41)$$

where

$$\overline{T} = \frac{3}{R_d^3}\int_0^{R_d} R^2 T(R)\mathrm{d}R \qquad (3.42)$$

is the average temperature of the droplet. Note that (3.41) is identical to (3.34), describing droplet heating in isothermal (infinite liquid conductivity) approximation, if h is replaced by h^* found from the modified Nusselt number

$$\mathrm{Nu}^{**} = \mathrm{Nu}\frac{T_g - T_s}{T_g - \overline{T}}. \qquad (3.43)$$

Having substituted (3.39) into (3.42) and keeping in mind (3.40), we obtain

$$\overline{T} = \frac{2T_c + 3T_s}{5} = T_s - 0.2\zeta(T_g - T_s). \tag{3.44}$$

Substitution of (3.44) into (3.43) gives

$$\text{Nu}^{**} = \frac{\text{Nu}}{1 + 0.2\zeta}. \tag{3.45}$$

On the other hand, the combination of (3.40) and (3.44) gives [44]

$$T_s = \frac{\overline{T} + 0.2\zeta T_g}{1 + 0.2\zeta}, \tag{3.46}$$

$$T_c = \frac{(1 + 0.5\zeta)\overline{T} - 0.3\zeta T_g}{1 + 0.2\zeta}. \tag{3.47}$$

The combination of (3.45)–(3.47) gives the full solution to the problem of convective droplet heating as predicted by (3.39). First, (3.34) is solved with h replaced by h^* from (3.45). Then the values of T_s and T_c are obtained from (3.46) and (3.47). These two parameters are then substituted into (3.39) to give the radial distribution of temperature inside droplets. For practical engineering applications, we are primarily interested in the values of T_s, which determine the rate of evaporation and break-up of droplets. The model based on (3.39) was called the parabolic temperature profile model in [44].

As follows from the comparison of this model with the numerical solution to (3.1), the predictions of this model show good accuracy at large times, but can differ considerably from numerical results for small times. The period of time when the latter happens is usually reasonably short and can be ignored in practical calculations. Actual implementation of this model into a customised version of the CFD code VECTIS is described in [120]. The predictions of the model were reasonably close to those predicted by (3.37), and its additional CPU requirements are very small and should be acceptable in most practical applications. The parabolic temperature profile model was generalised to take into account the initial heating of droplets in [44], but the implementation of this generalised model into CFD codes has not yet been investigated.

Note that the parabolic temperature profile model can be developed more rigorously based on (3.37) if only the first term in this series is taken into account and the initial temperature inside droplets is assumed to be constant. This approach was suggested in [152]. In the limit $t \to \infty$, equation (18) of [152] is identical to (3.44) in this paper. For $t \to 0$, the accuracy of equation (18) of [152] becomes questionable, since in this case all terms in (3.37) become comparable.

The analysis of [22] is based on the generalised form of (3.39) in which the constant power 2 is replaced with a time-dependent factor n_T. This allowed the author of this paper to widen the range of applicability of the model while retaining its analytical simplicity.

The applicability of the model based on the parabolic approximation of temperature distribution and its generalisations (power law and polynomial approximations, heat balance integral model) to modelling the heating and evaporation of various droplets has been investigated in [130]. Note that models based on the power law (with power not equal to 2) polynomial approximations do not have clear physical meaning, since they cannot be considered the limiting cases of the more general model based on the analytical solution to the heat conduction equation inside the droplet (in contrast to the conventional parabolic model). The same comment applies to the model described in [22].

The analysis of (3.1) presented in this section was essentially based on the assumption that thermal conductivities, specific heat capacities, and densities of gas and liquid are constant. This assumption is reasonable for CFD applications, where the calculations are performed over small time steps in which the variations of these parameters can be ignored. They can be updated from one step to another. Attempts to take into account the temperature dependence of these parameters (i.e., to solve nonlinear heat conduction equations) began more than 50 years ago and are still continuing (e.g., [13, 23, 24, 50, 62, 66, 76, 151]).

3.1.2 Moving Droplets

The analysis of heat exchange between gas and stationary droplets was simplified by the fact that this problem is essentially one-dimensional in space. The complexity of the problem of heat transfer between gas and moving droplets lies in the fact that this problem is at least two-dimensional in space. This leads to the need to replace (3.1) with a more general equation for the temperatures of both liquid and gas and represent it in the form [93]

$$\frac{\partial T}{\partial t} + \nabla \cdot (\mathbf{u}(t, \mathbf{x})T) = \kappa \nabla^2 T, \tag{3.48}$$

where $\mathbf{u}(t, \mathbf{x})$ is the fluid velocity, which depends on time t and position in space $\mathbf{x}$ in the general case. The Laplacian ∇^2 is three-dimensional in the general case; its two dimensional approximation is almost universally used. In most practical applications, the time-dependence of $\mathbf{u}$ is ignored, even if the problem of transient heat transfer is considered. In the general case, this equation is solved both in droplets and in the surrounding gas. Equation (3.48) does not contain the so-called history terms. These will be briefly discussed later.

Application of (3.48) instead of (3.1) leads to qualitative differences between the mechanisms of heat transfer in the cases of stationary and moving droplets. In the case of stationary droplets, the heat transfer takes place via conduction, and is described by the heat conduction equation in both liquid and gas. In the case of moving droplets, however, convection heat transfer takes place, which incorporates effects of bulk fluid motion (advection) and diffusion effects (conduction). The modelling of heat transfer in this case needs to be based on the Navier–Stokes equations for enthalpy and momentum [67]. For stationary droplets, the thickness of the boundary layer around them can be infinitely large. In the case of moving droplets, however, this thickness is always finite (see (3.14)). All these differences between the heat transfer processes in the case of stationary and moving droplets require the development of different methods of analysis. While in the case of stationary droplets, we started with the general transient solution and ended up with the analysis of the limiting steady-state case, in the case of moving droplets, the starting point will be the simplest steady-state case.

3.1.2.1 Steady-State Heating

In the case of steady-state droplet heating, the first term in (3.48) can be assumed equal to zero, and this leads to considerable simplification of the solution of this equation. In some practically important cases, this solution can be avoided altogether, and the analysis can be performed at a semiqualitative level, backed by experimental observations.

The first simplification of the problem is based on the assumption that there is no spatial gradient along a droplet's surface. This allows us to separate the analysis of the gas and liquid phases. If the surface temperature of the droplet is fixed, then using dimensional analysis, one can show that Nusselt number (Nu) can depend only on the Reynolds and Prandtl numbers (Re and Pr). If $\mathrm{Re} = 0$ (stagnant droplet), then $\mathrm{Nu} = 2$. The qualitative analysis of the problem for large Re, but in the laminar boundary layer region, shows that Nu is expected to be proportional to $\mathrm{Re}^{1/2}\mathrm{Pr}^{1/3}$ [58]. Thus the general formula for Nu can be written as

$$\mathrm{Nu} = 2 + \beta_c \mathrm{Re}^{1/2}\mathrm{Pr}^{1/3}, \tag{3.49}$$

where the coefficient β_c cannot be derived from this simplified analysis. It should be obtained either from experimental observations or from more rigorous numerical analysis of the basic equations. The most widely used value of this coefficient, supported by experimental observations, is $\beta_c = 0.6$ (see, e.g., [17]). In a number of papers (see, e.g., [6, 128]) the value $\beta_c = 0.552$ is used. Sometimes, the power 1/3 of the Prandtl number is replaced by 0.4 [93].

In the case of stationary droplets heated or cooled due to natural convection, the value of Nu can be estimated as [34, 150]

$$\mathrm{Nu} = 2 + 0.43\ (\mathrm{Gr}\,\mathrm{Pr})^{1/4}, \tag{3.50}$$

where the Grashof number is defined as

$$\mathrm{Gr} = \frac{g\beta_G(T_s)(2R_d)^3|T_g - T_s|}{\left[\nu_g(T_s)\right]^2}, \tag{3.51}$$

$$\beta_G = \frac{\rho_g(T_g) - \rho_g(T_s)}{\rho_g(T_s)[T_s - T_g]}$$

is the coefficient of expansion, approximated for the ideal gas as $\beta_G = 1/T_s$. Formula (3.51) is valid in the range $1 < \mathrm{Gr\,Pr} < 100,000$.

Whitaker [143] drew attention to the fact that in the wake region, for large Re, Nu is expected to be proportional to $\mathrm{Re}^{2/3}\mathrm{Pr}^{1/3}$. Also, the difference in gas dynamic viscosities in the vicinity of the droplet surface (μ_{gs}) and in the bulk flow ($\mu_{g\infty}$) had to be taken into account. Hence, he looked for correlation for Nu in the form

$$\mathrm{Nu} = 2 + \left(a_\mathrm{W}\mathrm{Re}^{1/2} + b_\mathrm{W}\mathrm{Re}^{2/3}\right)\mathrm{Pr}^{1/3}\left(\frac{\mu_{g\infty}}{\mu_{gs}}\right)^{c_\mathrm{W}}, \qquad (3.52)$$

where the constants a_W, b_W, and c_W had to be calculated based on available experimental data. The analysis of these data led to the following result: $a_\mathrm{W} = 0.4$, $b_\mathrm{W} = 0.06$, and $c_\mathrm{W} = 0.25$. Equation (3.52) might be expected to be more accurate than (3.49) (at least due to the larger number of fitting constants), but it has rarely been used in practical applications. Note that the ratio $\mu_{g\infty}/\mu_{gs}$ is expected to be very close to unity.

As noticed in [6, 31], Eq. (3.49) overestimates the heat transfer rate at low Reynolds numbers ($\mathrm{Re} \leq 10$). Also, it and (3.52) predict the physically incorrect supersensitivity of the heat transfer rate to the small-velocity fluctuations near $\mathrm{Re} = 0$, since $\partial\mathrm{Nu}/\partial\mathrm{Re} \to \infty$ when $\mathrm{Re} \to 0$ [128].

As an alternative to (3.49), the following correlation was recommended in [31, 128]:

$$\mathrm{Nu} = 1 + (1 + \mathrm{RePr})^{1/3} f_c(\mathrm{Re}), \qquad (3.53)$$

where $f_c(\mathrm{Re}) = 1$ at $\mathrm{Re} \leq 1$ and $f_c(\mathrm{Re}) = \mathrm{Re}^{0.077}$ at $1 < \mathrm{Re} \leq 400$. Equation (3.53) approximates the numerical results obtained by a number of authors for $0.25 < \mathrm{Pr} < 100$ with an error less than 3 % [128].

Analyses of the process of heating of a spherical body in a flow of fluid were often restricted to the case of small Re when the flow can be considered Stokesian. In this case, the radial and azimuthal components of flow velocity around a spherical solid body can be found from the following formulas [30, 108]:

$$\begin{aligned} u_r = u_0 \Bigg\{ & \frac{1}{2}\left(1 + \frac{3\mathrm{Re}}{16}\right)\left[2 - \frac{3R_d}{R} + \left(\frac{R_d}{R}\right)^3\right]\cos\theta \\ & - \frac{3\mathrm{Re}}{64}\left[2 - \frac{3R_d}{R} + \left(\frac{R_d}{R}\right)^2 - \left(\frac{R_d}{R}\right)^3 + \left(\frac{R_d}{R}\right)^4\right]\left(3\cos^2\theta - 1\right) \\ & + O\left(\mathrm{Re}^2\log\mathrm{Re}\right)\Bigg\}, \end{aligned} \qquad (3.54)$$

$$u_\theta = u_0 \left\{ -\frac{1}{4}\left(1+\frac{3\mathrm{Re}}{16}\right)\left[4-\frac{3R_d}{R}-\left(\frac{R_d}{R}\right)^3\right]\sin\theta \right.$$
$$+\frac{3\mathrm{Re}}{64}\left[4-\frac{3R_d}{R}+\left(\frac{R_d}{R}\right)^3-2\left(\frac{R_d}{R}\right)^4\right]\cos\theta\sin\theta$$
$$\left. +O\left(\mathrm{Re}^2\log\mathrm{Re}\right)\right\}, \tag{3.55}$$

where u_0 is the flow velocity unperturbed by the sphere, u_r and u_θ are the radial (away from the centre of the sphere) and circumferential (angle θ is estimated relative to the direction of the velocity in the unperturbed flow) flow velocity components.

As follows from (3.54) and (3.55), both components of flow velocity are zero at $R = R_d$, as expected (no slip condition).

In the limit Re→ 0, Eqs. (3.54) and (3.55) can be simplified to

$$u_r = u_0\left[1-\frac{3R_d}{2R}+\frac{1}{2}\left(\frac{R_d}{R}\right)^3\right]\cos\theta, \tag{3.56}$$

$$u_\theta = -u_0\left[1-\frac{3R_d}{4R}-\frac{1}{4}\left(\frac{R_d}{R}\right)^3\right]\sin\theta. \tag{3.57}$$

Note that the main advantages of (3.54) and (3.55) compared with (3.56) and (3.57) can be observed at R compared with R_d. In the limit $R \to \infty$, Eqs. (3.56) and (3.57) predict the unperturbed flow velocity.

In the case of droplets in which recirculation is allowed, (3.56) and (3.57) can be generalised to [101, p. 697], [54]

$$u_r = u_0\left[1-\frac{A_v R_d}{R}+2B_v\left(\frac{R_d}{R}\right)^3\right]\cos\theta, \tag{3.58}$$

$$u_\theta = -u_0\left[1-\frac{A_v R_d}{2R}-B_v\left(\frac{R_d}{R}\right)^3\right]\sin\theta, \tag{3.59}$$

where

$$A_v = \frac{3\lambda_v+1}{2(\lambda_v+1)}, \qquad B_v = \frac{\lambda_v}{4(\lambda_v+1)},$$

$\lambda_v = \mu_l/\mu_g$ is the ratio of dynamic viscosities (note that this definition of λ_v is different from the one used in [54, 101] but consistent with the definition of the parameter used later). In the case $\lambda_v = \infty$ (solid body), (3.58) and (3.59) refer to the case $R \geq R_d$, and they reduce to (3.56) and (3.57). Similar equations for the velocities inside the droplets could be derived from equation (21.14.5b) of [101]. These describe the well-known Hill's spherical vortex [101]. A more in-depth analysis of recirculation within a fluid sphere is given in [11].

If the effects of viscosity are ignored, then for both solid and liquid moving spheres, the components of velocity can be represented in particularly simple forms valid for $R \geq R_d$ (see [101, p. 562]):

$$u_r = u_0 \left[1 - \left(\frac{R_d}{R} \right)^3 \right] \cos\theta, \tag{3.60}$$

$$u_\theta = -u_0 \left[1 + \frac{1}{2} \left(\frac{R_d}{R} \right)^3 \right] \sin\theta. \tag{3.61}$$

The values of the components of fluid velocity predicted by the above-mentioned systems of equations are substituted into (3.48). All these systems of equations have actually been used in analyses of steady-state and transient heating of droplets or, more generally, spherical bodies. These are reviewed below.

Equations (3.56) and (3.57) were used in [8] for the asymptotic analysis of the process of heating a spherical body in a flow of fluid. The authors of this paper assumed that both Reynolds number and Peclet number ($\mathrm{Pe} \equiv \mathrm{Re} \cdot \mathrm{Pr}$) are small, and they used the technique originally developed in [108], which led to the derivation of (3.54) and (3.55). Their final formula was represented in the form

$$\mathrm{Nu} = 2 \left(1 + \frac{1}{4}\mathrm{Pe} + \frac{1}{8}\mathrm{Pe}^2 \ln \mathrm{Pe} + 0.01702\mathrm{Pe}^2 + \frac{1}{32}\mathrm{Pe}^3 \ln \mathrm{Pe} + O(\mathrm{Pe}^3) \right). \tag{3.62}$$

As follows from this equation, $\mathrm{Nu} \to 2$ when $\mathrm{Pe} \to 0$. Although this equation was derived under the assumption that $\mathrm{Re} \ll 1$ and $\mathrm{Pe} < 1$, it can still be applied for Re and Pe in the range from 0 to 0.7 [93]. Sometimes, Re is defined based on R_d and not $2R_d$ (as in our case). This led to a different form of (3.62) [112].

Using (3.54) and (3.55) but assuming that $\mathrm{Pe} \to \infty$, an alternative expression for Nu was derived in [8]:

$$\mathrm{Nu} = 0.991\mathrm{Pe}^{1/3} \left(1 + \frac{1}{16}\mathrm{Re} + \frac{3}{160}\mathrm{Re}^2 \ln \mathrm{Re} + O(\mathrm{Re}^2) \right). \tag{3.63}$$

A generalisation of the analysis by [8] for the case $\mathrm{Re} < 1$ and $\mathrm{Pe} < 1$ by taking extra terms in the velocity field (that is, replacing (3.56) and (3.57) with (3.54) and (3.55)) was reported in [112]. Generalizations of this equation to the case of bodies of arbitrary shape were reported in [7, 12, 21].

Based on the numerical study of the transient heat transfer from a sphere at high Reynolds and Peclet numbers, the following steady-state correlation was suggested in [56]:

$$\mathrm{Nu} = 0.922 + \mathrm{Pe}^{1/3} + 0.1\mathrm{Re}^{1/3}\mathrm{Pe}^{1/3}. \tag{3.64}$$

If Re = 0, then (3.64) reduces to the solution by Acrivos [7]. Also, the prediction of this equation is close to that of (3.63) at large Pe but relatively small Re.

In [92], a mathematical model was developed to describe the heat transfer process when a melting sphere is immersed in a moving fluid. Based on this model, the following correlation for Nu was obtained:

$$\mathrm{Nu} = 2 + 0.47\mathrm{Re}^{0.5}\mathrm{Pr}^{0.36}, \tag{3.65}$$

where

$$0.003 \leq \mathrm{Pr} \leq 10, \qquad 10^2 \leq \mathrm{Re} \leq 5 \times 10^4.$$

This model was validated against various experimental results involving metals and water. The value of Nu predicted by (3.65) is reasonably close to the value predicted by (3.49).

None of (3.62)–(3.65) take into account recirculation inside the droplets. The latter could be accounted for based on asymptotic or rigorous numerical analysis of coupled fluid dynamics and heat transfer equations. Levich [84] (see p. 408) was perhaps the first to provide an asymptotic solution for small Re but large Pe in the form (see also [93])

$$\mathrm{Nu} = \sqrt{\frac{4}{3\pi}\frac{1}{1+\lambda_v}\mathrm{Pe}}. \tag{3.66}$$

Since this formula is valid for very large Pr, its practical applicability to the analysis of droplet heating in a gas is expected to be limited.

Earlier studies in this direction were reviewed by Feng and Michaelides [57]. Also, in that paper, a comprehensive numerical analysis of heat and mass transfer coefficients of viscous moving spheres (droplets) was reported. Their analysis was based on a number of conventional assumptions. Namely, it was assumed that droplets retain their spherical forms and there are no temperature gradients along the droplet surfaces. In what follows, their main results are summarised.

One of the aims of [57] was to investigate the dependence of Nu on Re, Pe $\equiv$ Re$\cdot$Pr, $\lambda_v = \mu_l/\mu_g$, and ρ_l/ρ_g, where $\mu_{l(g)}$ and $\rho_{l(g)}$ are dynamic viscosities and densities of liquid (gas). As follows from their analysis, the dependence of Nu on ρ_l/ρ_g was negligibly small for ρ_l/ρ_g in the range 0.1–10 (bubbles and droplets), and the range of Pe from 1 to 500, Re= 10, and $\lambda_v = 1$. It was not expected to be important for other ranges and values of parameters. Hence the analysis was focused on a study of the dependence of Nu on Re, Pe, and λ_v. The results were presented in a table, covering the range of Re from 0 to 500, the range of Pe from 1 to 1000, and of λ_v from 0 to ∞. Based on the information presented in this table, the following working correlations were developed [57].

At small but finite Re ($0 < \mathrm{Re} < 1$) and $10 \leq \mathrm{Pe} \leq 1000$, the following expression for Nu was derived [57]:

$$\mathrm{Nu}(\lambda_v, \mathrm{Pe}, \mathrm{Re}) = \left[\frac{0.651}{1+0.95\lambda_v}\mathrm{Pe}^{1/2} + \frac{0.991\lambda_v}{1+\lambda_v}\mathrm{Pe}^{1/3}\right][1+\alpha_v(\mathrm{Re})] + \left[\frac{1.65(1-\alpha_v(\mathrm{Re}))}{1+0.95\lambda_v} + \frac{\lambda_v}{1+\lambda_v}\right], \tag{3.67}$$

where

$$\alpha_v(\mathrm{Re}) = \frac{0.61\mathrm{Re}}{\mathrm{Re}+21} + 0.032.$$

Note that in the limit $\mathrm{Re} \to 0$ and $\lambda_v \to \infty$, Nu given by expression (3.67) is not reduced to 2, as predicted for stationary droplets. In this limit, Pr should be infinitely large to satisfy the condition that $\mathrm{Pe} > 10$. At $1 \le \mathrm{Re} \le 500$ and $10 \le \mathrm{Pe} \le 1000$, the following expressions for Nu were suggested [57]:

$$\mathrm{Nu}(\lambda_v, \mathrm{Pe}, \mathrm{Re}) = \frac{2-\lambda_v}{2}\mathrm{Nu}(0, \mathrm{Pe}, \mathrm{Re}) + \frac{4\lambda_v}{6+\lambda_v}\mathrm{Nu}(2, \mathrm{Pe}, \mathrm{Re}) \tag{3.68}$$

for $0 \le \lambda_v \le 2$, and

$$\mathrm{Nu}(\lambda_v, \mathrm{Pe}, \mathrm{Re}) = \frac{4}{2+\lambda_v}\mathrm{Nu}(2, \mathrm{Pe}, \mathrm{Re}) + \frac{\lambda_v - 2}{\lambda_v + 2}\mathrm{Nu}(\infty, \mathrm{Pe}, \mathrm{Re}) \tag{3.69}$$

for $2 \le \lambda_v \le \infty$, where

$$\mathrm{Nu}(0, \mathrm{Pe}, \mathrm{Re}) = 0.651\mathrm{Pe}^{1/2}\left[1.032 + \frac{0.61\mathrm{Re}}{\mathrm{Re}+21}\right] + \left[1.60 + \frac{0.61\mathrm{Re}}{\mathrm{Re}+21}\right],$$

$$\mathrm{Nu}(2, \mathrm{Pe}, \mathrm{Re}) = 0.64\mathrm{Pe}^{0.43}\left[1 + 0.233\mathrm{Re}^{0.287}\right] + 1.41 - 0.15\mathrm{Re}^{0.287},$$

$$\mathrm{Nu}(\infty, \mathrm{Pe}, \mathrm{Re}) = 0.852\mathrm{Pe}^{1/3}\left[1 + 0.233\mathrm{Re}^{0.287}\right] + 1.3 - 0.182\mathrm{Re}^{0.355}.$$

These equations could be potentially incorporated into any CFD code and used in engineering applications.

A more recent and detailed analysis of heat and mass exchange between a moving sphere and ambient fluid is presented in [14]. While the analysis in the latter paper was formulated in terms of mass transfer, it is also applicable to heat transfer, with concentration replaced by temperature and the Sherwood number by the Nusselt number. The authors of that paper investigated the impact of viscosity on the steady-state mass transfer from a sphere at low Peclet numbers, based on singular perturbation techniques, to find the approximate asymptotic solution for the temperature profiles. They provided simple analytical expressions that can be used to calculate the local and average Nusselt numbers.

3.1.2.2 Transient Heating

The complexity of the problem of transient heating of moving droplets lies in the fact that variations in both temperature and flow velocity need to be accounted for in the general case. This is something that is performed in most commercial CFD codes. This analysis, however, is usually case-dependent, and the results might have limited general application. At the same time, in many applications, the characteristic times of variation of flow parameters are much longer than the characteristic times of droplet heating. This allows us to consider the problem of droplet heating assuming that the flow is fixed. This assumption is almost universally used in the analysis of this problem. In the case of small Re, we can further assume that flow is Stokesian, with the velocities described by (3.54) and (3.55), or their simplified versions (3.56) and (3.57).

Any effects of droplet oscillations on the heating process are generally ignored. This could be justified when the droplet Weber number $\mathrm{We} = 2u_0 R_d \rho_l / \sigma_s$ (where σ_s is the interfacial surface tension) is <4 [144]. This result was obtained in [144] based on experimental studies of falling droplets at Re in the range 138–971.

Most approaches to investigating the problem of transient heating of moving droplets performed so far can be presented in two groups. The first approach is based on the assumption that the temperature of the droplet surface remains constant throughout the whole heating process. The second approach allows for changes of droplet temperature with time. In what follows, these two approaches are considered separately.

Fixed surface temperature

Choudhoury and Drake [30] and Konopliv and Sparrow [79] were perhaps the first to solve the problem of transient heating of a solid sphere at small Reynolds and Peclet numbers, assuming that the surface temperature of the sphere is fixed. In the first paper, the steady-state velocity distribution in fluid described by (3.54) and (3.55) was used, and it was assumed that initially, the temperatures of the sphere and surrounding fluid were equal. Then at $t = 0$, the temperature of the sphere abruptly changed to another constant value. The solution obtained for small t in the limit of very small Re reduced to that predicted by (3.27). In the limit of large t, the solution predicts values of Nu close to 2. This is as expected, since this solution was based on the assumption of small Re and Pe.

Konopliv and Sparrow [80] generalised the analysis of their earlier paper [79] to the case of large Pe. As in [79], the analysis was based on (3.56) and (3.57). They drew attention to the fact that for high Pe, the temperature field is confined to a boundary layer which is very thin compared with the velocity field boundary layer. This allowed these authors to simplify both the temperature Eq. (3.48) (assuming that heat conduction takes place in the radial direction only) and Eqs. (3.56) and (3.57) (assuming that R is close to R_d). Two methods were employed to obtain the solution to the simplified temperature equation. One method involved numerical inversion of an integral transform, while the second was based on an asymptotic expansion in the Laplace transform s-plane. For small t, both methods predicted similar results, while for larger t, the first method was recommended. In the limit of large times, the results of [80] reduced to those predicted by (3.63) with Re = 0.

A similar problem, but for the heating of a liquid droplet, was considered in [54]. These authors assumed that both Re and Pe are small, based their analysis on (3.58) and (3.59), and obtained analytical expressions for Nu, valid in short and long time limits. In the short time limit, their expression reduced to (3.27). In the long time limit ($\mathrm{Fo} = O(\mathrm{Pe}^{-2})$), they obtained the following expression:

$$\mathrm{Nu} = 2\left\{1 + \frac{\mathrm{Pe}}{2}\left[\frac{1}{2}\mathrm{erf}\left(\mathrm{Pe}\frac{\sqrt{\mathrm{Fo}}}{4}\right) + \frac{2}{\mathrm{Pe}\sqrt{\pi \mathrm{Fo}}}\exp\left(-\frac{\mathrm{Pe}^2\mathrm{Fo}}{16}\right)\right] + \frac{3+2\lambda_v}{24(1+\lambda_v)}\mathrm{Pe}^2\ln\left(\frac{\mathrm{Pe}}{2}\right)\right\} + O(\mathrm{Pe}^{1+}), \tag{3.70}$$

where Fo is the Fourier number introduced in (3.5) and (3.6).

In [56], it was shown that the results predicted by (3.70) for a solid sphere, Pe = 0.5, and long times differ by less than 3 % from the results predicted by numerical analyses for Stokesian flow.

In the same paper [56], it was shown that at sufficiently small times, the predicted values of Nu practically do not depend on Re. This dependence was shown to be rather weak for relatively small Pe (Pe = 2) at all times, but significant for Pe = 200 at large times.

The models described so far have focused predominantly on spherical droplets, while the actually observed droplets in sprays are typically far from being spherical [33, 85]. Several papers have considered the problem of transient heat transfer between aspherical bodies and ambient fluid. In [55], a model for the heat transfer processes from an oblate and a prolate spheroid at the limit of very small Pe was developed. Other approaches to this and similar problems are discussed in [53, 70, 75, 77, 99, 131]. In [105], a more general problem of heat transfer from a particle with an arbitrary shape suspended in a fluid was considered.

Variable surface temperature

Chao [26] was perhaps the first to solve the problem of conjugate heat transfer between a translating droplet and ambient gas. He took into account the recirculation inside a droplet, but ignored the effects of viscosity. This allowed him to approximate the velocity profile outside the droplet by (3.60) and (3.61). A similar set of equations for the internal flow (Hill vortex) was used. This approximation is expected to lead to substantial errors at small Re (Stokesian flow), but could be justified for Re > 200. Also, the assumption concerning a thin thermal boundary layer was made, which could be justified for large Pe. This allowed further simplification of (3.60) and (3.61), assuming that $|R - R_d| \ll R_d$ and using (3.48) (assuming that heat conduction takes place in the radial direction only (cf. [80])). The simplified temperature equation was solved outside and inside the droplet, assuming that the temperature and heat fluxes on both sides of the interface are equal. The solution for the average value of the Nusselt number was presented in an explicit analytical form. As in the case of stationary droplets, this solution predicted large Nu for small t. For large times, it asymptotically approached a steady-state solution in the form

$$\mathrm{Nu} = \frac{2\mathrm{Pe}^{1/2}}{\sqrt{\pi}(1+\beta_v)}, \tag{3.71}$$

where $\beta_v = \sqrt{k_g \rho_g c_{pg}/(k_l \rho_l c_l)}$. In the case of fuel droplets moving in gas, β_v is expected to be small.

Chao and Chen [28] solved the same problem as in [26] using a different analytical technique originally developed in [27]. The results were almost identical to those predicted by the analytical solution [26]. The previously mentioned results obtained in [80] for a sphere with fixed temperature were generalised in the same paper to the case that sphere heating is taken into account, provided that the thermal conductivity of this sphere is infinitely large.

Abramzon and Borde [2] presented the results of numerical analysis of transient heat transfer to slowly moving droplets (Stokesian approximation) in a wide range of Peclet numbers $0 \leq \mathrm{Pe} \leq 1000$. They studied the conjugate problem, so that the heat transfer processes in gas and droplets were taken into account. They used the flow velocity distribution predicted by (3.58) and (3.59) (note a typographical error in their equation (2)). It was shown that the droplet heating process can be divided into two stages. In the first stage, the heat is transferred primarily by conduction, and the isotherms have almost spherical shapes. In the second stage, convective heat transfer predominates. During this latter stage, the internal isotherms were shown to be close to streamlines. At high Pe, Nu was shown to oscillate in time about its asymptotic value. The fact that initially, the droplet is heated predominantly by conduction agrees with the result reported in [56], where it was shown that Nu practically does not depend on Re at the initial stage of heating (see the discussion in the previous subsection). It was shown that for large Pe ($100 \leq \mathrm{Pe} \leq 1000$), the heating can be regarded as quasistationary after a short initial time lapse. During this initial time period, the analytical solution of [26] was shown to give satisfactory results. A detailed comparison of the asymptotic values of Nu, predicted by the numerical analysis, and previously reported results of asymptotic analyses were presented for $\lambda_v = 1$ (fluids with similar hydrodynamic properties) and $\lambda_v = \infty$ (solids).

Results of further analysis of transient heat transfer between an ambient fluid and a single sphere were reported in [3]. This analysis was restricted to solid spheres, but a wide range of Pe ($1 \leq \mathrm{Pe} \leq 10000$) and the ratios of volume heat capacities of the fluid to that of a sphere in the range $0 \leq (\rho c)_{12} \equiv \rho_g c_{pg}/(\rho_l c_l) \leq 2$ were considered. The droplet thermal conductivity was assumed to be large, so that the temperature gradients inside the sphere could be ignored. The flow field was assumed to be Stokesian, and the velocity distribution outside the sphere was taken in the form given by (3.56) and (3.57). It was pointed out that for $(\rho c)_{12} = 0$ and Pe close to zero, the values of Nu asymptotically approach 2, as predicted by (3.27). However, for $(\rho c)_{12} \geq 0.2$ and the same Pe, the asymptotic values of Nu were shown to be close to zero. This is consistent with the prediction of Fig. 3.4. Note that this figure is presented for $(\rho c)_{12} = 0.0156$, for which the calculations were not performed in [3]. The author anticipates that one of the main conclusions made in this paper that 'at $(\rho c)_{12} \geq 0.2$ the asymptotic values of the Nusselt number appears to be considerably less than the corresponding steady-state value for the case of constant sphere temperatures' could be generalised to any $(\rho c)_{12} > 0$. In [74], the analysis presented in [3] was generalised to take into account the flow recirculation in droplets, although the range of Pe was reduced to $10 \leq \mathrm{Pe} \leq 1000$. Whenever the results of [3] and [74] have been compared, the difference between them has been less than 1%.

The energy equation of a rigid sphere of infinitely large thermal conductivity in a viscous fluid subject to an unsteady flow and temperature field under the assumption of small Pe was developed in [94]. The values of the sphere's temperature were shown to depend on instantaneous values of fluid temperature and the 'histories' of both fluid and sphere temperatures. These histories were accounted for via the so-called history integral, which, in the energy equation, is analogous to the history force (Basset force) in the equation for the motion of a sphere. The presence of the history integral effectively turns the differential equation for the sphere temperature into an integrodifferential equation, the analysis of which can be computationally intensive. The authors of [59] presented a more detailed analysis of the effect of the history term on the transient energy equation of a sphere. More specifically, they attempted to clarify when this term can be ignored if a predetermined calculation error can be tolerated. They considered the transient heat transfer cases related to three different fluid temperature processes: step temperature change, ramp temperature change, and sinusoidal temperature change. It was shown that the importance of the history term is mainly controlled by the above-mentioned parameter $(\rho c)_{12}$. In the case $0.002 \leq (\rho c)_{12} \leq 0.2$ (the range of typical values for fuel droplets in air), this term is important and needs to be retained. More detailed analysis of the history term is given in [93].

Talley and Yao [134] drew attention to the fact that heat transfer to a moving droplet can still be correctly predicted by (3.1), with appropriate boundary conditions, if the liquid thermal conductivity k_l is replaced by the so-called effective thermal conductivity k_{eff}. This approach was further developed by Abramzon and Sirignano [6], who defined k_{eff} by the following equation:

$$k_{\mathrm{eff}} = \chi_T k_l, \tag{3.72}$$

where the coefficient χ_T varies from about 1 (at droplet Peclet number $\mathrm{Pe}_{d(l)} = \mathrm{Re}_{d(l)}\mathrm{Pr}_l < 10$) to 2.72 (at $\mathrm{Pe}_{d(l)} > 500$) and can be approximated as

$$\chi_T = 1.86 + 0.86 \tanh\left[2.225 \log_{10}\left(\mathrm{Pe}_{d(l)}/30\right)\right]. \tag{3.73}$$

Liquid fuel transport properties were used for calculating $\mathrm{Pe}_{d(l)}$, in which liquid transport properties and the maximum surface velocity inside droplets were used. The latter velocity was calculated as [6]

$$U_s = \frac{1}{32} \Delta U \left(\frac{\mu_g}{\mu_l} \right) \mathrm{Re}_d C_F, \tag{3.74}$$

where $\Delta U \equiv U_g - U_d$ is the relative velocity between ambient gas and droplets, $\mu_{g(l)}$ is the dynamic viscosity of gas (liquid), Re_d is the droplet Reynolds number based on the droplet diameter (denoted by Re_∞ in [6]), C_F is the friction drag coefficient [6]. This model can predict reasonably accurately (especially at small and large $\mathrm{Re}_{d(l)}$) the droplet average surface temperature, but not the distribution of temperature inside droplets. In our case, however, we are primarily interested in the prediction of the surface temperature, which controls droplet evaporation. Hence, the applicability of this model can be justified. It is widely known as the 'effective thermal conductivity (ETC) model'. The simplicity of this model makes it particularly attractive for application in CFD codes. For example, it allows the application of solution (3.37) to the case of moving droplets by replacing k_l by k_{eff}, assuming that $h = \mathrm{const}$.

More recent studies of the problem of transient heating of moving droplets were more case-specific, and in most cases they were linked with evaporation processes. They will be reviewed later, in Chap. 4.

3.2 Radiative Heating

3.2.1 Basic Equations and Approximations

Since thermal radiation propagates with a velocity close to the velocity of light, droplet motion does not influence the process of radiative heating. Hence, the analysis in this section is equally applicable to stationary and moving droplets.

The simplest model for radiative heating of droplets can be based on the assumption that droplets are opaque grey spheres, characterised by emissivity ε. In this case, the effect of radiative heating of droplets could be described by replacing the heat flux given by (3.28) with the following equation:

$$|\dot{q}| = h(T_{g\infty} - T_s) + \sigma\epsilon \left(\theta_R^4 - T_s^4 \right), \tag{3.75}$$

where θ_R is the so-called radiative temperature, σ is the Stefan–Boltzmann constant. For optically thick gas, θ_R can be identified with the ambient gas temperature $T_{g\infty}$, while for optically thin gas, it can be identified with the external temperature T_{ext} (e.g., temperature of a remote flame) [98, 127]. The simplicity of this model makes it particularly attractive for applications, including application in CFD codes (see, e.g., [91, 113, 114, 117, 129, 133, 141, 147, 153]). The value of ϵ in this equation could be considered a free parameter, which could be specified based on more rigorous calculations or obtained from the analysis of experimental data.

Assuming that there are no temperature gradients inside droplets and $\theta_R = T_{\mathrm{ext}}$, Eq. (3.75) allows us to generalise (3.34) and its solution (3.35) to account for the effect of thermal radiation by replacing $T_{g\infty}$ with

$$T_{g\infty} + \frac{\sigma \epsilon \left(T_{\mathrm{ext}}^4 - T_s^4 \right)}{h}.$$

In a similar way, radiative heating could be accounted for in solution (3.37). In both cases, it is implicitly assumed that the external radiation is that of a grey body.

In most practically important cases of fuel droplet heating, $T_s \ll T_{\mathrm{ext}}$, and the contribution of the terms proportional to T_s^4 can be safely ignored. If this is not the case (T_{ext} is of order less than T_s), then T_s is still below the boiling temperature of fuel (about 600 K), and the contribution of radiative heating can almost always be ignored altogether.

The approach based on the assumption that droplets are opaque grey spheres, however, overlooks the fact that droplet radiative heating takes place not at their surface (as in the case of convective heating) but via the absorption of thermal radiation penetrating inside the droplets. Thus the droplets should be considered semitransparent rather than grey opaque bodies [4, 5, 40, 43, 45–49, 60, 82, 87, 95, 96, 121, 138]. The focus of this section will be on analysis of the models of thermal radiation absorption in these semitransparent droplets, without taking into account the contribution of radiation from the droplets (see [87, 95, 96] for an analysis of the latter effect). Also, the analysis of thermal radiation propagation in ambient nongrey gas or gas–droplet mixture is beyond the scope of this book (see, e.g., [137, 149]).

Equation (3.1) for semitransparent droplets in the presence of thermal radiation can be generalised to

$$\frac{\partial T}{\partial t} = \kappa \left(\frac{\partial^2 T}{\partial R^2} + \frac{2}{R} \frac{\partial T}{\partial R} \right) + P(R), \tag{3.76}$$

where $P(R)$ accounts for the radiative heating of droplets. This equation is to be solved subject to the same boundary conditions (3.36) as before. Spherical symmetry of the problem is assumed. Note that the units of $P(R)$ are K/s.

The solution to (3.76) is a straightforward generalisation of solution (3.37) and can be written as [122] (see Appendix A):

$$T(R,t) = \frac{R_d}{R}\sum_{n=1}^{\infty}\left\{\frac{p_n}{\kappa_R\lambda_n^2} + \exp\left[-\kappa_R\lambda_n^2 t\right]\left(q_n - \frac{p_n}{\kappa_R\lambda_n^2}\right)\right.$$
$$- \frac{\sin\lambda_n}{\| v_n \|^2 \lambda_n^2}\mu_0(0)\exp\left[-\kappa_R\lambda_n^2 t\right]$$
$$\left. - \frac{\sin\lambda_n}{\| v_n \|^2 \lambda_n^2}\int_0^t \frac{\mathrm{d}\mu_0(\tau)}{\mathrm{d}\tau}\exp\left[-\kappa_R\lambda_n^2(t-\tau)\right]\mathrm{d}\tau\right\}$$
$$\times \sin\left[\lambda_n\left(\frac{R}{R_d}\right)\right] + T_g(t), \tag{3.77}$$

where

$$p_n = \frac{1}{R_d^2 \| v_n \|^2}\int_0^{R_d}(RP(R))\sin(\lambda_n R/R_d)\,\mathrm{d}R.$$

Other notation is the same as in (3.37).

If we ignore the effect of a temperature gradient inside droplets, then (3.34) for semitransparent droplets in the presence of thermal radiation can be generalised to

$$\frac{4}{3}\pi R_d^3\rho_l c_l\frac{\mathrm{d}T_d}{\mathrm{d}t} = 4\pi R_d^2 h(T_{g\infty} - T_d) + \rho_l c_l P_{\mathrm{total}}, \tag{3.78}$$

where

$$P_{\mathrm{total}} = 4\pi\int_0^{R_d} P(R)R^2\mathrm{d}R \tag{3.79}$$

is the total amount of radiation absorbed in a droplet.

The practical application of (3.77) and (3.78) is complicated by the fact that we do not know the value of $P(R)$. Estimates of this function, using various approximations, will be one of the main focuses of the rest of this section.

3.2.2 Mie Theory

The most rigorous approach to the calculation of $P(R)$ is based on the solution of the Maxwell equations, with boundary conditions at the droplet's surface. These boundary conditions are continuity of the normal component of the wave electric field and the jump in its tangential components controlled by the complex index of refraction of the liquid [18, 140]. This is the well-known Mie theory. In the case of interaction between a plane linearly polarised electromagnetic wave and a semitransparent sphere (droplet), this theory gives the following relations for the complex amplitudes of the components of the wave electric field inside the droplet [18, 39, 140, 145]:

$$E_r = \frac{E_0 \cos\varphi}{m_\lambda^2 \rho_\lambda^2} \sum_{k=1}^{\infty} \mathrm{i}^k (2k+1) d_k \psi_k(m_\lambda \rho_\lambda) P_k^1(\mu_\theta), \tag{3.80}$$

$$E_\theta = \frac{E_0 \cos\varphi}{m_\lambda \rho_\lambda} \sum_{k=1}^{\infty} \frac{\mathrm{i}^k (2k+1)}{k(k+1)} \left[c_k \psi_k(m_\lambda \rho_\lambda) \pi_k(\mu_\theta) - \mathrm{i} d_k \psi_k'(m_\lambda \rho_\lambda) \tau_k(\mu_\theta) \right], \tag{3.81}$$

$$E_\varphi = \frac{E_0 \sin\varphi}{m_\lambda \rho_\lambda} \sum_{k=1}^{\infty} \frac{\mathrm{i}^{k+1} (2k+1)}{k(k+1)} \left[\mathrm{i} c_k \psi_k(m_\lambda \rho_\lambda) \tau_k(\mu_\theta) + d_k \psi_k'(m_\lambda \rho_\lambda) \pi_k(\mu_\theta) \right], \tag{3.82}$$

where

$$c_k = \frac{m_\lambda \mathrm{i}}{m_\lambda \zeta_k(x_\lambda) \psi_k'(m_\lambda x_\lambda) - \zeta_k'(x_\lambda) \psi_k(m_\lambda x_\lambda)},$$

$$d_k = \frac{m_\lambda \mathrm{i}}{\zeta_k(x_\lambda) \psi_k'(m_\lambda x_\lambda) - m_\lambda \zeta_k'(x_\lambda) \psi_k(m_\lambda x_\lambda)},$$

$$\pi_k(\mu_\theta) = \frac{P_k^1(\mu_\theta)}{\sqrt{1-\mu_\theta^2}}, \qquad \tau_k(\mu_\theta) = -\sqrt{1-\mu_\theta^2} \frac{\mathrm{d} P_k^1(\mu_\theta)}{\mathrm{d}\mu_\theta},$$

$m_\lambda = n_\lambda - \mathrm{i}\kappa_\lambda$ is the complex index of refraction, $\rho_\lambda = 2\pi R/\lambda = R x_\lambda / R_d$ (x_λ is size parameter defined as $x_\lambda = 2\pi R_d/\lambda$), $\mu_\theta = \cos\theta$, the prime denotes differentiation with respect to the argument, $\theta = 0$ is the direction of wave propagation, φ is the azimuthal angle measured from the plane of electric field oscillations of the incident wave (as viewed along the direction of wave propagation), E_0 is the amplitude of the electric field in the incident wave, ψ_k and ζ_k are the Riccati–Bessel functions defined as [39]

$$\psi_k(z) = \left(\frac{\pi z}{2}\right)^{1/2} J_{k+\frac{1}{2}}(z), \qquad \zeta_k(z) = \left(\frac{\pi z}{2}\right)^{1/2} H_{k+\frac{1}{2}}^{(2)}(z),$$

$J_{k+\frac{1}{2}}$ are the Bessel functions, $H_{k+\frac{1}{2}}^{(2)}$ are the Hankel functions of the second kind, $P_k^1(\mu_\theta)$ are the associated Legendre polynomials. Note that there is a typographical error in the expression for E_φ given in [40] (cf. equation (5)). Expressions for E_r, E_θ, and E_φ given in [82] differ from those given above by the signs and the powers of i and signs of κ_λ. This difference is related to the fact that we follow the definitions of complex amplitudes and the azimuthal angle given in [106, 140], while Lage and Rangel [82] followed the definitions given in [18, 51]. This obviously does not influence the values of $|E_r|$, $|E_\theta|$, and $|E_\varphi|$ used in the following analysis. Equations similar to (3.80)–(3.82) could be obtained for the components of the wave magnetic field [51], but these will not be needed for our analysis.

The distribution of thermal radiation power density absorbed inside the droplet for a given wavelength of incident radiation can be found from the equation [51]

$$p_\lambda(R,\mu_\theta,\varphi) = \frac{4\pi n_\lambda \kappa_\lambda}{\lambda} I_\lambda^{\mathrm{ext}} \frac{|E_r|^2 + |E_\theta|^2 + |E_\varphi|^2}{|E_0|^2}, \tag{3.83}$$

where I_λ^{ext} is the intensity of external thermal radiation in a given direction. This equation was derived under the assumption that magnetic permeability of liquid is equal to 1 (this is a natural assumption for fuel droplets).

In the case of unpolarised external thermal radiation coming from one direction $\theta = 0$, we should set $\varphi = \pi/4$ on the right-hand side of (3.83) [51]. If the illumination of the droplet by unpolarised thermal radiation is spherically symmetric, then the radial profile of the absorbed power can be found via integration of (3.83) over all angles φ and θ, i.e., from the equation [46, 89, 106, 139]

$$p_\lambda(R) = 2\pi \int_{-1}^{1} p_\lambda(R,\mu_\theta,\varphi)\mathrm{d}\mu_\theta = \frac{16\pi^2 n_\lambda \kappa_\lambda}{\lambda} I_\lambda^{\mathrm{ext}} S(R), \tag{3.84}$$

where

$$\begin{aligned} S(R) =& \frac{1}{2|m_\lambda|^4 \rho_\lambda^4} \sum_{k=1}^{\infty} (2k+1) \Big[k(k+1)\, |d_k \psi_k(m_\lambda \rho_\lambda)|^2 \\ &+ |m_\lambda|^2 \rho_\lambda^2 \left(|c_k \psi_k(m_\lambda \rho_\lambda)|^2 + \left| d_k \psi_k'(m_\lambda \rho_\lambda) \right|^2 \right) \Big]. \end{aligned} \tag{3.85}$$

The required value of $P(R)$ can be obtained from (3.84) via integration of the whole spectrum of thermal radiation:

$$P(R) = \frac{1}{\rho_l c_l} \int_0^\infty p_\lambda(R)\mathrm{d}\lambda, \tag{3.86}$$

where the division of the integral by $\rho_l c_l$ converts the units of the radiation power absorbed in droplets from W/m^3 to K/s. The total power absorbed inside droplets ($\rho_l c_l P_{\mathrm{total}}$) is calculated from (3.79).

The calculations of $P(R)$ from (3.86) and P_{total} from (3.79) require knowledge of I_λ^0, the spectral index of absorption, and the index of refraction. If we assume that the external radiation is that of a black body (as was done in deriving (3.75)), then we can write [98, 127]

$$I_\lambda^{\mathrm{ext}} = B_\lambda(T_{\mathrm{ext}}) \equiv \frac{C_1}{\pi\lambda^5 \left[\exp\left(C_2/(\lambda T_{\mathrm{ext}})\right) - 1\right]}, \tag{3.87}$$

$$C_1 = 3.742 \times 10^8 \ \frac{\mathrm{W}\mu\mathrm{m}^4}{\mathrm{m}^2}, \qquad C_2 = 1.439 \times 10^4 \ \ \mu\mathrm{m}\cdot\mathrm{K},$$

where λ is the wavelength in μm. The function $B_\lambda(T_{\mathrm{ext}})$ is known as the Planck function. In the case of optically thick gas, the external temperature T_{ext} responsible for radiative heating should be replaced with the ambient gas temperature.

The index of absorption κ_λ can only be measured experimentally (see, e.g., [10]) or calculated on the basis of measured values of spectral reflectance at near normal incidence [102]. Examples of such measurements for Diesel fuels are presented in a number of papers including [48, 49, 121, 122, 124]. In all cases, the dependence of κ_λ on the type of Diesel fuel was shown to be noticeable, especially in the ranges of semitransparency $\lambda < 3$ μm and 4 μm $< \lambda < 6$ μm. Peaks of absorption for all types of Diesel fuel practically coincided.

The results of measurements of κ_λ could allow us to calculate the spectral distribution of the index of refraction n_λ, using Kramers–Kronig equations [98]. This, however, is expected to lead to substantial errors due to the limited range of wavelengths for which the measurements of κ_λ are usually performed. This problem can be overcome if we use subtractive Kramers-Kronig analysis, based on the measurements of n_λ at one particular wavelength, and its calculation for other wavelengths [9, 49]. For Diesel fuels, it was shown that the results of these measurements and calculations can be accurately approximated by the following relationship [47]:

$$n_\lambda = n_0 + 0.02\frac{\lambda - \lambda_m}{(\lambda - \lambda_m)^2 + 0.001}, \tag{3.88}$$

where $n_0 = 1.46$, $\lambda_m = 3.4$ μm, λ is the wavelength in μm. For practical calculations of κ_λ, the dependence of n_λ on λ can be ignored and n_λ can be made equal to 1.46 [49].

Formally, (3.86), combined with the measured values of κ_λ and measured or calculated values of n_λ, gives the required input parameters for (3.77) and (3.80) describing droplet heating in the presence of thermal radiation. Direct application of (3.86) and (3.79) is limited by the complexity of relevant calculations, and their incorporation into CFD codes is not feasible. Simpler but practically important models are considered in the next sections.

3.2.3 Integral Absorption of Radiation in Droplets

A convenient and widely used approach to characterise the integral absorption of thermal radiation in droplets is based on the application of the efficiency factor of absorption Q_a. This is defined as the ratio of radiation power absorbed in a droplet to the power of thermal radiation illuminating this droplet. In view of (3.84), an explicit expression for Q_a can be represented as [46]

$$Q_a = \frac{P_{\lambda(\mathrm{abs})}}{P_{\lambda(\mathrm{ins})}} = \frac{\int_0^{R_d} 4\pi R^2 p_\lambda(R)\mathrm{d}R}{4\pi^2 R_d^2 I_\lambda^0} = \frac{8 n_\lambda \kappa_\lambda}{x_\lambda^2}\int_0^{x_\lambda} S(\rho_\lambda)\rho_\lambda^2 \mathrm{d}\rho_\lambda, \tag{3.89}$$

where S is defined by (3.85).

As follows from (3.89) and (3.85), for a given value of λ, Q_a depends on n_λ, κ_λ, and R_d. Based on the fact that n_λ remains almost constant over the whole range of λ under consideration, $n \equiv n_\lambda \approx 1.46$, it was shown that Q_a is most sensitive towards the optical thickness of droplets $\tau_0 = 2\kappa_\lambda x_\lambda = a_\lambda R_d$, where a_λ is the absorption coefficient [48]. The results of calculations based on (3.89) were approximated by a particularly simple formula [41]:

$$Q_a = \frac{4n}{(n+1)^2}\left[1 - \exp(-2\tau_0)\right], \tag{3.90}$$

where $\tau_0 = a_\lambda R_d$ (hereinafter, the subscript $_\lambda$ for n is omitted). This is a slight improvement of the formula used in [48] ($Q_a = 1 - \exp(-2\tau_0)$). Equation (3.90) is much simpler than (3.89), and this is expected to be particularly useful for application in CFD codes. The additional errors introduced by this equation are generally less than the errors introduced by other approximations used in the model (say, the assumption about the sphericity of droplets).

Assuming that the thermal radiation illuminating the droplet is that of a black body and n is constant, the average efficiency factor of absorption of thermal radiation in the range of λ from λ_1 to λ_2 was found to be [48, 121]

$$\begin{aligned}\overline{Q_a} =& \frac{4n}{(n+1)^2}\\ &\times\left\{1 - \left[\int_{\lambda_1}^{\lambda_2} \frac{\exp\left(-\frac{8\pi\kappa_\lambda R_d}{\lambda}\right)}{\lambda^5\left[\exp\left(C_2/(\lambda T_{\text{ext}})\right) - 1\right]}\mathrm{d}\lambda \Big/ \int_{\lambda_1}^{\lambda_2} \frac{\mathrm{d}\lambda}{\lambda^5\left[\exp\left(C_2/(\lambda T_{\text{ext}})\right) - 1\right]}\right]\right\}.\end{aligned} \tag{3.91}$$

Taking into account the experimentally measured values of κ_λ, it was found that the best approximation for $\overline{Q_a}$ in the ranges $5 \le R_d \le 50\,\mu\text{m}$ and $1000 \le T_{\text{ext}} \le 3000\,\text{K}$ is provided by the function [48, 121]:

$$\Lambda_0 = aR_d^b, \tag{3.92}$$

where a and b are quadratic functions of T_{ext} approximated as

$$\left.\begin{aligned} a &= a_0 + a_1\left(T_{\text{ext}}/1000\right) + a_2\left(T_{\text{ext}}/1000\right)^2 \\ b &= b_0 + b_1\left(T_{\text{ext}}/1000\right) + b_2\left(T_{\text{ext}}/1000\right)^2 \end{aligned}\right\}. \tag{3.93}$$

The values of the coefficients in these expressions for a and b depend on the type of Diesel fuel under consideration. If unboiled yellow Diesel fuel is taken (similar results, but with deferent numerical values of the coefficients, were obtained for yellow boiled Diesel fuel, imitating the aging process due to long storage, and pink Diesel fuel (unboiled and boiled), used in off-road equipment, in which dye was added

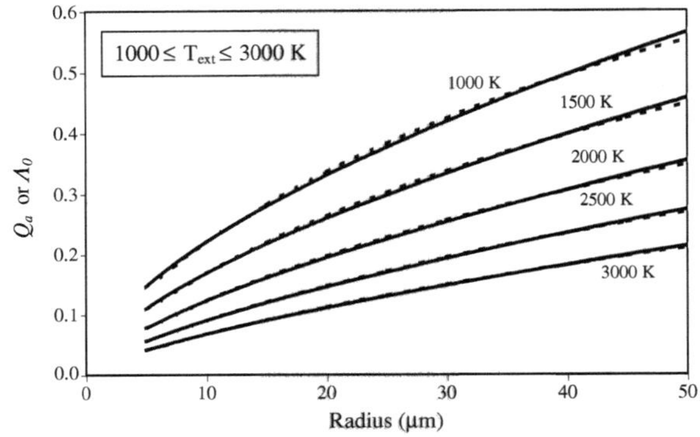

Fig. 3.5 Plots of $\overline{Q_a}$, defined by (3.91), versus R_d (*solid*) and Λ_0, defined by (3.92), versus R_d (*dashed*) for $T_{\text{ext}} = 1000$ K, $T_{\text{ext}} = 1500$ K, $T_{\text{ext}} = 2000$ K, $T_{\text{ext}} = 2500$ K and $T_{\text{ext}} = 3000$ K (indicated near the *curves*) for yellow unboiled Diesel fuel. Plots are based on the values of the coefficient found for external gas temperatures in the range $1000 \le T_{\text{ext}} \le 3000$ K and droplet radii in the range $5 \le R_d \le 50$ µm. Reproduced from Fig. 2a of [121] with permission of ASME

for legislative purposes) and R_d is measured in µm, then (3.93) can be represented in a more explicit form [121]:

$$a = 0.10400 - 0.054320T_{\text{ext}}/1000 + 0.008\,(T_{\text{ext}}/1000)^2\,,$$

$$b = 0.49162 + 0.098369T_{\text{ext}}/1000 - 0.007857\,(T_{\text{ext}}/1000)^2\,.$$

Typical plots of $\overline{Q_a}$ and Λ_0 versus R_d for a range of external temperatures from 1000 to 3000 K are shown in Fig. 3.5. As follows from this figure, the values of $\overline{Q_a}$ and Λ_0 are surprisingly close to each other in the whole range of temperatures under consideration. This justifies the application of approximation (3.92) for practical applications in CFD codes. When the range of temperatures is extended to $500 \le T_{\text{ext}} \le 3000$ K, then the values of a and b should be approximated by more complicated polynomials of the fourth degree [121, 124]. These, however, seem to be of limited practical importance, since the contribution of radiation at external temperatures <1000 K is generally small.

Based on (3.92), an explicit expression for $P(R)$ in (3.76) can be represented as

$$P(R) = 3a\sigma R_d^{b-1}\theta_R^4 \Big/ c_b\rho_b. \tag{3.94}$$

Formula (3.94) allows us to generalise solution (3.8), taking into account the contribution of thermal radiation. This was done by replacing (3.1) with (3.76) and keeping the same boundary and initial conditions as in the case of (3.1). The final expression for $T(R,t)$ was represented in the form (see Appendix B)

$$
\begin{aligned}
T(R,t) = {} & T_{g0}(R_g) \\
& + \frac{1}{R}\sum_{n=1}^{\infty}\left[\exp\left(-\lambda_n^2 t\right)\frac{1}{||v_n||^2}\left(\int\limits_0^{R_b}(-(T_{g0}(R_g)-T_{b0}(R))Rv_n(R)c_l\rho_l \mathrm{d}R\right.\right. \\
& \left. + \int\limits_{R_b}^{R_g}(-(T_{g0}(R_g)-T_{g0}(R))Rv_n(R)c_{pg}\rho_g \mathrm{d}R\right) \\
& \left. + \int\limits_0^t \exp\left(-\lambda_n^2(t-\tau)\right)p_n(\tau)\mathrm{d}\tau\right]v_n(R), \qquad (3.95)
\end{aligned}
$$

where

$$
p_n = \frac{c_l\rho_l}{||v_n||^2}\int\limits_0^{R_b} RP(t,R)v_n(R)\mathrm{d}R = \frac{3a\sigma R_d^{b-1}\theta_R^4}{||v_n||^2\lambda_n^2 a_d^2}\left[1-\lambda_n a_b R_b \cot(\lambda_n a_b R_b)\right]. \qquad (3.96)
$$

All other notation is the same as in solution (3.8).

The effects of thermal radiation predicted by solution (3.95) on nonevaporating droplet heating in the limit when $T_{g0}(R) = \mathrm{const}$ at $R_d < R \leq R_g$ were investigated in [125]. The authors of that paper concluded that the radiative effects are negligible at small Fourier numbers (Fo), but can be dominant at Fo > 50.

3.2.4 Geometric Optics Analysis

So far, two extreme cases have been considered: the distribution of thermal radiation absorption in droplets as predicted by Mie theory (Sect. 3.2.2) and a simplified model predicting the overall absorption of thermal radiation in droplets (Sect. 3.2.3). In this section, we return to the problem of finding the distribution of thermal radiation absorption in droplets, but this problem will be solved using a much simpler model than the one used in Sect. 3.2.2.

Firstly, we assume that the size parameter $x_\lambda = 2\pi R_d/\lambda$ is much greater than unity. This enables us to replace analysis based on Mie theory by analysis based on the geometric optics approximation. As in the previous sections, we assume that the illumination of the droplet by external radiation is spherically symmetric, the angular distribution of this radiation is known, and the droplet's shape can be approximated by a sphere. In this case, the radiation transfer equation in the droplet can be represented as [100]

$$
\mu_\theta\frac{\partial I_\lambda}{\partial R} + \frac{1-\mu_\theta^2}{R}\frac{\partial I_\lambda}{\partial \mu_\theta} + a_\lambda I_{a\lambda} = 0, \qquad (3.97)
$$

where $I_\lambda(r, \mu_\theta)$ is the spectral radiation intensity at a given point integrated along the azimuthal angle, θ is measured from the R-direction. This equation does not take into account the effects of scattering and the contribution of the internal source of thermal radiation (see [63, 104]).

The boundary conditions for $I_\lambda(R, \mu_\theta)$ can be represented as [100]

$$\left.\begin{aligned} I_\lambda(0, -\mu_\theta) &= I_\lambda(0, \mu_\theta), \\ I_\lambda(R_d, -\mu_\theta) &= R_{\mathrm{ref}}(n, \mu_\theta) I_\lambda(R_d, \mu_\theta) + \left[1 - R_{\mathrm{ref}}(1/n, -\mu'_\theta)\right] n^2 I_\lambda^{\mathrm{ext}}(-\mu'_\theta) \end{aligned}\right\}, \tag{3.98}$$

where $\mu'_\theta = \sqrt{1 - n^2(1 - \mu_\theta^2)}$, R_{ref} is the reflection coefficient [19], I_λ^{ext} is the spectral intensity of external radiation. Note that $R_{\mathrm{ref}}(1/n, -\mu'_\theta) = R_{\mathrm{ref}}(n, \mu_\theta)$.

The first boundary condition in (3.98) is the symmetry condition at the droplet center. The second boundary condition in (3.98) indicates that the value of $I_\lambda(R_d, -\mu_\theta)$ at the surface of the droplet is the sum of the intensity of the reflected radiation (the first term on the right-hand side of this equation) and the intensity of the refracted radiation (the second term on the right-hand side of this equation).

The radiation power absorbed per unit volume of the droplet is determined as

$$p(R) = \int_0^\infty p_\lambda(R) \mathrm{d}\lambda, \tag{3.99}$$

where

$$p_\lambda(R) = a_\lambda I_\lambda^0(R), \qquad I_\lambda^0(R) = \int_{-1}^{1} I_\lambda(R, \mu_\theta) \mathrm{d}\mu_\theta,$$

$I_\lambda^0(R)$ is the spectral radiation power density. Note that $p(R)$ (in contrast to $P(R)$) has units of W/m^3.

Calculation of $p(R)$ based on (3.97)–(3.99) is rather difficult. Following [40, 42], this problem is simplified using the so-called MDP$_0$ approximation. In this approximation, it is assumed that in the droplet core ($R \leq R_* \equiv R_d/n$), radiation intensity is constant in the angular ranges $-1 \leq \mu_\theta < 0$ and $0 < \mu_\theta \leq 1$. At the droplet periphery ($R_* < R \leq R_d$), however, constant values of the radiation intensity are assumed when $-1 \leq \mu_\theta < -\mu_*$ and $\mu_* < \mu_\theta \leq 1$, where $\mu_* = \sqrt{1 - (R_*/R)^2}$. External radiation cannot penetrate into the droplet peripheral zone at $-\mu_* < \mu_\theta < \mu_*$. Numerical solutions to (3.97) with boundary conditions (3.98) have shown that MDP$_0$ adequately predicts the angular dependence of the radiation intensity [40]. This approximation is based on the analysis of the following function:

$$g_0(R) = \begin{cases} I_\lambda^0(R), & R \le R_* \\ I_\lambda^0(R)/(1-\mu_*), & R_* < R \le R_d. \end{cases} \quad (3.100)$$

Integration of (3.97) and (3.98) over μ_θ in the ranges $-1 \le \mu_\theta < 0$ and $0 < \mu_\theta \le 1$ (droplet core) and in the ranges $-1 \le \mu_\theta < -\mu_*$ and $\mu_* < \mu_\theta \le 1$ (droplet periphery) leads to the following boundary value problem for $g_0(R)$ [40, 42, 45]:

$$\frac{1}{4R^2}\left(R^2 g_0'\right)' + \frac{C_{g1}-1}{2R} g_0' = C_{g2} a_\lambda^2 g_0, \quad (3.101)$$

where

$$C_{g1} = \begin{cases} 1 & \text{when } R \le R_* \\ (1-\mu_*)/2\mu_* & \text{when } R_* < R \le R_d, \end{cases}$$

$$C_{g2} = \begin{cases} 1 & \text{when } R \le R_* \\ (1+\mu_*)^{-2} & \text{when } R_* < R \le R_d, \end{cases}$$

where the differentiation is over R.

The boundary conditions for (3.101) are written as

$$\left.\begin{array}{ll} g_0' = 0 & \text{when } R = 0 \\ \frac{1+\mu_c}{2} g_0' = \frac{2na_\lambda}{n^2+1}\left(n^2 I_\lambda^{0(\text{ext})} - g_0\right) & \text{when } R = R_d, \end{array}\right\} \quad (3.102)$$

where

$$\mu_c = \sqrt{1-(1/n^2)}, \qquad I_\lambda^{0(\text{ext})} = \int_{-1}^{1} I_\lambda^{\text{ext}}(\mu)\mathrm{d}\mu.$$

In deriving (3.101) and (3.102), it was assumed that a_λ does not depend on R and the average values of $R_{\text{ref}}(n, \mu_\theta)$ in the ranges $-1 \le \mu_\theta < -\mu_c$ and $\mu_c < \mu_\theta \le 1$ are equal to $R_{\text{ref}}(n, 1)$.

The spectral power of the radiation absorbed per unit volume inside the droplet is determined as

$$p_\lambda(R) = a_\lambda \left[1 - \mu_* \Theta(R - R_*)\right] g_0(R), \quad (3.103)$$

where Θ is the Heaviside unit step function:

$$\Theta(x) = \begin{cases} 0 & \text{when } x < 0 \\ 1 & \text{when } x \ge 0. \end{cases}$$

Presentation of the results for the differential absorption of thermal radiation is simplified if the following normalised function is introduced:

$$w_\lambda(R) = p_\lambda(R) \Bigg/ \left[\frac{3}{R_d^3} \int_0^{R_d} p_\lambda(R) R^2 \mathrm{d}R \right]. \tag{3.104}$$

Equation (3.104) and the fact that the total radiation power absorbed in the droplet is given by the expression

$$\rho_l c_l P_{\mathrm{total}} = \pi R_d^2 \int_0^\infty Q_a I_\lambda^{0(\mathrm{ext})} \mathrm{d}\lambda$$

allows us to calculate the radiation power absorbed per unit volume inside the droplet as

$$p(R) = \frac{0.75}{R_d} \int_0^\infty Q_a w(R) I_\lambda^{0(\mathrm{ext})} \mathrm{d}\lambda. \tag{3.105}$$

If the external thermal radiation is that of a black body at temperature T_{ext}, then $I_\lambda^{0(\mathrm{ext})} = 4\pi B_\lambda(T_{\mathrm{ext}})$, and (3.105) is rewritten as

$$P(R) = \frac{3\pi}{c_l \rho_l R_d} \int_0^\infty Q_a w(R) B_\lambda(T_{\mathrm{ext}}) \mathrm{d}\lambda, \tag{3.106}$$

where B_λ is the Planck function introduced earlier. Note that $P(R)$ in (3.106) is in K/s, as in (3.76).

As follows from (3.106), the problem of the approximate calculation of the radiation power absorbed per unit volume inside droplets reduces to the problem of finding an approximation for $w_\lambda(R)$.

Following [45], $w(R)$ can be approximated as

$$w_\lambda(R) = \frac{[1 - \mu_* \Theta(R - 1/n)] \left(R^2 + \overline{\gamma}\right)}{\left[0.6\left(1 - \mu_c^5\right) - \mu_c^3/n^2\right] + \overline{\gamma}\left(1 - \mu_c^3\right)}, \tag{3.107}$$

where $\overline{\gamma} = (1.5/\tau_0^2) - (0.6/n^2)$ and

$$w(R) = \frac{\xi^2 \tau_0^3}{3} \frac{\exp\left[-\xi\left(\tau_0 - \tau_\lambda\right)\right]}{\tau_0(\xi\tau_0 - 2) + (2/\xi)\left[1 - \exp\left(-\xi\tau_0\right)\right]}, \tag{3.108}$$

where $\xi = 2/(1 + \mu_c)$.

Equation (3.107) was used when $\tau_0 < n\sqrt{2.5}$; otherwise, (3.108) was used. Equations (3.107) and (3.108) were used for the analysis of Diesel fuel droplet heating and evaporation in the presence of thermal radiation in [4, 5]. An alternative formula for $w_\lambda(R)$, valid in the whole range of τ_0, was suggested in [43].

Some authors have combined the geometric optics approximation with the Monte Carlo approach. This turned out to be particularly useful for studying the anisotropic emission characteristics of a semitransparent spherical body with spherically asymmetric temperature distribution [86].

References

1. Abarham, M., & Wichman, I. S. (2011). Mono-component fuel droplet evaporation in the presence of background fuel vapor. *International Journal of Heat and Mass Transfer*, *54*, 4090–4098.
2. Abramzon, B., & Borde, I. (1980). Conjugate unsteady heat transfer from a droplet in creeping flow. *AIChE Journal*, *26*, 536–544.
3. Abramzon, B., & Elata, C. (1984). Heat transfer from a single sphere in Stokes flow. *International Journal of Heat and Mass Transfer*, *27*, 687–695.
4. Abramzon, B., & Sazhin, S. (2005). Droplet vaporization model in the presence of thermal radiation. *International Journal of Heat and Mass Transfer*, *48*, 1868–1873.
5. Abramzon, B., & Sazhin, S. (2006). Convective vaporization of fuel droplets with thermal radiation absorption. *Fuel*, *85*, 32–46.
6. Abramzon, B., & Sirignano, W. A. (1989). Droplet vaporization model for spray combustion calculations. *International Journal of Heat and Mass Transfer*, *32*, 1605–1618.
7. Acrivos, A. (1980). A note on the rate of heat and mass transfer from a small particle freely suspended in linear shear field. *Journal of Fluid Mechanics*, *98*, 299–304.
8. Acrivos, A., & Taylor, T. E. (1962). Heat and mass transfer from single spheres in stokes flow. *Physics of Fluids*, *5*, 387–394.
9. Ahrenkiel, R. K. (1971). Modified Kramers-Kronig analysis of optical spectra. *Journal of the Optical Society of America*, *61*, 1651–1655.
10. Aliaev, V.A., & Panfilovich, K, B. (2003). *Radiation and conduction heat transfer in semi-transparent organic liquids*. Kazan: Kazan State University Publishing House.
11. Barry, D. A., & Parlange, J.-Y. (2002). Recirculation within a fluid sphere at moderate Reynolds numbers. *Journal of Fluid Mechanics*, *465*, 293–300.
12. Batchelor, G. K. (1979). Mass transfer from a particle suspended in a fluid with a steady linear ambient velocity distribution. *Journal of Fluid Mechanics*, *95*, 369–400.
13. Becker, M. (1999). Nonlinear transient heat conduction using similarity groups. *ASME Journal of Heat Transfer*, *122*, 33–39.
14. Bell, C. G., Byrne, H. M., Whiteley, J. P., & Waters, S. L. (2014). Heat or mass transfer at low Peclet number for Brinkman and Darcy flow round a sphere. *International Journal of Heat and Mass Transfer*, *68*, 247–258.
15. Bertoli, C., & Migliaccio, M. (1999). A finite conductivity model for diesel spray evaporation computations. *International Journal of Heat and Fluid Flow*, *20*, 552–561.
16. Bilger, R. W. (2011). A mixture fraction framework for the theory and modeling of droplets and sprays. *Combustion and Flame*, *158*, 191–202.
17. Bird, R. B., Stewart, W. E., & Lightfoot, E. N. (2002). *Transport phenomena*. New York: John Wiley & Sons.
18. Bohren, C. F., & Huffman, D. R. (1983). *Absorption and scattering of light by small particles*. New York: John Wiley & Sons.
19. Born, M., & Wolf, E. (1975). *Principles of optics*. London: Pergamon Press.
20. Brenn, G., Deviprasath, L. J., Durst, F., & Fink, C. (2007). Evaporation of acoustically levitated multi-component liquid droplets. *International Journal of Heat and Mass Transfer*, *50*, 5073–5086.
21. Brenner, H. (1963). Forced convection heat and mass transfer at small Peclet numbers from a particle of arbitrary shape. *Chemical Engineering Science*, *18*, 109–122.

22. Brereton, G. J. (2013). A discrete multicomponent temperature-dependent model for the evaporation of spherical droplets. *International Journal of Heat and Mass Transfer, 60*, 512–522.
23. Cai, R., & Zhang, N. (2001). Some algebraically explicit analytical solutions of unsteady non-linear conduction. *ASME Journal of Heat Transfer, 123*, 1189–1191.
24. Carslaw, H. S., & Jaeger, J. C. (1986). *Conduction of heat in solids*. Oxford: Clarendon Press.
25. Castanet, G., Lavieille, P., Lemoine, F., Lebouché, M., Atthasit, A., Biscos, Y., et al. (2002). Energetic budget on an evaporating monodisperse droplet stream using combined optical methods. Evaluation of the convective heat transfer. *International Journal of Heat and Mass Transfer, 45*, 5053–5067.
26. Chao, B. T. (1969). Transient heat and mass transfer to a translating droplet. *ASME Journal of Heat Transfer, 91*, 273–281.
27. Chao, B. T., & Cheema, L. S. (1968). Unsteady heat transfer in laminar boundary layer over a flat plate. *International Journal of Heat and Mass Transfer, 11*, 1311–1324.
28. Chao, B. T., & Chen, L. S. (1970). Series solution of unsteady heat or mass transfer to a translating fluid sphere. *International Journal of Heat and Mass Transfer, 13*, 359–367.
29. Cheng, L., Xu, M., & Wang, L. (2008). From Boltzmann transport equation to single-phase-lagging heat conduction. *International Journal of Heat and Mass Transfer, 51*, 6018–6023.
30. Choudhoury, P. N., & Drake, D. G. (1971). Unsteady heat transfer from a sphere in a low Reynolds number flow. *Quarterly Journal of Mechanics and Applied Mathematics, 24*, 23–36.
31. Clift, R., Grace, J. R., & Weber, M. E. (1978). *Bubbles, drops and particles*. New York: Academic Press.
32. Cooper, F. (1977). Heat transfer from a sphere to an infinite medium. *International Journal of Heat and Mass Transfer, 20*, 991–993.
33. Crua, C., Shoba, T., Heikal, M., Gold, M., & Higham, C. (2010). High-speed microscopic imaging of the initial stage of diesel spray formation and primary breakup. SAE, International 2010-01-2247.
34. Daho, T., Vaitilingom, G., Sanogo, O., Ouiminga, S. K., Segda, B. G., Valette, J., et al. (2012). Model for predicting evaporation characteristics of vegetable oils droplets based on their fatty acid composition. *International Journal of Heat and Mass Transfer, 55*, 2864–2871.
35. Davidzon, M. I. (2012). Newton's law of cooling and its interpretation. *International Journal of Heat and Mass Transfer, 55*, 5397–5402.
36. De, S., & Kim, S. H. (2013). Large eddy simulation of dilute reacting sprays: droplet evaporation and scalar mixing. *Combustion and Flame, 160*, 2048–2066.
37. Deepu, P., Basu, S., & Kumar, R. (2013). Vaporization dynamics of functional droplets in a hot laminar air jet. *International Journal of Heat and Mass Transfer, 56*, 69–79.
38. Dirbude, S., Eswaran, V., & Kushari, A. (2012). Droplet vaporization modeling of rapeseed and sunflower methyl esters. *Fuel, 92*, 171–179.
39. Dombrovsky, L. A. (1996). *Radiation heat transfer in disperse systems*. New York, Wallingford: Begell House.
40. Dombrovsky, L. A. (2000). Thermal radiation from nonisothermal spherical particles of a semitransparent material. *International Journal of Heat and Mass Transfer, 43*, 1661–1672.
41. Dombrovsky, L. A. (2002). Spectral model of absorption and scattering of thermal radiation by diesel fuel droplets. *High Temperature, 40*, 242–248.
42. Dombrovsky, L. A. (2002). A modified differential approximation for thermal radiation of semitransparent nonisothermal particles: application to optical diagnostics of plasma spraying. *Journal of Quantitative Spectroscopy and Radiative Transfer, 73*, 433–441.
43. Dombrovsky, L. A. (2004). Absorption of thermal radiation in large semi-transparent particles at arbitrary illumination of a polydisperse system. *International Journal of Heat and Mass Transfer, 47*, 5511–5522.
44. Dombrovsky, L. A., & Sazhin, S. S. (2003). A parabolic temperature profile model for heating of droplets. *ASME Journal of Heat Transfer, 125*, 535–537.
45. Dombrovsky, L.A., & Sazhin, S.S. (2003). Absorption of thermal radiation in a semi-transparent spherical droplet: a simplified model. *International Journal of Heat and Fluid Flow 24*, 919–927.

46. Dombrovsky, L. A., & Sazhin, S. S. (2004). Absorption of external thermal radiation in asymmetrically illuminated droplets. *Journal of Quantitative Spectroscopy and Radiative Transfer*, *87*, 119–135.
47. Dombrovsky, L.A., Sazhin, S.S., & Heikal, M.R. (2000). A model for computation of radiative characteristics of diesel fuel droplets. In A Leontiev & A. Klimenko (Eds.), *Proceedings of 3rd Russian national heat and mass transfer conference* (Vol. 6, pp. 262–265). Moscow: Moscow Power Engineering Institute Publishing House (in Russian).
48. Dombrovsky, L. A., Sazhin, S. S., Sazhina, E. M., Feng, G., Heikal, M. R., Bardsley, M. E. A., et al. (2001). Heating and evaporation of semi-transparent diesel fuel droplets in the presence of thermal radiation. *Fuel*, *80*, 1535–1544.
49. Dombrovsky, L. A., Sazhin, S. S., Mikhalovsky, S. V., Wood, R., & Heikal, M. R. (2003). Spectral properties of diesel fuel droplets. *Fuel*, *82*, 15–22.
50. Dowding, K. J., & Blackwell, B. F. (2001). Sensitivity analysis for nonlinear heat conduction. *ASME Journal of Heat Transfer*, *123*, 1–10.
51. Dusel, P. W., Kerker, M., & Cooke, D. D. (1979). Distribution of absorption centres within irradiated spheres. *Journal of the Optical Society of America*, *69*, 55–59.
52. Eyssartier, A., Cuenot, B., Gicquel, L. Y. M., & Poinsot, T. (2013). Using LES to predict ignition sequences and ignition probability of turbulent two-phase flames. *Combustion and Flame*, *160*, 1191–1207.
53. Feng, Z.-G. (2013). Forced heat and mass transfer from a slightly deformed sphere at small but finite Peclet numbers in Stokes flow. *ASME J Heat Transfer*, *135*, 081702-1.
54. Feng, Z.-G., & Michaelides, E. E. (1996). Unsteady heat transfer from a sphere at small Peclet numbers. *ASME Journal of Fluids Engineering*, *118*, 96–102.
55. Feng, Z.-G., & Michaelides, E. E. (1997). Unsteady heat and mass transfer from a spheroid. *AIChE Journal*, *43*, 609–614.
56. Feng, Z.-G., & Michaelides, E. E. (2000). A numerical study of the transient heat transfer from a sphere at high Reynolds and Peclet numbers. *International Journal of Heat and Mass Transfer*, *43*, 219–229.
57. Feng, Z.-G., & Michaelides, E. E. (2001). Heat and mass transfer coefficients of viscous spheres. *International Journal of Heat and Mass Transfer*, *44*, 4445–4454.
58. Fuchs, N. A. (1959). *Evaporation and droplet growth in gaseous media*. London: Pergamon Press.
59. Gay, M., & Michaelides, E. E. (2003). Effect of the history term on the transient energy equation for a sphere. *International Journal of Heat and Mass Transfer*, *46*, 1575–1586.
60. Godoy, W. F., & DesJardin, P. E. (2009). Radiation driven evaporation for polydisperse water sprays. *International Journal of Heat and Mass Transfer*, *52*, 2893–2901.
61. Goldstein, P. P. (2012). Van Kampen's relativistic heat transport revisited. *International Journal of Heat and Mass Transfer*, *55*, 4285–4290.
62. Goodman, T. R. (1964). Application of integral methods to transient nonlinear heat transfer. In T. F. Irvine Jr & J. P. Hartnett (Eds.), *Advances in heat transfer* (Vol. 1, pp. 51–122). New York: Academic Press.
63. Gusarov, A. V. (2011). Differential approximations to the radiation transfer equation by Chapman-Enskog expansion. *ASME Journal of Heat Transfer*, *133*, 082701.
64. Gouesbet, G., & Berlemont, A. (1999). Eulerian and Lagrangean approaches for predicting the behaviour of discrete particles in turbulent flows. *Progress in Energy Combustion Science*, *25*, 133–159.
65. Gröber, H., & Erk, S. (1961). Die Grudgesetze der Wärmeübertragung (3rd ed. pp. 57–59). Berlin: Springer
66. Haq, S., & Ishaq, M. (2012). Solution of strongly nonlinear ordinary differential equations arising in heat transfer with optimal homotopy asymptotic method. *International Journal of Heat and Mass Transfer*, *55*, 5737–5743.
67. Harada, T., Watanabe, H., Suzuki, Y., Kamata, H., Matsushita, Y., Aoki, H., et al. (2011). A numerical investigation of evaporation characteristics of a fuel droplet suspended from a thermocouple. *International Journal of Heat and Mass Transfer*, *54*, 649–655.

68. Honnery, D., Nguyen, D., & Soria, J. (2013). Microdroplet evaporation under increasing temperature conditions: experiments and modelling. *Fuel*, *105*, 247–257.
69. Incropera, F. P., & DeWitt, D. P. (1996). *Fundamentals of heat and mass transfer*. New York: John Wiley & Sons.
70. Jain, P. K., Singh, S., & Rizwan-uddin (2010). An exact analytical solution for two-dimensional, unsteady, multilayer heat conduction in spherical coordinates. *International Journal of Heat and Mass Transfer*, *53*, 2133–2142.
71. Jin, H., Song, S., Chen, Y., Wang, C., & Fan, J. (2014). Numerical investigation of evaporating isopropyl alcohol spray with LES-FDF model. *Fuel*, *119*, 98–105.
72. Jones, W. P., Lyra, S., & Marquis, A. J. (2010). Large eddy simulation of evaporating kerosene and acetone sprays. *International Journal of Heat and Mass Transfer*, *53*, 2491–2505.
73. Jones, W. P., Marquis, A. J., & Vogiatzaki, K. (2014). Large-eddy simulation of spray combustion in a gas turbine combustor. *Combustion and Flame*, *161*, 222–239.
74. Juncu, Gh. (2001). Unsteady heat and/or mass transfer from a fluid sphere in creeping flow. *International Journal of Heat and Mass Transfer*, *44*, 2239–2246.
75. Juncu, Gh. (2010). Unsteady heat transfer from an oblate/prolate spheroid. *International Journal of Heat and Mass Transfer*, *53*, 3483–3494.
76. Kartashov, E. M. (2001). *Analytical methods in the heat transfer theory in solids*. Moscow: Vysshaya Shkola. (in Russian).
77. Kishore, N., & Gu, S. (2011). Momentum and heat transfer phenomena of spheroid particles at moderate Reynolds and Prandtl numbers. *International Journal of Heat and Mass Transfer*, *54*, 2595–2601.
78. Kollár, L. E., & Farzaneh, M. (2007). Modeling the evolution of droplet size distribution in two-phase flows. *International Journal of Multiphase Flow*, *33*, 1255–1270.
79. Konopliv, N., & Sparrow, E. M. (1971). Temperature and heat transfer history of a solid body in a forced convection flow. *Quarterly of Applied Mathematics*, *29*, 225–235.
80. Konopliv, N., & Sparrow, E. M. (1972). Unsteady heat transfer and temperature for stokesian flow about a sphere. *ASME Journal of Heat Transfer*, *94*, 266–272.
81. Kuang, Z.-B. (2014). Discussions on the temperature wave equation. *International Journal of Heat and Mass Transfer*, *71*, 424–429.
82. Lage, P. L. C., & Rangel, R. H. (1993). Total thermal radiation absorption by a single spherical droplet. *Journal of Thermophysics and Heat Transfer*, *7*, 101–109.
83. Lam, T. T. (2013). A unified solution of several heat conduction models. *International Journal of Heat and Mass Transfer*, *56*, 653–666.
84. Levich, V. G. (1962). *Physicochemical hydrodynamics*. Englewood Cliffs NJ: Prentice Hall.
85. Linne, M. (2013). Imaging in the optically dense regions of a spray: a review of developing techniques. *Progress in Energy Combustion Science*, *39*, 403–440.
86. Liu, L. H. (2004). Anisotropic emission characteristics of semi-transparent spherical particle with spherically asymmetric temperature distribution. *Journal of Quantitative Spectroscopy and Radiative Transfer*, *85*, 135–143.
87. Liu, L. H., Li, B. X., Tan, H. P., & Yu, Q. Z. (2002). Emissive power of semitransparent spherical particles with nonuniform temperature. *International Journal of Heat and Mass Transfer*, *45*, 4907–4910.
88. Loth, E. (2000). Numerical approaches for motion of dispersed particles, droplets and bubbles. *Progress in Energy Combustion Science*, *26*, 161–223.
89. Mackowski, D. W., Altenkirch, R. A., & Menguç, M. P. (1990). Internal absorption cross sections in a stratified sphere. *Applied Optics*, *29*, 1551–1559.
90. Maqua, C., Castanet, G., Grisch, F., Lemoine, F., Kristyadi, T., & Sazhin, S. S. (2008). Monodisperse droplet heating and evaporation: experimental study and modelling. *International Journal of Heat and Mass Transfer*, *51*, 3932–3945.
91. Marchese, A. J., & Dryer, F. L. (1997). The effect of non-luminous thermal radiation in microgravity droplet combustion. *Combustion Science and Technology*, *124*, 371–402.
92. Melissari, B., & Argyropoulos, S. A. (2005). Development of a heat transfer dimensionless correlation for spheres immersed in a wide range of Prandtl number fluids. *International Journal of Heat and Mass Transfer*, *48*, 4333–4341.

93. Michaelides, E. E. (2003). Hydrodynamic force and heat/mass transfer from particles, bubbles, and drops—the Freeman scholar lecture. *ASME Journal of Fluid Engineering, 125*, 209–238.
94. Michaelides, E. E., & Feng, Z.-G. (1994). Heat transfer from a rigid sphere in a nonuniform flow and temperature field. *International Journal of Heat and Mass Transfer, 37*, 2069–2076.
95. Miliauskas, G. (2001). Regularities of unsteady radiative-conductive heat transfer in evaporating semitransparent liquid droplets. *International Journal of Heat and Mass Transfer, 44*, 785–798.
96. Miliauskas, G. (2003). Interaction of the transfer processes in semitransparent liquid droplets. *International Journal of Heat and Mass Transfer, 46*, 4119–4138.
97. Mishra, S. C., & Sahai, H. (2014). Analysis of non-Fourier conduction and volumetric radiation in a concentric spherical shell using lattice Boltzmann method and finite volume method. *International Journal of Heat and Mass Transfer, 68*, 51–66.
98. Modest, M. F. (2003). *Radiative heat transfer* (2nd ed.). Amsterdam: Academic Press.
99. de Monte, F., Beck, J. V., & Amos, D. E. (2011). A heat-flux based 'building block' approach for solving heat conduction problems. *International Journal of Heat and Mass Transfer, 54*, 2789–2800.
100. Özişik, M. N. (1973). *Radiative transfer and interaction with conduction and convection*. New York: John Wiley & Sons.
101. Panton, R. L. (1996). *Incompressible flow*. New York, Chichester: John Wiley & Sons.
102. Pinkley, L. W., Sethna, P. P., & Williams, D. (1978). Infrared band intensities of saturated hydrocarbons. *Journal of Physical Chemistry, 82*, 1532–1537.
103. Polyanin, A.D., Kutepov, A.M., Vyazmin, A.V., & Kazenin, D.A. (2002). *Hydrodynamics, mass and heat transfer in chemical engineering* (pp. 149–214). London and New York:Taylor & Francis.
104. Pomraning, G. C., & Siewert, C. E. (1982). On the integral form of the equation of transfer for a homogeneous sphere. *Journal of Quantitative Spectroscopy and Radiative Transfer, 28*, 503–506.
105. Pozrikidis, C. (1997). Unsteady heat or mass transport from a suspended particle at low Peclet numbers. *Journal of Fluid Mechanics, 334*, 111–133.
106. Prishivalko, A. P. (1983). *Optical and thermal fields in light scattering particles*. Minsk: Nauka. (in Russian).
107. Protheroe, M. D., Al-Jumaily, A., & Nates, R. J. (2013). Prediction of droplet evaporation characteristics of nebuliser based humidification and drug delivery devices. *International Journal of Heat and Mass Transfer, 60*, 772–780.
108. Proudman, I., & Pearson, J. R. A. (1956). Expansion at small Reynolds numbers for a flow past a sphere and a circular cylinder. *Journal of Fluid Mechanics, 2*, 237–262.
109. Rao, P. M., Raghavan, V., Velusamy, K., Sundararajan, T., & Shet, U. S. P. (2012). Modeling of quasi-steady sodium droplet combustion in convective environment. *International Journal of Heat and Mass Transfer, 55*, 734–743.
110. Rahbari, I., Mortazavi, F., & Rahimian, M. H. (2014). High order numerical simulation of non-Fourier heat conduction: an application of numerical Laplace transform inversion. *International Communications in Heat and Mass Transfer, 51*, 51–58.
111. Reitz, R. D., & Rutland, C. J. (1995). Development and testing of diesel engine CFD models. *Progress in Energy and Combustion Science, 21*, 173–196.
112. Rimmer, P. L. (1968). Heat transfer from a sphere in a stream of small Reynolds number. *Journal of Fluid Mechanics 32*, 1–7 (Erratum *35*, 827–828 (1969)).
113. Sahu, V. K., Raghavan, V., Pope, D. N., & Gogos, G. (2011). Numerical modeling of steady burning characteristics of spherical ethanol particles in a spray environment. *ASME Journal of Heat Transfer, 133*, 094502-1.
114. Saitoh, T., Yamazaki, K., & Viscanta, R. (1993). Effect of thermal radiation on transient combustion of a fuel droplet. *Journal of Thermophysics and Heat Transfer, 7*, 94–100.

115. Sanjosé, M., Senoner, J. M., Jaegle, F., Cuenot, B., Moreau, S., & Poinsot, T. (2011). Fuel injection model for Euler-Euler and Euler-Lagrange large-eddy simulations of an evaporating spray inside an aeronautical combustor. *International Journal of Multiphase Flow*, *37*, 514–529.
116. Sazhin, S. S., & Krutitskii, P. A. (2003). A conduction model for transient heating of fuel droplets. In H. G. W. Begehre, R. P. Gilber, & M. W. Wong (Eds.), *Progress in analysis* (Vol. II, pp. 1231–1239): *proceedings of the 3rd international ISAAC (international society for analysis, applications and computations) congress* (August 20–25, 2001, Berlin). Singapore: World Scientific.
117. Sazhin, S. S., Sazhina, E. M., & Heikal, M. R. (2000). Modelling of the gas to fuel droplets radiative exchange. *Fuel*, *79*, 1843–1852.
118. Sazhin, S. S., Feng, G., Heikal, M. R., Goldfarb, I., Goldshtein, V., & Kuzmenko, G. (2001). Thermal ignition analysis of a monodisperse spray with radiation. *Combustion and Flame*, *124*, 684–701.
119. Sazhin, S. S., Goldshtein, V., & Heikal, M. R. (2001). A transient formulation of Newton's cooling law for spherical bodies. *ASME Journal of Heat Transfer*, *123*, 63–64.
120. Sazhin, S. S., Dombrovsky, L. A., Krutitskii, P. A., Sazhina, E. M., & Heikal, M. R. (2002), Analytical and numerical modelling of convective and radiative heating of fuel droplets in diesel engines. In J. Taine (Ed.), *Proceedings of the twelfth international heat transfer conference*, V. 1. Grenoble (August 18–23, 2002) (pp. 699–704). Paris: Éditions Scientifique et Médicale Elsevier SAS.
121. Sazhin, S. S., Abdelghaffar, W. A., Sazhina, E. M., Mikhalovsky, S. V., Meikle, S. T., & Bai, C. (2004). Radiative heating of semi-transparent diesel fuel droplets. *ASME Journal of Heat Transfer 126*, 105–109. Erratum *126*, 490–491.
122. Sazhin, S. S., Krutitskii, P. A., Abdelghaffar, W. A., Mikhalovsky, S. V., Meikle, S. T., & Heikal, M. R. (2004). Transient heating of diesel fuel droplets. *International Journal of Heat and Mass Transfer*, *47*, 3327–3340.
123. Sazhin, S. S., Abdelghaffar, W. A., Krutitskii, P. A., Sazhina, E. M., & Heikal, M. R. (2005). New approaches to numerical modelling of droplet transient heating and evaporation. *International Journal of Heat and Mass Transfer*, *48*, 4215–4228.
124. Sazhin, S. S., Kristyadi, T., Abdelghaffar, W. A., Begg, S., Heikal, M. R., Mikhalovsky, S. V., et al. (2007). Approximate analysis of thermal radiation absorption in fuel droplets. *ASME Journal of Heat Transfer*, *129*, 1246–1255.
125. Sazhin, S. S., Krutitskii, P. A., Martynov, S. B., Mason, D., Heikal, M. R., & Sazhina, E. M. (2007). Transient heating of a semitransparent spherical body. *International Journal of Thermal Sciences*, *46*(5), 444–457.
126. Sazhin, S. S., Gusev, I. G., Krutitskii, P. A., & Heikal, M. (2011). Transient heating of a semitransparent spherical body immersed into a gas with inhomogeneous temperature distribution. *International Journal of Thermal Sciences*, *50*, 1215–1222.
127. Siegel, R., & Howell, J. P. (1992). *Thermal radiation heat transfer*. Washington: Hemisphere Publishing Corporation.
128. Sirignano, W. A. (1999). *Fluid dynamics and transport of droplets and sprays*. Cambridge: Cambridge University Press.
129. Sleicher, C. A., & Churchill, S. W. (1956). Radiant heating of dispersed particles. *Industrial and Engineering Chemistry*, *48*, 1819–1824.
130. Snegirev, A Yu. (2013). Transient temperature gradient in a single-component vaporizing droplet. *International Journal of Heat and Mass Transfer*, *65*, 80–94.
131. Sreenivasulu, B., Srinivas, B., & Ramesh, K. V. (2014). Forced convection heat transfer from a spheroid to a power law fluid. *International Journal of Heat and Mass Transfer*, *70*, 71–80.
132. Stamoudis, N., Chryssakis, C., & Kaiktsis, L. (2014). A two-component heavy fuel oil evaporation model for CFD studies in marine diesel engines. *Fuel*, *115*, 145–153.
133. Su, J. (2004). Improved lumped models for transient radiative cooling of a spherical body. *International Communications in Heat and Mass Transfer*, *31*, 85–94.

134. Talley, D. G., & Yao, S. C. (1986). A semi-empirical approach to thermal and composition transients inside vaporizing fuel droplets. *Proceedings of Twenty-first Symposium (International) on Combustion/The Combustion Institute*, pp. 609–616.
135. Tillou, J., Michel, J.-B., Angelberger, C., & Veynante, D. (2014). Assessing LES models based on tabulated chemistry for the simulation of diesel spray combustion. *Combustion and Flame, 161*, 525–540.
136. Todes, O. M. (1968). Quasi-stationary regimes of mass and heat transfer between a spherical body and ambient medium. In V. A. Fedoseev (Ed.), *Problems of evaporation, combustion and gas dynamics in disperse systems: Proceedings of the Sixths Conference on Evaporation, Combustion and Gas Dynamics in Disperse Systems (October 1966)*. Odessa: Odessa University Publishing House; pp. 151–159 (in Russian).
137. Trivic, D. N. (2014). 3-D radiation modeling of nongray gases-particles mixture by two different numerical methods. *International Journal of Heat and Mass Transfer, 70*, 298–312.
138. Tseng, C. C., & Viskanta, R. (2005). Effect of radiation absorption on fuel droplet evaporation. *Combustion Science and Technology, 177*, 1511–1542.
139. Tuntomo, A., Tien, C. L., & Park, S. H. (1991). Internal distribution of radiant absorption in a spherical particle. *ASME Journal of Heat Transfer, 113*, 402–412.
140. Van de Hulst, H. C. (1957). *Light scattering by small particles*. New York: Dover Publications Inc.
141. Viscanta, R., & Merriam, R. L. (1968). Heat transfer by combined conduction and radiation between concentric spheres separated by radiating medium. *ASME Journal of Heat Transfer, 90*, 248–256.
142. Wadewitz, A., & Specht, E. (2001). Limit value of the Nusselt number for particles of different shape. *International Journal of Heat and Mass Transfer, 44*, 967–975.
143. Whitaker, S. (1972). Forced convection heat transfer correlations for flow in pipes, past flat plates, single cylinders, single spheres, and for flow in packed beds and tube bundles. *AIChE Journal, 18*, 361–371.
144. Winnikov, S., & Chao, B. T. (1966). Droplet motion in purified systems. *Physics of Fluids, 9*, 50–61.
145. Xu, H. (2004). Electromagnetic energy flow near nanoparticles - i: Single spheres. *Journal of Quantitative Spectroscopy and Radiative Transfer, 87*, 53–67.
146. Xu, M., & Li, X. (2012). The modeling of nanoscale heat conduction by Boltzmann transport equation. *International Journal of Heat and Mass Transfer, 55*, 1905–1910.
147. Yang, J.-R., & Wong, S.-C. (2001). On the discrepancies between theoretical and experimental results for microgravity droplet evaporation. *International Journal of Heat and Mass Transfer, 44*, 4433–4443.
148. Ye, H., & Ma, Y.-L. (2011). Combined radiation-evaporation model of a liquid droplet layer in space. *ASME Journal of Heat Transfer, 133*, 111502-1.
149. Yuen, W. W. (2014). Development of the concept of mean temperatures in the analysis of radiative heat transfer in an inhomogeneous non-isothermal non-gray medium. *International Journal of Heat and Mass Transfer, 68*, 259–268.
150. Yuge, T. (1960). Experiments on heat transfer from spheres including combined natural and forced convection. *ASME Journal of Heat Transfer, 82*, 214–220.
151. Zeldovich, Ya. B., & Kompaneets, A.S. (1950). *On the theory of propagation of heat with temperature dependent thermal conductivity*. Moscow: Academy of Science of USSR Publishing House (in Russian).
152. Zeng, Y., & Lee, C. F. (2002). A preferential vaporization model for multicomponent droplets and sprays. *Atomization and Sprays, 12*, 163–186.
153. Zeoli, N., Gu, S., & Kamnis, S. (2008). Numerical modelling of metal droplet cooling and solidification. *International Journal of Heat and Mass Transfer, 51*, 4121–4131.
154. Zhang, L., & Kong, S.-C. (2010). Vaporization modeling of petroleum-biofuel drops using a hybrid multi-component approach. *Combustion and Flame, 157*, 2165–2174.
155. Zhang, S., Zhang, H. W., Yang, D. S., & Bi, J. Y. (2011). Multiple spatial and temporal scales method for numerical simulation of non-Fourier heat conduction problems: multidimensional case. *International Journal of Heat and Mass Transfer, 54*, 863–873.

Chapter 4
Heating and Evaporation of Monocomponent Droplets

Abstract The models used for the analysis of monocomponent droplet heating and evaporation in this chapter are based on the assumption that fuel vapour in the vicinity of the droplet surface is always saturated. Hence, the rate of droplet evaporation is equal to the rate of vapour diffusion from the droplet surface to ambient gas. These are known as the hydrodynamic models of droplet evaporation. The analysis starts with empirical correlations which are not directly linked with any evaporation model. Then classical hydrodynamic models of droplet evaporation are reviewed. All of these models are based on the assumption that the droplet's radius remains constant during the time step but changes from one time step to another due to droplet swelling and evaporation. Then the effects of changes in droplet radii during individual time steps on heating and evaporation processes are investigated. Finally, comparisons between the results of modelling and experimental data are presented and discussed.

In the most general case, the droplet evaporation process includes two main phases: (1) detachment of molecules from the surface of the droplet into gas in the immediate vicinity of the droplets (evaporation proper) and (2) diffusion of vapour from the surface of the droplet into the ambient gas. The mathematical modelling of the first process is far more complicated than the modelling of the second process. Hence, in most practical applications, researchers have tried to focus on just the second process, ignoring the details of the first one. The models which they used were based on the assumption that vapour in the vicinity of the droplet surface is always saturated. Hence, the rate of evaporation is equal to the rate of vapour diffusion from the droplet surface to ambient gas. These are known as hydrodynamic models of droplet evaporation. In contrast to these, models taking into account the details of detachment of fuel molecules are known as kinetic models (if they are based on the kinetic Boltzmann equation) or molecular dynamics models (if they are based on modelling the dynamics of individual molecules). This and the following chapters will be focused only on the hydrodynamic models. Kinetic and molecular dynamics models are discussed in Chap. 6.

S. Sazhin, *Droplets and Sprays*, DOI: 10.1007/978-1-4471-6386-2_4,

In the recent population balance model described in [75, 76], the spray evaporation processes are linked with the atomization processes within the Eulerian–Eulerian multiphase framework. In our analysis these two processes are modelled separately.

The most general approach to the problem of droplet heating and evaporation in the hydrodynamic limit should be based on the solution of the equations for conservation of mass, momentum, and energy in the liquid and gas phases with the corresponding boundary and initial conditions. Although this approach is widely used for the analysis of individual droplets (e.g., [8, 91]), it would not be feasible to apply it to modelling realistic sprays in complex geometries, in which the effects of turbulence and chemical transformations cannot be ignored. This and the following chapters are focused mainly on simplified models which can be potentially implemented into commercial and research CFD codes.

The analysis will start with empirical correlations which are not directly linked with any evaporation model (Sect. 4.1). Then classical hydrodynamic models of droplet evaporation are reviewed (Sect. 4.2). All of these models are based on the assumption that the droplet's radius remains constant during the time step but changes from one time step to another due to droplet swelling and evaporation. In Sect. 4.4, the effects of changes in droplet radii during individual time steps on heating and evaporation processes are investigated. A comparison between the results of modelling and experimental data is presented and discussed in Sect. 4.5. The analysis of this chapter is focused only on monocomponent droplets. It is generalised to the case of multicomponent droplets in Chap. 5.

4.1 Empirical Correlations

A brief summary of empirical correlations for the values of Nusselt numbers of nonevaporating moving droplets was given in Sect. 3.1.2. The most widely used of these correlations are those given by Equations (3.49) and (3.53). A detailed analysis of the early suggested correlations for evaporating droplets is given in [78]. Also in that paper, an analysis of the experimental data on droplets suspended at the end of a tube in a wind tunnel is presented. This analysis eventually led to the following correlation [20, 40, 78]:

$$\mathrm{Nu}_f = \left[2 + 0.57\mathrm{Re}_f^{1/2}\mathrm{Pr}_f^{1/3}\right] \Big/ \left(1 + B_f\right)^{0.7}, \tag{4.1}$$

where

$$B_f = \frac{c_{pv}(T_g - T_s)}{L}\left(1 - \frac{|\dot{q}_d|}{|\dot{q}_c|} + \frac{|\dot{q}_R|}{|\dot{q}_c|}\right), \tag{4.2}$$

$|\dot{q}_d|$ is the rate at which heat is spent on raising the temperature of a liquid droplet, $|\dot{q}_c|$ and $|\dot{q}_R|$ are the heat rates supplied to droplets by convection and radiation respectively, $\mathrm{Re}_f = 2R_d\rho_{g\infty}|\mathbf{v}_g - \mathbf{v}_d|/\mu_{gf}$, where the subscript $_f$ indicates that the values of parameters are taken inside the film surrounding the droplet, while the subscript $_\infty$ in the term T_g is hereinafter omitted. It was assumed that inside this film, the temperature is equal to $(T_g + T_s)/2$ and the vapour mass fraction is equal to $(Y_{v\infty}+Y_{vs})/2$, and Y_{vs} and $Y_{v\infty}$ are mass fractions of vapour in the vicinity of droplets and away from them. It was assumed that droplets can be treated as black bodies with emissivities equal to 0.95. This seems to be acceptable in the case of relatively large droplets used in the analysis of [78] (R_d was between about 0.5 mm and 3 mm), but is likely to be highly questionable in the case of smaller droplets (see Sect. 3.2). In the case of $|\dot{q}_d| = |\dot{q}_R| = 0$, B_f is identical to the Spalding heat transfer number introduced later. Equation (4.1) was obtained in the following range of parameters: $0.07 \leq B_f \leq 2.79$ and $24 \leq \mathrm{Re}_f \leq 1974$. It can be extrapolated beyond this range. A modified version of (4.1), with B_f replaced with the Spalding heat transfer number (introduced later in this chapter), was used in the analysis presented in [66].

Note that in the absence of evaporation, $|\dot{q}_d| = |\dot{q}_c| + |\dot{q}_R|$ and $B_f = 0$. In this case, (4.1) is identical to the corresponding equation for nonevaporating droplets (Eq. (3.49)). The predictions of (4.1) were shown to be in agreement with the results of numerical analysis [40, 79].

Similarly to how the convection heat transfer coefficient was introduced to describe droplet heating, the mass transfer coefficient h_m is used to describe droplet evaporation. This coefficient can be defined via the following equation:

$$m''_d = h_m \left(\rho_{vs} - \rho_{v\infty}\right), \tag{4.3}$$

where

$$m''_d = \frac{|\dot{m}_d|}{4\pi R_d^2},$$

ρ_{vs} and $\rho_{v\infty}$ are densities of fuel vapour in the vicinity of droplets and away from them (ambient gas), $|\dot{m}_d|$ is the absolute value of the rate of droplet evaporation. For stationary droplets evaporating into vacuum (no Stefan flow; see later in this chapter),

$$h_m = \frac{D_g}{R_d},$$

where D_g is the binary diffusion coefficient of vapour in gas. In the general case, the dimensionless mass transfer coefficient was introduced via the Sherwood number defined as [42]

$$\mathrm{Sh} = \frac{2R_d h_m}{D_g} \tag{4.4}$$

(cf. the introduction of the dimensionless heat transfer coefficient via Nu; see the discussion following (3.29)).

In the case of stationary droplets evaporating into vacuum or when the Stefan flow is not important (see the discussion later in this chapter), Sh = 2. Similarly to Eq. (4.1), the correlation for Sh for moving evaporating droplets was obtained in the form [40, 71]

$$\mathrm{Sh}_f = \left[2 + 0.87\mathrm{Re}_f^{1/2}\mathrm{Sc}_f^{1/3}\right] \Big/ (1 + B_M)^{0.7}, \tag{4.5}$$

where

$$B_M = \frac{\rho_{vs} - \rho_{v\infty}}{\rho_{gs}} = \frac{Y_{vs} - Y_{v\infty}}{1 - Y_{vs}} \tag{4.6}$$

is the Spalding mass transfer number, $\mathrm{Sc}_f \equiv \nu_f / D_f$ is the Schmidt number determined in the film region. Equation (4.5) was obtained for $20 \leq \mathrm{Re}_f \leq 2000$. The term $(1 + B_M)^{0.7}$ implicitly takes into account the contribution of the Stefan flow (see the discussion later in this chapter).

The value of ρ_{vs} is linked with the saturation vapour pressure p_{vs} and surface temperature by the ideal gas law and the Clausius–Clapeyron equation. Assuming that vapour leaving the droplet surface is saturated n-dodecane vapour, and using data presented in [64], the following approximation can be obtained [56]:

$$p_{vs} = p_{\mathrm{ref}} \exp\left[\frac{T_s - T_{\mathrm{ref}}}{a_{\mathrm{pres}}}\right], \tag{4.7}$$

where

$$\left.\begin{array}{ll} p_{\mathrm{ref}} = 70.44\ \mathrm{Pa};\ T_{\mathrm{ref}} = 300.18\ \mathrm{K};\ a_{\mathrm{pres}} = 22.37\ \mathrm{K} & \text{when } T_s \leq 440.00\ \mathrm{K} \\ p_{\mathrm{ref}} = 46204.48\ \mathrm{Pa};\ T_{\mathrm{ref}} = 449.87\ \mathrm{K};\ a_{\mathrm{pres}} = 56.97\ \mathrm{K} & \text{when } T_s > 440.00\ \mathrm{K}. \end{array}\right\}$$

An alternative approximation for p_{vs}, which turned out to be more accurate than (4.7), was used in [3]:

$$p_{vs} = \exp\left[8.1948 - 7.8099\left(\frac{300}{T_s}\right) - 9.0098\left(\frac{300}{T_s}\right)^2\right] \quad \text{(bar)}. \tag{4.8}$$

Note that the mass fraction of fuel vapour can be estimated as

$$Y_{vs} = \left[1 + \left(\frac{p}{p_{vs}} - 1\right)\frac{M_a}{M_v}\right]^{-1}, \tag{4.9}$$

where p is the total pressure of the mixture of vapour and air.

The prediction of (4.5) is consistent with the results of numerical analysis [40] (cf. the case of the Nusselt number). Although correlations (4.1) and (4.5) were suggested about 20 years ago, they are still widely used in practical applications (e.g., [103]).

An alternative correlation for Sh was suggested in [52] in the form

$$\mathrm{Sh} = \left[2.009 + 0.514\mathrm{Re}^{1/2}\mathrm{Sc}^{1/3}\right]. \tag{4.10}$$

This correlation was obtained for the reference temperature T_r using the $1/3$ rule:

$$T_r = T_s + \frac{1}{3}\left(T_g + T_s\right), \tag{4.11}$$

where T_s and T_g are the temperatures at the surface of the droplet and ambient gas respectively (the correlation (4.10) is valid only when the contribution of the Stefan flow is small). Equation (4.11) is widely used in most other models for Nu and Sh.

A number of more recent experimental correlations have been suggested [1, 11, 98–100] (see [49] for a comparative analysis of earlier correlations). These were mainly focused on the analysis of the effects of turbulence on droplet evaporation.

Results of the study of the dependence of Sh on Re, Pe, and viscosity ratio λ_v based on the numerical analysis of the transient mass transfer process from a single moving droplet were reported in [73]. The influence of Sh on Re was found to be small for $\mathrm{Re} < 2$, in agreement with (4.5). A sensitivity of Sh to λ_v was shown to exist only for $0.1 < \lambda_v < 100$. Below this range, the interface moves with the same velocity as the surrounding fluid, while above this range, the droplet behaves similarly to a rigid sphere. This is consistent with the prediction of the dependence of Nu on λ_v as given by (3.67)–(3.69).

4.2 Classical Models

4.2.1 Maxwell and Stefan–Fuchs Models

The simplest model for droplet evaporation was suggested by Maxwell back in 1877 [35]. According to this model, the rate of droplet evaporation is controlled exclusively by the diffusion process and is given by the equation

$$\dot{m}_d = 4\pi R^2 D_v \frac{\mathrm{d}\rho_v}{\mathrm{d}R}, \tag{4.12}$$

where D_v is the diffusion coefficient of the vapour and ρ_v is its density. Since $\dot{m}_d < 0$ does not depend on R in the steady-state case, this equation can be integrated from $R = R_d$ to $R = \infty$ to give

$$\dot{m}_d = -4\pi R_d D_v \left(\rho_{vs} - \rho_{v\infty}\right), \tag{4.13}$$

where ρ_{vs} and $\rho_{v\infty}$ are the same as in (4.6).

Equation (4.13) is known as the Maxwell equation [35]. Its limitation lies in the fact that it takes into account only the diffusion process, but ignores the effect of convective flow of the mixture of gas (air) and vapour away from the surface of the droplet (Stefan flow). Since the net gas flux towards the droplet is equal to zero, we can write

$$U\rho_g = D_g \frac{\mathrm{d}\rho_g}{\mathrm{d}R}, \tag{4.14}$$

where D_g is the diffusion coefficient of the gas, ρ_g is its density, and U is the value of the net velocity of the mixture of gas and fuel vapour away from the droplet.

Using (4.14), the droplet evaporation rate in the presence of vapour diffusion and Stefan flow can be found from the equation

$$\dot{m}_d = 4\pi R^2 \left(D_v \frac{\mathrm{d}\rho_v}{\mathrm{d}R} - \rho_v U \right) = 4\pi R^2 \left(D_v \frac{\mathrm{d}\rho_v}{\mathrm{d}R} - D_g \frac{\rho_v}{\rho_g} \frac{\mathrm{d}\rho_g}{\mathrm{d}R} \right). \tag{4.15}$$

For the diffusion of a single-component vapour in gas, we expect that $D_g = D_v$. Further assumptions are needed to solve (4.15). Following [35], we could assume that the total pressure and overall molar concentration of the mixture of gas and vapour is constant. A more widely used approach is based on the assumption that the total density of the mixture of gas and vapour is constant (see, e.g., [29]). None of these assumptions can be rigorously justified. Following widely used practice, we base our further analysis on the second assumption and assume that $\rho_{\text{total}} = \rho_g + \rho_v = \text{const}$. In this case, the rearrangement and integration of (4.15) from $R = R_d$ to $R = \infty$ gives the following well-known expression for $\dot{m}_d$:

$$\dot{m}_d = -4\pi R_d D_v \rho_{\text{total}} \ln\left(1 + B_M\right), \tag{4.16}$$

where B_M is the Spalding mass transfer number defined by (4.6). The evaporation model based on (4.16) is sometimes referred to as the Stefan–Fuchs model (e.g., [94]) to recognise the facts that this model was first introduced by Fuchs [35] and that it is based on the concept of the Stefan flow. If $Y_{vs} \ll 1$ and $Y_{v\infty} \ll 1$, then $B_M \ll 1$, and (4.16) reduces to Maxwell's equation (4.13). These conditions are typically satisfied except when the droplet surface temperature approaches the boiling temperature. This can justify the application of (4.13) in asymptotic studies of droplet evaporation and the ignition of fuel vapour/air mixture (see [13, 36–38, 65, 83] and Chap. 7).

If we assume that the droplet surface temperature is constant, then B_M would be constant is well. Making an additional assumption that D_v and ρ_{total} are also constant, one can show that expression (4.16) reduces to the statement that $\mathrm{d}R_d^2/\mathrm{d}t$ is constant. This is the well-known D^2-law ($D = 2R_d$), widely referred to in the literature (e.g., [6, 19, 43, 47, 58]).

Note that D_v in (4.16) is a mass diffusion coefficient. This should not be confused with the molar diffusion coefficient (see [26] for details).

An alternative expression for $\dot{m}_d$ can be obtained based on an analysis of the energy balance equation. Assuming that the evaporating droplet is stationary, this equation can be represented in the form [92]

$$4\pi R^2 k_g \frac{\mathrm{d}T}{\mathrm{d}R} = -\dot{m}_d c_{pv}(T - T_s) - \dot{m}_d L(T_s) + |\dot{q}_d|, \tag{4.17}$$

where $R > R_d$. As in the previous equations, $\dot{m}_d \leq 0$. The left-hand side of this equation shows the heat supplied from the surrounding gas to the droplet. The first term on the right-hand side shows the heat required to increase vapour temperature from T_s to $T = T(R)$ (gas temperature at the distance R from the centre of the droplet). The second and third terms on the right hand-side show the heat spent on droplet evaporation and raising its temperature respectively.

Equation (4.17) can be rearranged to yield

$$4\pi k_g \frac{\mathrm{d}T}{c_{pv}(T - T_s) + L(T_s) - (|\dot{q}_d|/\dot{m}_d)} = -\frac{\dot{m}_d \mathrm{d}R}{R^2}. \tag{4.18}$$

Integration of the left- and right-hand sides of this equation from $T = T_s$ to $T = T_g$ and from $R = R_d$ to $R = \infty$ respectively gives

$$\dot{m}_d = -\frac{4\pi k_g R_d}{c_{pv}} \ln(1 + B_T), \tag{4.19}$$

where

$$B_T = \frac{c_{pv}(T_g - T_s)}{L(T_s) - (|\dot{q}_d|/\dot{m}_d)} \tag{4.20}$$

is the Spalding heat transfer number [4]. From (4.16) and (4.19) follows the relation

$$B_T = (1 + B_M)^{\varphi} - 1 \tag{4.21}$$

between B_T and B_M [4], where

$$\varphi = \left(\frac{c_{pv}}{c_{pg}}\right)\frac{1}{\mathrm{Le}}, \tag{4.22}$$

$\mathrm{Le} = k_g/(c_{pg}\rho_{\mathrm{total}} D_v)$ is the Lewis number.

If $\mathrm{Le} = 1$ and $c_{pv} = c_{pg}$, then $B_T = B_M$. This condition is widely used in the analysis of droplet evaporation (e.g., [56]). Note that in some papers, (4.21) is used with $\varphi = \frac{1}{\mathrm{Le}}$ (see, e.g., [5, 59, 72]), which cannot be justified unless $c_{pv} = c_{pg}$. This difference in the presentation of these equations can be traced to the form of energy conservation equation (Eqs. (4.17) or (4.18)) used in some reviews (e.g., Eq. (3.4) in [30]), where c_{pv} was replaced by c_p, and the latter was implicitly identified with the heat capacity of the ambient gas.

The generalisation of formula (4.22) to the case of moving droplets is discussed in Section 4.2.2.

From (4.19) and (4.20) follows the relation

$$|\dot{q}_s| = |\dot{q}_{\mathrm{ins}}| \frac{z}{\mathrm{e}^z - 1}, \tag{4.23}$$

where $|\dot{q}_s| = -\dot{m}_d L + |\dot{q}_d|$ is the heat which reaches the surface of the droplet, $|\dot{q}_{\mathrm{ins}}| = 4\pi R_d^2 (k_g/R_d)(T_g - T_s)$ is the heat transferred from gas to droplets in the absence of evaporation, $z = -\dot{m}_d c_{pv}/(4\pi k_g R_d) = \dot{m}_d^{''} R_d c_{pv}/k_g$. Equation (4.23) was first obtained in [28] and is widely used in CFD applications (e.g., [48]).

Having introduced the convective heat transfer coefficients describing droplet heating in the presence and absence of evaporation as h and h_0 and the corresponding Nusselt numbers as Nu and Nu_0, we can rearrange (4.23) as

$$|\dot{q}_s| = 4\pi R_d^2 \frac{\mathrm{Nu}}{2} \frac{k_g}{R_d} (T_g - T_s) = \frac{z}{\mathrm{e}^z - 1} 4\pi R_d^2 \frac{\mathrm{Nu}_0}{2} \frac{k_g}{R_d} (T_g - T_s). \tag{4.24}$$

It follows from this equation that

$$\mathrm{Nu} = \frac{z}{\mathrm{e}^z - 1} \mathrm{Nu}_0 = \frac{\ln(1 + B_T)}{B_T} \mathrm{Nu}_0. \tag{4.25}$$

Recalling that in the case of stationary nonevaporating droplets, $\mathrm{Nu}_0 = 2$, we see that the combination of Eqs. (4.25) and (4.19) gives

$$\dot{m}_d^{''} = \frac{k_g}{2 c_{pv} R_d} \mathrm{Nu}\, B_T, \tag{4.26}$$

Comparing (4.3) and (4.13), one can see that when the contribution of Stefan flow can be ignored and the droplets are stationary, then $\mathrm{Sh} = \mathrm{Sh}_0 = 2$. In the presence of this flow, the analysis needs to be based on (4.16). This equation can be rewritten as

$$\dot{m}_d^{''} = \frac{D_v \rho_{\mathrm{total}}}{2 R_d} \mathrm{Sh}\, B_M, \tag{4.27}$$

where

$$\mathrm{Sh} = \frac{\ln(1 + B_M)}{B_M} \mathrm{Sh}_0. \tag{4.28}$$

Both Nu_0 and Sh_0 can take into account the movement of droplets (see Eqs. (3.49)–(3.53) for Nu_0 and similar equations for Sh_0, with Pr replaced by Sc).

Equation (4.26) can be applied to modelling evaporating moving droplets if the dependence of Nu on Re and Pr is taken into account. Various models for this were discussed in Sect. 3.1.2. Models which are most widely used in practical applications in CFD codes are usually the simplest ones, i.e., those based on (3.49) or (3.53). All equations presented in Sect. 3.1.2 are equally applicable to the analysis of Sherwood number Sh for moving droplets if we replace Nu with Sh, and the Prandtl number Pr with Sc. In this case, the Peclet number is defined as Pe = ReSc. This would allow us to use (4.27) for the analysis of evaporating moving droplets alongside (4.26).

These comments about the values of Sh for moving droplets suggest why (4.27) is almost universally used for the analysis of droplet evaporation processes in the CFD framework. In most cases, it is linked with the equation describing droplet heating in the absence of internal temperature gradients (see Chap. 3). The former equation can be generalised to the case in which the effects of droplet evaporation are taken into account:

$$\frac{\mathrm{d}T_s}{\mathrm{d}t} = \frac{3}{R_d \rho_l c_l} (q_s - j_v L), \tag{4.29}$$

where ρ_l, c_l, and L are liquid density, liquid specific heat capacity, and latent heat of evaporation j_v is vapour mass flux.

As mentioned earlier, in Sect. 3.1.2, ignoring temperature gradients inside droplets cannot always be tolerated, and a more general model taking into account these gradients would have to be used. Such a model could be based on direct numerical solutions of (3.1) (in the absence of thermal radiation), (3.76) (in the presence of thermal radiation), or analytical solutions to these equations (Eqs. (3.37) and (3.77)). To take into account the effect of droplet evaporation in analytical solutions (3.37) and (3.77), gas temperature was replaced by the so-called effective temperature, defined as

$$T_{\mathrm{eff}} = T_g + \frac{\rho_l L \dot{R}_d}{h}, \tag{4.30}$$

where the value of $\dot{R}_d$ (the derivative of the droplet radius with respect to time) can be taken from the previous time step. This approach is based on the observation that gas temperature changes more quickly than droplet radius during the evaporation process. The latter is assumed constant in the analytical solutions, but is updated at the end of each time step. Also, the modelling of droplet heating in the presence of evaporation can be based on the parabolic model described in Sect. 3.1.1, as discussed later in this chapter. An alternative version of the parabolic model specifically designed for the analysis of droplet evaporation was suggested in [25]. The latter model was based on the assumption that the droplet surface temperature is fixed. This can be justified when the heat-up period has been completed, which makes this model rather restrictive for practical engineering applications.

Results of detailed comparison between the performances of the algorithms based on the analytical solution (3.77), the algorithm based on numerical solution of the discretised Eq. (3.76), and the parabolic model (Eqs. (3.39)–(3.47)) are presented and discussed in [85]. As shown in that paper, the algorithm based on the analytical solution is more effective (from the points of view of accuracy and CPU time requirement) than the algorithm based on the numerical solution of the discretised heat conduction equation inside the droplet and more accurate than the algorithm based on the parabolic temperature profile model. Thermal radiation makes a relatively small contribution to droplet heating and evaporation in most cases. This allows us to describe the effects of this radiation using a simplified model, which takes into account the semitransparency of droplets but does not consider the spatial variations of radiation absorption inside them (see Sect. 3.2). This result was confirmed in [2, 3], where the predictions of the models taking and not taking into account the distribution of radiative heating inside droplets were compared. Note that a model for droplet heating and evaporation rather similar to the one developed in [84, 85] (that is, based on the analytical solution to the heat transfer equation inside a droplet over the time step) but without taking into account the effect of thermal radiation was suggested in [68].

In realistic CFD codes designed for modelling spray combustion (e.g., in Diesel engines) the above-mentioned models for droplet heating and evaporation would have been complemented by the models of droplet dynamics, breakup, gas dynamics, and heating and ignition of fuel vapour (see Chaps. 2 and 7).

Although the models described above have been successfully applied to many engineering problems, there is still much scope for their further refinement. Possible approaches to this refinement are considered in the following sections.

4.2.2 Abramzon and Sirignano Model

The classical model described in the previous section was substantially refined by Abramzon and Sirignano [4]. The main ideas of the model developed in that paper are summarised below.

To take into account the effect of convective transport caused by the droplet motion relative to the gas, they employed the so called 'film' theory, discussed in several well-known monographs (e.g., [10, 33]) and previously used in Sect. 3.1.1. The key concepts of this theory are film thicknesses δ_T and δ_M, the expressions for which are derived from the requirements that the rates of a purely molecular transport by thermal conduction or diffusion through the film must be equal to the actual intensity of the convective heat or mass transfer between the droplet surface and the external flow. For the case of heat conduction at the surface of a nonevaporating spherical droplet, this requirement was presented in the form of (3.13). This equation led to expression (3.14) for δ_{T0}. A similar analysis of the mass transfer equation at the surface of a spherical droplet leads to the following expression for δ_{M0}:

$$\delta_{M0} = \frac{2R_d}{\mathrm{Sh}_0 - 2}, \tag{4.31}$$

where the index $_0$ indicates that the effects of the Stefan flow are ignored. The latter effects were taken into account by introducing the correction factors

$$F_T = \delta_T/\delta_{T0}; \qquad F_M = \delta_M/\delta_{M0}, \tag{4.32}$$

which represented the relative change of the film thickness due to the Stefan flow. To find the correction factors F_M and F_T, a model problem of the laminar boundary layer of the flow past a vaporizing wedge was considered. The range of parameters $0 \le (B_M,\ B_T) \le 20$; $1 \le (\mathrm{Sc},\ \mathrm{Pr}) \le 3$ was considered with the wedge angle in the range $(0,\ 2\pi)$. It was shown that in the case of an isothermal surface and constant physical properties of the fluid, the problem has a self-similar solution, and F_M and F_T do not depend on the local Re and are practically insensitive to Sc, Pr, and wedge angle. They were approximated as

$$F_{M(T)} = \left(1 + B_{M(T)}\right)^{0.7} \frac{\ln(1 + B_{M(T)})}{B_{M(T)}}. \tag{4.33}$$

The value of $F_{M(T)}$ predicted by this expression increases from 1 to 1.285 as $B_{M(T)}$ grows from 0 to 8. For $B_{M(T)}$ in the range (8, 20), $F_{M(T)}$ remains practically constant.

In [4], it was assumed that (3.14) and (4.31) are valid for evaporating droplets, provided that $\mathrm{Nu_0}$ is replaced with the 'modified' Nusselt number Nu^*, $\mathrm{Sh_0}$ is replaced with the 'modified' Sherwood number Sh^*, δ_{T0} is replaced with δ_T, and δ_{M0} is replaced with δ_M. This leads to the following expression for Nu^*:

$$\mathrm{Nu}^* = 2 + \frac{2R_d}{\delta_T} = 2 + \frac{\mathrm{Nu_0} - 2}{F_T}. \tag{4.34}$$

Similarly, the modified Sherwood number was obtained in the form

$$\mathrm{Sh}^* = 2 + \frac{2R_d}{\delta_M} = 2 + \frac{\mathrm{Sh_0} - 2}{F_M}. \tag{4.35}$$

We remark that Nu^* and Sh^* should substitute $\mathrm{Nu_0}$ and $\mathrm{Sh_0}$ in (4.25) and (4.28), respectively. The term 'modified' was introduced for Nu^* and Sh^*, since $\mathrm{Nu}^* \to \mathrm{Nu_0}$ as $F_T \to 1$, and $\mathrm{Sh}^* \to \mathrm{Sh_0}$ as $F_M \to 1$. The terms modified Nusselt and Sherwood numbers, however, might be misleading, since the actual Nusselt and Sherwood numbers are calculated from (4.25) and (4.28) with $\mathrm{Nu_0}$ and $\mathrm{Sh_0}$ replaced by Nu^* and Sh^* respectively (see [4]).

The introduction of parameters Nu^* and Sh^* allows us to generalise expression (4.22), to take into account the effect of the moving droplets, to [4]

$$\varphi = \left(\frac{c_{pv}}{c_{pg}}\right)\left(\frac{\mathrm{Sh}^*}{\mathrm{Nu}^*}\right)\frac{1}{\mathrm{Le}}. \tag{4.36}$$

Expression (4.36) is much more complex than (4.22), and its application would require an iteration process to solve the problem of droplet heating and evaporation [4]. In many practical applications, including Diesel engine conditions, the results of modelling based on (4.36) and (4.22) are practically indistinguishable [27].

Table 4.1 The values of Nu predicted by various models and various B_T

Model	Equations	$B_T = 0.1$	$B_T = 1$	$B_T = 3$
Experimental correlation	(4.1)	Nu = 6.605	4.347	2.676
Stefan–Fuchs model	(4.25)	Nu = 6.732	4.894	3.263
Abramzon and Sirignano	(4.25, 4.33, 4.34)	Nu = 6.641	4.501	2.975

This justifies the applications of expression (4.22) for the moving droplets, as has been done in a number of papers, including [84–86].

The predictions of the experimental correlation for Nu (Eq. (4.1)), the Stefan–Fuchs model (Eq. (4.25)), and the Abramzon and Sirignano [4] model (Eqs. (4.25), (4.33), and (4.34)) were compared. To achieve consistency between these models, it was assumed that the heat-up period was completed ($|\dot{q}_d| = 0$), effects of thermal radiation can be ignored, and Nu_0 is given by (3.49) with $\beta_c = 0.57$. It was assumed that Pr $= 0.7$ and Re $= 100$, and three values of B_T were considered. The results of calculations of Nu are presented in Table 4.1.

As follows from this table, the values of Nu predicted by the Abramzon and Sirignano model are noticeably closer to those predicted by the experimental correlations than by the Stefan–Fuchs model. A more detailed comparison between the models in a wider range of parameters would be required to further support this conclusion. Note that empirical correlation (4.1) was obtained in a rather limited range of parameters (droplet radii between 0.5 and 3 mm), and its application to modelling the heating of small droplets with radii less than 10 μm, as observed in Diesel engines, is not at first evident.

The effect of internal circulation on heat transfer in a droplet was simulated in [4] using two models: the 'extended model', which directly solves the convective energy equation inside the droplet, and the 'effective conductivity model', described earlier. In [2, 3], these models were generalised to take into account the contribution of thermal radiation and the temperature-dependence of liquid properties. In both models, the contribution of thermal radiation was taken into account based on the models for thermal radiation absorption described in Sect. 3.2.4, both taking and not taking into account the spatial distribution of radiation absorption in droplets. In the latter case, the analysis was based on the total radiation heat absorption in a droplet divided by its volume. These models were applied to the analysis of Diesel fuel and n-decane droplet heating and evaporation in Diesel engine-like conditions. Physical properties of Diesel fuel were approximated by those for n-dodecane. It was pointed out that the radiation absorption in Diesel fuel is generally stronger than in n-decane, and it needs to be taken into account in modelling the combustion processes in Diesel engines. The weak effect of thermal radiation in n-decane droplets, however, may be related to the fact that due to lack of experimental data, the absorption coefficient was assumed to be zero at $\lambda < 2.6\,\mu$m. When data were available, the absorption of radiation of n-decane was generally less than that of Diesel fuel, especially in the regions of semitransparency (λ not close to 3.4 μm). Comparison between the calculations performed using the 'extended model' with distributed radiation absorption heat source and those based on the 'effective-conductivity' model with the uniform distribution of the internal heat source showed exceptionally good agreement between the results. This allowed the authors of [2, 3] to recommend using the 'effective-conductivity' model with uniform radiation absorption (see Sect. 3.2.4) for spray combustion calculations, including applications in Diesel engines.

Note that in the absence of thermal radiation, the droplet temperature approaches an equilibrium or 'wet-bulb' temperature. At that temperature, all of the heat coming to the droplet surface from the gas is spent on evaporation (latent heat), and the net heat penetrating to the liquid phase becomes zero $\dot{q}_d = 0$. In the presence of thermal radiation, however, the droplet surface temperature continues to rise above the wet-bulb temperature [2, 3]. As the surface droplet temperature increases, the heat coming to the droplet surface through convection decreases, but the heat used for vaporization increases. As a result, the value of $\dot{q}_d$ becomes negative, as confirmed by calculations presented in [2, 3]. At the end of the evaporation period, the total radiation absorption decreases very quickly with the droplet radius (see Sect. 3.2.3), while the heat transferred through the droplet surface, $|\dot{q}_d|$, decreases relatively slowly. At a certain moment, the net amount of heat going into the droplet's interior becomes negative, and $\mathrm{d}\overline{T}/\mathrm{d}t < 0$ (see [2, 3]). Therefore, the temperature of the irradiated droplet reaches its maximal value, and then decreases toward the wet-bulb temperature of the droplet predicted by the model without thermal radiation. Physically, this resembles the situation in which a droplet suspended at room temperature is heated by internal heat sources. Since the evaporation is relatively slow, the droplet temperature approaches some steady-state value, which is higher than the regular wet-bulb temperature. If the internal heat sources are suddenly 'turned off', the droplet temperature starts to decrease to the wet-bulb temperature [2, 3]. These results were confirmed by calculations reported in [86] (the analysis in the latter paper was based on the Stefan–Fuchs evaporation model). The importance of taking the radiation effect into account in modelling droplet evaporation was indicated earlier in [102].

4.2.3 Yao, Abdel-Khalik, and Ghiaasiaan Model

As in the case of the Abramzon and Sirignano model, thermal and mass thicknesses δ_T and δ_M were introduced by Yao, Abdel-Khalik, and Ghiaasiaan (see [104]) in such a way that $T = T_g$ at $R = R_d + \delta_T$ and $Y_F = Y_{F\infty}$ at $R = R_d + \delta_M$. The effect of Stefan flow was taken into account, but in a way different from that suggested in [4] (not via functions $F(B_M)$ and $F(B_T)$). The boundary conditions at the droplet surface were the same as in the two previous models. Although the key equations presented in this section are the same as in [104], some details of the analysis are slightly different from those given in the original paper [104].

As in the previous models, the analysis of [104] is based on (4.18), but the right-hand side of this equation was integrated not from $R = R_d$ to $R = \infty$, but from $R = R_d$ to $R = R_d + \delta_T$ to take into account the finite thickness of the thermal boundary layer. As a result, the following expression for B_T was obtained:

$$B_T = \exp\left[\frac{\Omega_Y \delta_T}{R_d + \delta_T}\right] - 1, \tag{4.37}$$

where

$$\Omega_Y = \frac{\dot{m}''_d c_{pv} R_d}{k_g}.$$

Recalling that heat reaching droplets is spent on their heating and evaporation, we can write

$$h\left(T_g - T_s\right) 4\pi R_d^2 = L|\dot{m}_d| + |\dot{q}_d|, \tag{4.38}$$

where h is the convection heat transfer coefficient in the presence of the Stefan flow, introduced earlier. From (4.37) and (4.38) and recalling the definition of B_T, we obtain

$$h = \frac{\dot{m}''_d c_{pv}}{\exp\left[\frac{\Omega_Y \delta_T}{R_d+\delta_T}\right] - 1}. \tag{4.39}$$

This expression for h is the same as that derived in [104], which followed a different route. In the limit $\dot{m}''_d \to 0$, Eq. (4.39) can be simplified to

$$h = \frac{k_g(R_d + \delta_T)}{R_d \delta_T}. \tag{4.40}$$

This equation is equivalent to (3.14), obtained in the limit that the contribution of the Stefan flow can be ignored. In the limit $\delta_T \to \infty$, Eq. (4.40) gives the trivial result $h = k_g/R_d$. Equation (4.37) can be rewritten in the form

$$\dot{m}''_d = \frac{k_g}{c_{pv} R_d}\left(1 + \frac{R_d}{\delta_T}\right)\ln(1 + B_T) = \frac{k_g}{2c_{pv} R_d}\mathrm{Nu} B_T, \tag{4.41}$$

where

$$\mathrm{Nu} = 2\left(1 + \frac{R_d}{\delta_T}\right)\frac{\ln(1 + B_T)}{B_T}. \tag{4.42}$$

Equation (4.42) is a straightforward generalisation of (4.25) to take into account the contribution of the finite thickness of the thermal boundary layer. It reduces to (4.25) if δ_T is replaced by δ_{T0} defined by (3.14).

The effect of the mass boundary layer was taken into account in analyzing (4.15) (similar to the effect of the thermal boundary layer). This equation was integrated not from $R = R_d$ to $R = \infty$, as in the classical model, but from $R = R_d$ to $R = R_d + \delta_M$ to take into account the finite thickness of the mass boundary layer. As a result, the following equation for $\dot{m}''_d$ was obtained:

$$\dot{m}_d'' = \frac{D_v \rho_{\text{total}}}{R_d} \left(1 + \frac{R_d}{\delta_M}\right) \ln(1 + B_M). \tag{4.43}$$

Alternatively, this equation could be represented in the form (4.27), with Sh defined as

$$\text{Sh} = 2\left(1 + \frac{R_d}{\delta_M}\right) \frac{\ln(1 + B_M)}{B_M}. \tag{4.44}$$

Equation (4.44) is a straightforward generalisation of (4.28) to take into account the contribution of the finite thickness of the mass boundary layer. It reduces to (4.28) if δ_M is replaced by δ_{M0} as defined by (4.31).

Equations (4.42) and (4.44) reduce to those predicted by the Abramzon and Sirignano [4] model if δ_T and δ_M are defined by (4.32).

Note that in (4.42) and (4.44), the dependence of Nu and Sh on Re, Pr, and Sc takes place via the dependence of δ_T and δ_M on Re, Pr, and Sc. The value of δ_M in [104] was determined via (4.44), assuming that Sh is known. The latter was taken from the empirical correlation (4.5). Once the mass evaporation flux has been found, the value of Ω_Y can be determined. In this case, (4.41) was rearranged to produce

$$B_T = \frac{2\Omega_Y}{\text{Nu}}. \tag{4.45}$$

Thus, using the empirical correlation for Nu (Eq. (4.1)), one can find B_T. Finally, using the definition of B_T (Eq. (4.20)), the value of $|\dot{q}_d|$ can found. The calculation of the values of δ_T is not needed in this approach. The dependence of the results on the separately measured or calculated values of Nu and Sh seems to constitute the main limitation of the model of [104].

4.2.4 Tonini and Cossali Model

The relative simplicity of the models of droplet heating and evaporation described so far was achieved by introducing a number of simplifying assumptions for the analysis of the underlying mass, momentum, and energy conservation equations. Namely, these models impose constant total density in the flow field and ignore the effects of gas temperature. The former condition is untenable when droplet and gas temperature are different, and it is inconsistent with the concept of diffusion of the evaporating species [94]. The limitations of the latter assumptions were discussed in detail in Sect. 3.1.2.

Tonini and Cossali [94] developed a new model for the heat and mass transfer from suspended monocomponent spherical droplets evaporating in stagnant air, which relaxes the above-mentioned assumptions. The focus of their model was on the vapour phase; they assumed that the thermal conductivity of the liquid phase is infinitely large and that there is no temperature gradient inside the droplets. Their analysis was based on steady-state mass, momentum, and energy balance equations for the vapour and gas (air) mixture surrounding a spherical liquid drop floating in a gaseous atmosphere:

$$\frac{\mathrm{d}}{\mathrm{d}R}\left(R^2\rho_v U - R^2 D_v \rho_{\mathrm{total}}\frac{\mathrm{d}Y_v}{\mathrm{d}R}\right) = 0, \tag{4.46}$$

$$\frac{\mathrm{d}}{\mathrm{d}R}\left(R^2\rho_a U - R^2 D_v \rho_{\mathrm{total}}\frac{\mathrm{d}Y_a}{\mathrm{d}R}\right) = 0, \tag{4.47}$$

$$\rho_{\mathrm{total}} U \frac{\mathrm{d}U}{\mathrm{d}R} = \frac{\mathrm{d}p_{\mathrm{total}}}{\mathrm{d}R} + \mu_{\mathrm{mix}}\left(\frac{\mathrm{d}^2U}{\mathrm{d}R^2} + \frac{2}{R}\frac{\mathrm{d}U}{\mathrm{d}R}\right), \tag{4.48}$$

$$\rho_{\mathrm{total}} U c_{p,\mathrm{mix}} \frac{\mathrm{d}T}{\mathrm{d}R} = k_{\mathrm{mix}}\left(\frac{\mathrm{d}^2T}{\mathrm{d}R^2} + \frac{2}{R}\frac{\mathrm{d}T}{\mathrm{d}R}\right), \tag{4.49}$$

where U is the Stefan flow velocity described in Sect. 4.2.1, all transport and thermodynamic properties refer to the mixture of vapour (v) and air (a), and they are assumed to be constant, $R \geq R_d$ is the distance from the droplet centre. In contrast to the previously described models, both ρ_{total} and T are assumed to be functions of R, partial (p_v and p_a) and total (p_{total}) pressures are calculated from the ideal gas and Dalton laws.

The boundary conditions used for solving system (4.46)–(4.49) are as follows: $T(R = R_d) = T_s$, $T(R = \infty) = T_{a,\infty}$, $Y_v(R = \infty) = Y_{v,\infty}$, $p_v(R = R_d) = p_{vs}(T_s)$.

Combining (4.46)–(4.47) yields the following relation:

$$\rho_{\mathrm{total}} U = \frac{\dot{m}_d}{4\pi R^2}, \tag{4.50}$$

where $\dot{m}_d$ is the droplet evaporation rate.

Equation (4.50) is a trivial statement of the conservation of mass flux. This equation allows us to decouple (4.49) for T from other variables and represent its solution in the form [94]

$$T = T_{a,\infty} A_{\mathrm{TC}}(R) + (1 - A_{\mathrm{TC}}(R))\, T_s, \tag{4.51}$$

where

$$A_{\mathrm{TC}}(R) = \frac{\exp\left(-\frac{\hat{m}_d R_d}{\mathrm{Le}R}\right) - \exp\left(-\frac{\hat{m}_d}{\mathrm{Le}}\right)}{1 - \exp\left(-\frac{\hat{m}_d}{\mathrm{Le}}\right)}, \tag{4.52}$$

$\mathrm{Le} \equiv \mathrm{Sc}/\mathrm{Pr} = k_{\mathrm{mix}}/(D_v \rho_{a,\infty} c_{p,\mathrm{mix}})$, $\hat{m}_d = \dot{m}_d/(4\pi R_d D_v \rho_{a,\infty})$ is the normalised mass flow rate.

Note that (4.51) describes a steady-state distribution of temperature, in contrast to the transient distributions considered in Sect. 3.1.1.

New variables were introduced for the further analysis of system (4.46)–(4.49):

$$\xi_{\mathrm{TC}} = R_d/R, \qquad G_{\mathrm{TC}} = \ln(Y_a).$$

These new variables allowed Tonini and Cossali [94] to present the expression for the earlier introduced normalised mass $\hat{m}_d$ in a particularly simple form:

$$\hat{m}_d = -\frac{\mathrm{d}G_{\mathrm{TC}}}{\mathrm{d}\xi_{\mathrm{TC}}}. \tag{4.53}$$

The problem was further simplified under the assumption that

$$\Lambda_{\mathrm{TB}} = \frac{R_u T_{a,\infty} R_d^2}{M_v D_v^2} \gg 1. \tag{4.54}$$

Condition (4.54), relevant to many practically important engineering applications, allowed Tonini and Cossali [94] to obtained an analytical solution for G_{TC}, which led to the following implicit equation for $\hat{m}_d$:

$$\hat{m}_d + \left(\frac{\hat{m}_d}{1-\exp\left(-\frac{\hat{m}_d}{\mathrm{Le}}\right)}\right)\left(\frac{T_s}{T_a}-1\right) = -\hat{p}_{v,\mathrm{cr}} \ln\left[\frac{\hat{p}_{v,\mathrm{cr}}-\hat{p}_{v,s}}{\hat{p}_{v,\mathrm{cr}}-Y_{v,\infty}}\right], \tag{4.55}$$

where

$$\hat{p}_{v,\,\mathrm{cr}} = 1+\frac{\theta_{\mathrm{TC}}}{1-Y_{v,\infty}}, \quad \theta_{\mathrm{TC}} = \frac{M_v - M_a}{M_a}, \quad \hat{p}_{v,s} = \frac{p_{v,s} M_v}{R_u T_{a,\infty} \rho_{a,\infty}},$$

$p_{v,s}$ is the saturation vapour pressure corresponding to T_s, R_u is the universal gas constant, M_v and M_a are molar masses of vapour and air.

The application of the model to unsteady droplet evaporation showed that the droplet lifetime is always larger than that predicted by Maxwell and Stefan–Fuchs models and that the difference increases with air temperature.

Note that in some applications, the analysis of the effects of droplet evaporation can be detached from the analysis of the effects of droplet heating by assuming that all heat supplied to the droplet is spent on its evaporation. This ensures that the droplet surface temperature remains constant (wet-bulb temperature); see, e.g., [54].

An alternative approach to fully coupled numerical simulation of droplet heating and evaporation, using the arbitrary Lagrangian–Eulerian formulation, is described in [101].

4.3 Effects of Real Gases

In most models of droplet evaporation, including those discussed in Sect. 4.2, it is assumed that ambient gas is ideal. This assumption becomes questionable when the pressures and temperatures are high enough, as observed in internal combustion engines. In this section, the main approaches to taking into account 'real gas' effects are briefly summarised.

Real gases show deviations from the perfect gas law due to molecular interactions via repulsive and attractive forces. The difference can be described in terms of a compression factor [7, 77]

$$Z = V_m / V_{m0},$$

where V_m is the actual molar volume of gas, and V_{m0} is the molar volume predicted by the ideal gas law. Since the molar volume of an ideal gas is equal to $R_u T/p$, this expression for Z can be rewritten as [7, 77]

$$Z = \frac{pV_m}{R_u T}. \tag{4.56}$$

In the limit $p \to 0$, $Z = 1$. For most substances, for small p, $Z < 1$ (attractive forces dominate when the molecules are well separated), and for large p, $Z > 1$ (repulsive forces dominate when the molecules are close enough). Various methods of estimation of Z have been described in a number of classical textbooks [7, 77]. The most widely used approach is based on the solution of the equations of state for real gases. A number of these equations have been discussed, the best known being the van der Waals and Redlich–Kwong equations [7, 77]. All these equations reduce to the ideal gas law in the limit $p \to 0$ and ideally should satisfy the thermodynamic stability criteria at the critical point [77]

$$\left(\frac{\partial p}{\partial V_m}\right)_{T_c} = 0, \tag{4.57}$$

$$\left(\frac{\partial^2 p}{\partial V_m^2}\right)_{T_c} = 0, \tag{4.58}$$

where T_c is the critical temperature.

In what follows, we discuss just one of these equations, originally suggested by Peng and Robinson [74]. This equation was found to be particularly convenient for application to modelling droplet evaporation [24, 41, 46]. It was presented in the following form [74]:

$$p = \frac{R_u T}{V_m - b} - \frac{a}{V_m(V_m + b) + b(V_m - b)}, \tag{4.59}$$

where a was assumed to be a function of T, while b was assumed to be constant. Explicit expressions for these coefficients can be found from the requirements (4.57) and (4.58). On substituting (4.59) into (4.57) and (4.58), we obtain the following equations:

$$a(T_c) = \frac{R_u T_c (\tilde{V}_m^2 + 2\tilde{V}_m - 1)^2}{2(\tilde{V}_m + 1)(\tilde{V}_m - 1)^2} b, \tag{4.60}$$

$$(\tilde{V}_m + 1)^3 = 2\left(\tilde{V}_m^3 - 1\right), \tag{4.61}$$

where $\tilde{V}_m = V_m/b$. Analytical or numerical solution to (4.61) gives $\tilde{V}_m = 3.946$. Then (4.59) and (4.60) give us the following explicit expressions [74]:

$$a(T_c) = 0.45724 \frac{R_u^2 T_c^2}{p_c}, \tag{4.62}$$

$$b = 0.0778 \frac{R_u T_c}{p_c}, \tag{4.63}$$

$$Z_c \equiv Z(T_c) = 0.307. \tag{4.64}$$

In the case $T \neq T_c$, the value of a was found from the relation [74]:

$$a(T) = a(T_c)\left[1 + \kappa_\omega \left(1 - \sqrt{\frac{T}{T_c}}\right)\right]^2, \tag{4.65}$$

where

$$\kappa_\omega = 0.37464 + 1.54226\omega - 0.26992\omega^2,$$

ω is the acentric factor, defined as [77]

$$\omega = -\log\left(\frac{p(T = 0.7T_c)}{p_c}\right) - 1. \tag{4.66}$$

For monoatomic gases, $\omega = 0$ and $a(T) = a(T_c)$. In the general case, this parameter can be obtained from tables provided by various authors (see [77] and the references therein). For example, $\omega = 0.562$ for n-dodecane, $\omega = 0.04$ for nitrogen, and $\omega = 0.021$ for oxygen.

Equation (4.59) can be rewritten in a form convenient for finding the above-mentioned parameter Z [74]:

$$Z^3 - (1 - B)Z^2 + (A - 3B^2 - 2B)Z - (AB - B^2 - B^3) = 0, \tag{4.67}$$

where

$$A = \frac{ap}{R_u^2 T^2}, \tag{4.68}$$

$$B = \frac{bp}{R_u T}. \tag{4.69}$$

The analysis of real gases is considerably simplified if a new variable f called *fugacity* is introduced via the following relation [7, 77]:

$$\mathrm{d}G = N R_u T \mathrm{d}(\ln f), \tag{4.70}$$

where G is the Gibbs function. Equation (4.70) is easily derived from the ideal gas theory when f is replaced by pressure p. The main attractive feature of (4.70) is that it is valid for real gases. The main problem, however, lies in the fact that the relation between fugacity and pressure, via the so-called fugacity coefficient

$$\varphi = f/p,$$

needs to be found.

It can be shown that (see page 129 in [7])

$$\ln \varphi = \int_0^p \frac{Z-1}{p} \mathrm{d}p. \tag{4.71}$$

The calculation of this integral requires knowledge of the function $Z(p)$. If the latter function is defined by Equation (4.67), then $\ln \varphi$ can be expressed in an explicit form [41, 74]:

$$\ln \varphi = Z - 1 - \ln(Z - B) - \frac{A}{2\sqrt{2}B} \ln\left(\frac{Z + (1+\sqrt{2})B}{Z - (1-\sqrt{2})B}\right). \tag{4.72}$$

In practical applications, it is useful to introduce fugacities and fugacity coefficients for individual components in the mixture:

$$\varphi_i = f_i / (X_i p),$$

where X_i are molar fractions of components. Equation (4.72) in this case can be generalised to [74]:

$$\begin{aligned} \ln \varphi_i = & \frac{b_i}{b}(Z-1) - \ln(Z-B) \\ & - \frac{A}{2\sqrt{2}B}\left(\frac{2\sum_j X_j a_{ji}}{a} - \frac{b_i}{b}\right) \ln\left(\frac{Z + (1+\sqrt{2})B}{Z - (1-\sqrt{2})B}\right), \end{aligned} \tag{4.73}$$

where the mixture parameters are defined by the following mixing rules:

$$a = \sum_i \sum_j X_i X_j a_{ij}, \tag{4.74}$$

$$b = \sum_i X_i b_i, \tag{4.75}$$

$$a_{ij} = (1 - \zeta_{ij})\sqrt{a_i a_j}, \tag{4.76}$$

ζ_{ij} are empirically determined binary interaction coefficients (they were assumed equal to zero in [41]).

The specific enthalpy of evaporation for each component $\Delta h_{v,i}$ can be calculated from the equation [41]

$$\Delta h_{v,i} = -\frac{R_u T^2}{M_i} \frac{\partial}{\partial T} \left(\ln \frac{f_{i,g}}{f_{i,l}} \right), \tag{4.77}$$

where M_i is the molar mass of the ith component, subscripts $_l$ and $_g$ refer to liquid and gas phases respectively. There seems to be a typographical error in Eq. (22) of [41], where the fugacity coefficients were used instead of fugacities.

In the case of ideal gases, when fugacities reduce to pressures, (4.77) reduces to the classical Clausius–Clapeyron equation (e.g., see Eq. (1–216) of [53]). In [41], the above model was applied to simulate droplet evaporation at high pressures. A noticeably better agreement of the predictions of this model with experimental data, when compared with the predictions of the ideal gas model, was demonstrated. Applications of the real gas model to modelling droplet evaporation under Rapid Compression Machine (RCM) and Diesel engine conditions were discussed in [44, 57], respectively.

Perhaps one of the most comprehensive analyses of the real gas models is given in the recent monograph [50].

None of the droplet evaporation models considered so far takes into account the motion of the droplet surface boundary (interface) during the evaporation process (it is assumed fixed during each time step). This effect is considered in the models discussed in the next section.

4.4 Effects of the Moving Interface

Taking into account the effect of receding droplet radius on droplet heating and evaporation leads to the well-known Stefan problem, which has been widely discussed in the literature (e.g., [12, 14, 15, 39, 45, 69, 70, 80, 97, 105]) but has rarely been applied to engineering sprays, due to their complex structure. Hence, a substantial gap has developed between mathematical and engineering research in this field. This gap was partly filled in recent publications [88, 90]. In what follows, the main results reported in these papers is summarised. To simplify the analysis, it is assumed that the ideal gas model is valid.

4.4.1 Basic Equations and Approximations

As in Sect. 4.2, let us assume that a stationary evaporating droplet is immersed in a homogeneous hot gas at temperature T_g. The droplet is heated by convection with convection heat transfer coefficient $h(t)$ depending on time t, and cooled due to evaporation. In contrast to Sect. 4.2, the droplet radius is allowed to change with time and can be represented as $R_d(t)$, where $R_d(t)$ is a continuously twice differentiable function of time in the range $0 \le t \le t_e$, where t_e is the evaporation time. Both $R_d(t)$ and $h(t)$ are assumed to be known. Effects of thermal radiation are ignored, which is justified for typical Diesel fuel droplets and gas temperatures about or less than 1,000 K, when the effect of radiation on droplet evaporation time is typically less than 1 % [3]. The changes in the droplet temperatures are described by the heat conduction Eq. (3.1), which is solved subject to boundary condition (3.36) with T_g replaced with T_{eff} defined by (4.30). The initial condition is taken in the form $T(t=0) = T_{d0}(R)$, where $0 \le R \le R_{d0} = R_d(t=0)$.

Let us rewrite boundary condition (3.36) with T_{eff} defined by (4.30) in the form

$$\left(\frac{\partial T}{\partial R} + \frac{h}{k_l} T\right)\Bigg|_{R=R_d(t)} = \frac{h}{k_l} T_g + \frac{\rho_l}{k_l} L \dot{R}_d(t) \equiv M(t), \tag{4.78}$$

and introduce the new variable $u = TR$. Using this new variable, we can rewrite (3.1) as

$$\frac{\partial u}{\partial t} = \kappa \frac{\partial^2 u}{\partial R^2} \tag{4.79}$$

for $t \in (0, t_e)$, $R \in (0, R_d(t))$ with the boundary conditions

$$\left(\frac{\partial u}{\partial R} + H(t) u\right)\Bigg|_{R=R_d(t)} = \mu(t), \tag{4.80}$$

$$u|_{R=0} = 0, \tag{4.81}$$

where

$$H(t) = \frac{h(t)}{k_l} - \frac{1}{R_d(t)}, \qquad \mu(t) = M(t) R_d(t)$$

for $t \in (0, t_e)$. The initial condition is

$$u(R)|_{t=0} = R T_{d0}(R) \tag{4.82}$$

for $R \in (0, R_{d0})$.

As in the case of models discussed in Sect. 4.2, Eq. (4.79) is solved together the relevant equation for droplet radius to model the process of droplet heating and evaporation.

In the following sections, several solutions to (4.79) for various $R_d(t)$ and $T_{d0}(R)$ are presented and discussed.

4.4.2 Solution When $R_d(t)$ Is a Linear Function

Let us assume that $R_d(t)$ is a linear function of t during the time step

$$R_d(t) = R_{d0}(1 + \alpha t), \tag{4.83}$$

where R_{d0} is the initial droplet radius. Given the physical background to the problem, α can be expressed as [81]

$$\alpha = -\frac{k_g \ln(1 + B_M)}{\rho_l c_{pg} R_{d0}^2}, \tag{4.84}$$

where $B_M = Y_{vs}/(1 - Y_{vs})$ is the Spalding mass transfer number, Y_{vs} is the mass fraction of fuel vapour near the droplet surface, k_g and c_{pg} are gas thermal conductivity and specific heat capacity respectively.

Note that (4.84) was derived under the assumption that Le $= 1$ (cf. (4.16)), and ignoring the effect of swelling. A more rigorous expression for α can be represented as

$$\begin{aligned} \alpha = \frac{\dot{R}_d}{R_{d0}} &\equiv \frac{1}{R_{d0}}\left(\dot{R}_{de} + \dot{R}_{ds}\right) \\ &= -\frac{1}{R_{d0}}\left[\frac{k_g \ln(1 + B_M)}{\rho_l c_{pg} R_d} + \frac{R_{d0}}{\Delta t}\left[\left(\frac{\rho(\overline{T}_0)}{\rho(\overline{T}_1)}\right)^{1/3} - 1\right]\right], \end{aligned} \tag{4.85}$$

where $\dot{R}_{de}$ and $\dot{R}_{ds}$ are the rates of change of droplet radius due to evaporation and swelling (or contraction), subscripts $_0$ and $_1$ indicate the values of average temperatures at the beginning and at the end of the time step.

The following analysis is based on (4.84), but it can be easily generalised to the general case of Le $\neq 1$ taking into account the effect of swelling described by (4.85).

For $R_d(t)$ defined by (4.83), the solution to (4.79), subject to the above boundary and initial conditions, leads to the following expression for $T(R)$ [88] (see Appendix D):

$$T(R) = \frac{1}{R\sqrt{R_d(t)}} \exp\left[-\frac{\alpha R_{d0} R^2}{4\kappa R_d(t)}\right] \times \left[\sum_{n=1}^{\infty} \Theta_n(t) \sin\left(\lambda_n \frac{R}{R_d(t)}\right) + \frac{\mu_0(t)}{1+h_0} \frac{R}{R_d(t)}\right], \tag{4.86}$$

where

$$\Theta_n(t) = \Theta_n(0) \exp\left[\frac{\kappa\lambda_n^2}{\alpha R_{d0}^2}\left(\frac{1}{1+\alpha t} - 1\right)\right] + f_n \int_0^t \frac{\mathrm{d}\mu_0(\tau)}{\mathrm{d}\tau} \exp\left[\frac{\kappa\lambda_n^2}{\alpha R_{d0}^2}\left(\frac{1}{1+\alpha t} - \frac{1}{1+\alpha\tau}\right)\right] \mathrm{d}\tau, \tag{4.87}$$

$$\mu_0(t) \equiv \tilde{\mu}(t)\sqrt{R_d(t)} \exp\left[\frac{R_d'(t) R_d(t)}{4\kappa}\right], \tag{4.88}$$

$$\tilde{\mu}(t) = M(t) R_d^2(t),\ M(t) = \frac{h}{k_l} T_g + \frac{\rho_l}{k_l} L \dot{R}_d(t),\ f_n = -\frac{\sin\lambda_n}{\|\, v_n \,\|^2 \lambda_n^2},$$

$$v_n(r) = \sin\lambda_n r \qquad (n = 1, 2, \ldots). \tag{4.89}$$

$$\|\, v_n \,\|^2 = \frac{1}{2}\left(1 - \frac{\sin 2\lambda_n}{2\lambda_n}\right) = \frac{1}{2}\left(1 + \frac{h_0}{h_0^2 + \lambda_n^2}\right), \tag{4.90}$$

$r = R/R_d$, λ_n are positive solutions to the equation

$$\lambda \cos\lambda + h_0 \sin\lambda = 0 \tag{4.91}$$

presented in ascending order,

$$h_0 = \frac{h(t)}{k_l} R_d(t) - 1 - \frac{R_d'(t) R_d(t)}{2\kappa}$$

is assumed to be constant during the time step,

$$\Theta_n(0) = q_n + \mu_0(0) f_n, \tag{4.92}$$

$$q_n = \frac{1}{\|\, v_n \,\|^2} \int_0^1 W_0(r) v_n(r) \mathrm{d}r,$$

$$W_0(r) = R_{d0}^{3/2} r T_0(r R_{d0}) \exp\left[\frac{R_d'(0) R_{d0}}{4\kappa} r^2\right]. \tag{4.93}$$

Note that the condition $h_0 = \text{const}$ can be violated for t in the immediate vicinity of t_e for finite time steps. This imposes a rather severe restriction on the choice of the time steps just before the droplet fully evaporates. Condition (4.83) allows us to use the above solution for individual time steps but not for the whole period of droplet heating and evaporation. In what follows, this assumption is relaxed, and we consider arbitrary $R_d(t)$, provided that this function is continuously twice differentiable (which is a natural condition given the physical background to the problem).

4.4.3 Solution for Arbitrary $R_d(t)$ but $T_{d0}(R) = \text{const}$

The analysis of this section is based on the assumption that $T_{d0}(R) = T_{d0} = \text{const}$. In this case, we can introduce the new variable $v = u - RT_{d0}$ and rearrange (4.79) as

$$\frac{\partial v}{\partial t} = \kappa \frac{\partial^2 v}{\partial R^2} \tag{4.94}$$

for $t \in (0, t_e)$ and $R \in (0, R_d(t))$ with the boundary conditions

$$\left(\frac{\partial v}{\partial R} + H(t)v \right) \Bigg|_{R=R_d(t)} = \mu_0(t), \tag{4.95}$$

$$v|_{R=0} = 0 \tag{4.96}$$

for $t \in (0, t_e)$ and the initial condition

$$v|_{t=0} = 0 \tag{4.97}$$

for $R \in (0, R_d(t))$, where

$$\mu_0(t) = -T_{d0} - H(t)R_d(t)T_{d0} + \mu(t) = -\frac{h(t)}{k_l} R_d(t)T_{d0} + \mu(t),$$

$\mu(t)$ is the same as introduced in (4.80).

We look for a solution to the problem (4.94)–(4.97) in the form

$$v(R,t) = \int_0^t \nu(\tau)G(t,\tau,R)\mathrm{d}\tau, \tag{4.98}$$

where

$$G(t,\tau,R) = G_0(t-\tau, R-R_d(\tau)) - G_0(t-\tau, R+R_d(\tau)),$$
$$G_0(t,x) = \frac{\sqrt{\kappa}}{2\sqrt{\pi t}} \exp\left[-\frac{x^2}{4\kappa t}\right].$$

The expression $G(t,\tau,R)$ can be represented in the alternative form

$$G(t,\tau,R) = \frac{\sqrt{\kappa}}{2\sqrt{\pi(t-\tau)}} \left\{\exp\left[-\frac{(R-R_d(\tau))^2}{4\kappa(t-\tau)}\right] - \exp\left[-\frac{(R+R_d(\tau))^2}{4\kappa(t-\tau)}\right]\right\}. \tag{4.99}$$

Note that $G(t,\tau,R=0)=0$, and $\nu(t)$ is an unknown continuous function to be determined later from one of the boundary conditions. The function $v(R,t)$ is known as a single-layer heat potential. It has the following properties for every continuous function $\nu(t)$ [45, 93]:

(1) It satisfies (4.94) for $0 < t < t_e$ and $0 < R < R_d(t)$.
(2) It satisfies conditions (4.96) and (4.97).
(3) It is continuous at $R \to R_d - 0$.
(4) For the derivative $\partial v(R,t)/\partial R$, the following limiting formula is valid:

$$\left.\frac{\partial v(R,t)}{\partial R}\right|_{R\to R_d(t)-0} = \frac{\nu(t)}{2} + \int_0^t \nu(\tau)\left[\frac{\partial G(t,\tau,R)}{\partial R}\right]\bigg|_{R=R_d(t)} \mathrm{d}\tau. \tag{4.100}$$

This means that for every continuous function $\nu(t)$, the potential $v(R,t)$ satisfies (4.94) and boundary and initial conditions (4.96) and (4.97). Choice of the function $\nu(t)$ satisfying the integral Eq. (4.100) should be made in such a way that the remaining boundary condition (4.95) is satisfied as well.

From Eq. (4.99), it follows that

$$\begin{aligned}\left.\frac{\partial G(t,\tau,R)}{\partial R}\right|_{R=R_d(t)} = &-\frac{1}{4\sqrt{\pi\kappa}(t-\tau)^{3/2}} \\ &\times\left\{(R_d(t)-R_d(\tau))\exp\left[-\frac{(R_d(t)-R_d(\tau))^2}{4\kappa(t-\tau)}\right]\right. \\ &\left.-(R_d(t)+R_d(\tau))\exp\left[-\frac{(R_d(t)+R_d(\tau))^2}{4\kappa(t-\tau)}\right]\right\}.\end{aligned} \tag{4.101}$$

Since $R_d(t)$ is a continuously differentiable function, we obtain, in the limit $\tau \to t-0$,

$$\left.\frac{\partial G(t,\tau,R)}{\partial R}\right|_{R=R_d(t)} \propto O\left(\frac{1}{\sqrt{t-\tau}}\right). \tag{4.102}$$

It follows from this equation that there is an improper integral on the right-hand side of (4.100).

In view of (4.100) and (4.98), we can rewrite the boundary condition (4.95) as

$$\frac{\nu(t)}{2}+\int_0^t \nu(\tau)\left[\frac{\partial G(t,\tau,R)}{\partial R}\right]\Bigg|_{R=R_d(t)} d\tau+H(t)\int_0^t \nu(\tau)G(t,\tau,R_d(t))d\tau=\mu_0(t),$$

or

$$\frac{\nu(t)}{2}+\int\limits_0^t \nu(\tau)\left\{\left[\frac{\partial G(t,\tau,R)}{\partial R}\right]\Bigg|_{R=R_d(t)}+H(t)G(t,\tau,R_d(t))\right\}d\tau=\mu_0(t), \tag{4.103}$$

where $G(t,\tau,R)$ and its derivative with respect to R are defined by (4.99) and (4.101).

As follows from the definition of $G(t,\tau,R_d(t))$ (see Eq. (4.99)), in the limit $\tau \to t-0$, this function has the singularity

$$G(t,\tau,R_d(t)) \propto O\left(\frac{1}{\sqrt{t-\tau}}\right)$$

(cf. Eq. (4.102)). Therefore, the integral in (4.103) is defined as an improper integral.

Equation (4.103) is an integral equation of Volterra type. It has a unique continuous solution. A scheme for its numerical solution is rather similar to one described in Appendix F (see also Appendix A in [90]). This solution is then substituted into (4.98). The final distribution of temperature inside the droplet can be calculated from the following expression:

$$\begin{aligned} T(t,R)=T_{d0}+\frac{\sqrt{\kappa}}{2R\sqrt{\pi}}\int\limits_0^t \frac{\nu(\tau)}{\sqrt{t-\tau}}\left\{\exp\left[-\frac{(R-R_d(\tau))^2}{4\kappa(t-\tau)}\right]\right.\\ \left.-\exp\left[-\frac{(R+R_d(\tau))^2}{4\kappa(t-\tau)}\right]\right\}d\tau. \end{aligned} \tag{4.104}$$

Details of the numerical calculation of the integral on the right-hand side of (4.104) are given in Appendix G.

4.4.4 Solution for Arbitrary $R_d(t)$ and $T_{d0}(R)$

Let us assume that an arbitrary continuously twice differentiable function $T_{d0}(R)$ is defined for $0 \le R \le R_{d0}$. This definition is extended for the whole range $0 \le R < \infty$:

$$T_{d0}(R) = \begin{cases} T_{d0}(R) & \text{when } 0 \le R \le R_{d0} \\ T_{\text{out}}(R) & \text{when } R_{d0} < R \le R_{\text{eff}} \\ 0 & \text{when } R > R_{\text{eff}}, \end{cases} \tag{4.105}$$

where

$$T_{\text{out}}(R) = \frac{1}{R}\left\{R_{d0}T_{d0}(R_{d0}) + (R - R_{d0})\left[\left.(RT_{d0}(R))'_R\right|_{R=R_{d0}}\right]\right\},$$

R_{eff} is the effective outer radius such that $R_{\text{eff}} > R_{d0}$. The function $T_{d0}(R)$ defined by (4.105) is continuously differentiable in the range $0 \le R \le R_{\text{eff}}$.

Let us now introduce a new function $U(t, R)$ defined as

$$U(t, R) = \int_0^{R_{\text{eff}}} (\zeta T_{do}(\zeta))G_1(t, R, \zeta)\mathrm{d}\zeta, \tag{4.106}$$

where

$$G_1(t, R, \zeta) = \frac{1}{\kappa}\left[G_0(t, R - \zeta) - G_0(t, R + \zeta)\right],$$

$G_0(t, x)$ is the same as in (4.98). Based on the latter equation, the expression $G_1(t, R)$ can be represented in an alternative form:

$$G_1(t, R) = \frac{1}{2\sqrt{\pi t \kappa}}\left\{\exp\left[-\frac{(R-\zeta)^2}{4\kappa t}\right] - \exp\left[-\frac{(R+\zeta)^2}{4\kappa t}\right]\right\}. \tag{4.107}$$

Note that $G_1(t, R = 0) = 0$.

The function $U(t, R)$ has the following properties [45, 93]:

(1) It satisfies (4.94) for $0 < t < t_e$ and $0 < R < \infty$.
(2) It satisfies the boundary condition (4.96) for $0 < t < t_e$.
(3) It satisfies the initial condition

$$U(t, R)|_{t=+0} = \begin{cases} RT_{d0}(R) & \text{when } 0 \le R \le R_{\text{eff}} \\ 0 & \text{when } R > R_{\text{eff}}. \end{cases} \tag{4.108}$$

The latter relation follows from the property of the delta function

$$\lim_{\alpha_{\text{delta}} \to \infty} \frac{\alpha_{\text{delta}}}{\sqrt{\pi}} \exp(-\alpha_{\text{delta}}^2 x^2) = \delta(x). \tag{4.109}$$

We look for the solution to (4.79) in the form

$$u(t, R) = U(t, R) + v(t, R). \tag{4.110}$$

Having substituted (4.110) into (4.79) and boundary and initial conditions (4.80)–(4.82), we obtain problem (4.94)–(4.97) for $v(t, R)$ in which

$$\mu_0(t) = -\left[U_R'(t, R) + H(t)U(t, R)\right]\Big|_{R=R_d(t)} + \mu(t). \tag{4.111}$$

The solution of the latter problem is similar to the one discussed in the previous section.

The expression for $\mu_0(t)$ contains

$$U_R'(t, R)\Big|_{R=R_d(t)} = \int_0^{R_{\text{eff}}} (\zeta T_{do}(\zeta)) \left.\frac{\partial G_1(t, R, \zeta)}{\partial R}\right|_{R=R_d(t)} \mathrm{d}\zeta, \tag{4.112}$$

where

$$\begin{aligned}\left.\frac{\partial G_1(t, R, \zeta)}{\partial R}\right|_{R=R_d(t)} &= \frac{1}{\kappa}\left[\frac{\partial}{\partial R}(G_0(t, R-\zeta) - G_0(t, R+\zeta))\right]\Big|_{R=R_d(t)} \\ &= \frac{1}{\kappa}\left[\frac{\partial}{\partial \zeta}(-G_0(t, R-\zeta) - G_0(t, R+\zeta))\right]\Big|_{R=R_d(t)},\end{aligned} \tag{4.113}$$

G_0 is the same as in (4.98). Using the latter equation, we can rewrite the expression for $\left.\frac{\partial G_1(t,R,\zeta)}{\partial R}\right|_{R=R_d(t)}$ in a more explicit form:

$$\begin{aligned}\left.\frac{\partial G_1(t, R, \zeta)}{\partial R}\right|_{R=R_d(t)} = &-\frac{1}{4\sqrt{\pi}(\kappa t)^{3/2}}\left\{(R-\zeta)\exp\left[-\frac{(R-\zeta)^2}{4\kappa t}\right]\right. \\ &\left.\left. - (R+\zeta)\exp\left[-\frac{(R+\zeta)^2}{4\kappa t}\right]\right\}\right|_{R=R_d(t)}.\end{aligned}$$

Hence, we obtain an explicit expression for $\mu_0(t)$ in the form

$$\begin{aligned}\mu_0(t) = &\frac{1}{4\sqrt{\pi}(\kappa t)^{3/2}}\int_0^{R_{\text{eff}}} (\zeta T_{d0}(\zeta))\left\{(R_d(t)-\zeta)\exp\left[-\frac{(R_d(t)-\zeta)^2}{4\kappa t}\right]\right. \\ &\left. - (R_d(t)+\zeta)\exp\left[-\frac{(R_d(t)+\zeta)^2}{4\kappa t}\right]\right\}\mathrm{d}\zeta \\ &- \frac{H(t)}{2\sqrt{\pi\kappa t}}\int_0^{R_{\text{eff}}} (\zeta T_{d0}(\zeta))\left\{\exp\left[-\frac{(R_d(t)-\zeta)^2}{4\kappa t}\right] - \exp\left[-\frac{(R_d(t)+\zeta)^2}{4\kappa t}\right]\right\}\mathrm{d}\zeta \\ &+ M(t)R_d(t).\end{aligned} \tag{4.114}$$

In the limit $t \to 0+$, the expression for $\mu_0(t)$ simplifies to (see [90])

$$\mu_0(0) = -\left[(\zeta T_{do}(\zeta))'_\zeta \Big|_{\zeta=R_{d0}} + H(0) R_{d0} T_{d0}(R_{d0}) \right] + \mu(0). \tag{4.115}$$

Combining (4.98) and (4.110), we can represent the final solution to our problem in the form

$$T(t,R) = \frac{1}{R}\left[U(t,R) + \frac{\sqrt{\kappa}}{2\sqrt{\pi}} \int\limits_0^t \frac{\nu(\tau)}{\sqrt{t-\tau}} \left\{ \exp\left[-\frac{(R - R_d(\tau))^2}{4\kappa(t-\tau)} \right] \right.\right.$$
$$\left.\left. - \exp\left[-\frac{(R + R_d(\tau))^2}{4\kappa(t-\tau)} \right] \right\} \mathrm{d}\tau \right], \tag{4.116}$$

where $\nu(\tau)$ is the solution to (4.103) with $\mu_0(t)$ given by (4.114), $U(t,R)$ is given by (4.106).

Note that taking into account the initial distribution of temperature along R is absolutely essential when the solution is applied to individual time steps. In the solution described in the last two sections, however, the same formulas describe the time evolution of droplet temperatures during the whole period of their evaporation. It is anticipated that in this case, the effect of the initial distribution of droplet temperatures is not important in most practical applications. Hence, the solution described in the previous section is expected to be more practically important than the solution described in this section.

Note that although the analysis presented so far refers to stationary droplets, it can be generalised in a straightforward way to the case of moving droplets, based on the effective thermal conductivity model (see Sect. 3.1.2).

In the solutions presented in the last two sections, it was assumed that $R_d(t)$ is known. From the point of view of the physical background to the problem, however, $R_d(t)$ depends on the time evolution of the droplet temperature $T(R,t)$, which is the solution to the problem. Hence, an iterative process is required. Firstly, the time evolution of the droplet radius $R_d(t)$ is obtained using the conventional approach, whereby it remains constant during the time step but changes from one time step to another due to the evaporation process. Then these values of $R_d(t)$ are used in the new solutions to obtain updated values of the time evolution of the distribution of temperatures inside the droplet and on its surface $T(R,t)$. These new values of droplet temperature are used to update the function $R_d(t)$. This process is continued until convergence is achieved, which typically takes place after about 15 iterations.

4.4.5 Results

In Figs. 4.1 and 4.2, we compared the results of calculations of droplet surface temperatures and radii, taking into account the effects of evaporation, using the integral

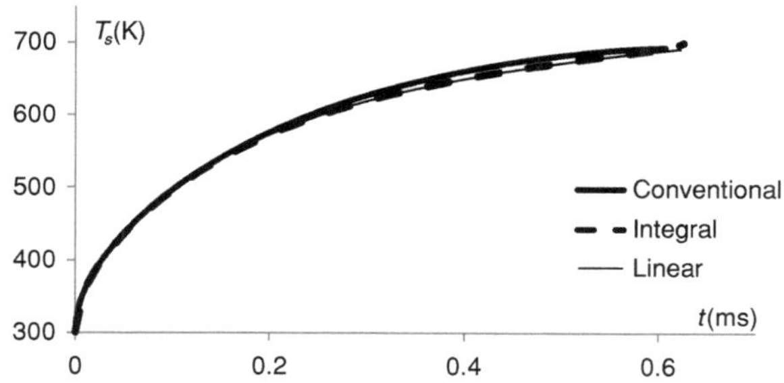

Fig. 4.1 Plots of T_s versus time using the conventional model (*thick solid*), integral model based on (4.104) (*dashed*) and linear model (*thin solid*) for a stationary n-dodecane ($M_f = 170$ kg/kmol) droplet with an initial radius 5 μm, evaporating in ambient air at a pressure of $p = 30$ bar $=$ 3,000 kPa and temperature 1,000K. Reprinted from [90], Copyright Elsevier (2011)

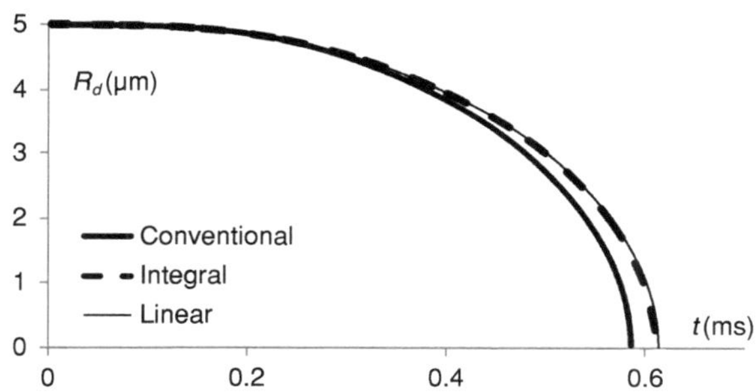

Fig. 4.2 The same as Fig. 4.1 but for the droplet radius versus time. Reprinted from [90], Copyright Elsevier (2011)

solution for arbitrary $R_d(t)$ but constant T_{d0} (Eq. (4.104)); the solution, based on the linear approximation of $R_d(t)$ (Eq. (4.86)); and the conventional approach, based on the assumption that droplet radius does not change during the time step. Droplets are assumed to be those of n-dodecane ($M_f = 170$ kg/kmol), and ambient air is assumed to be at the pressure of $p = 30$ bar $= 3{,}000$ kPa (typical values for Diesel engine-like conditions).

The results of calculations for $R_{d0} = 5\,\mu$m, and ambient air temperature 1,000 K are shown in Figs. 4.1 and 4.2 (these are reproduced with minor modifications from Fig. 1 of [90]). One thousand time steps were used in the conventional approach and the approach based on solution (4.86). In the integral solution based on (4.104), the integral (F.1) was approximated as the sum of 100 terms, and up to 15 iterations were used. At the first iteration, the time evolution of the droplet radius was assumed to be the same as predicted by the conventional model.

As follows from Figs. 4.1 and 4.2, the results predicted by the integral solution (4.104) and linear solution (4.86) practically coincide, which suggests that both approaches are correct and valid. Both these solutions predict lower droplet temperatures and longer evaporation times, in agreement with the results reported in [88].

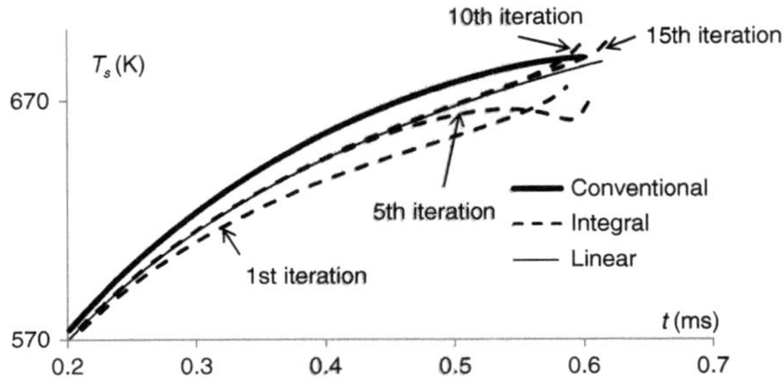

Fig. 4.3 The same as Fig. 4.1 but for different numbers of iterations in the integral solution. Reprinted from [90], Copyright Elsevier (2011)

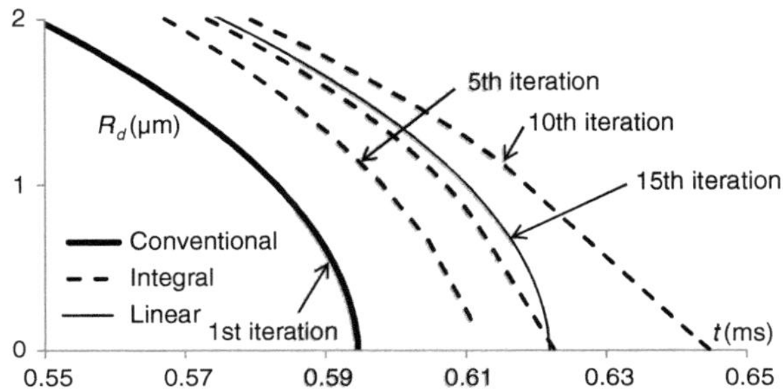

Fig. 4.4 The same as Fig. 4.3 but for the droplet radius versus time. Reprinted from [90], Copyright Elsevier (2011)

Note that deviations between the predictions of the integral and linear solutions were observed in the immediate vicinity of the instant of time when the droplet completely evaporates.

There were obvious numerical problems when we approached this instant of time, due to the fact that the time derivative of R_d becomes infinitely large. In practice, an extrapolation based on the assumption that the second derivative of $R_d(t)$ is constant was used for these times. This leads to small deviations between the predicted evaporation times. In the case shown in Figs. 4.1 and 4.2, the evaporation times predicted by the conventional model, linear solution, and integral solution were 0.595, 0.622, and 0.628 ms respectively. This means that the difference between the evaporation times predicted by the linear and integral solutions was less than 1 % and can be safely ignored in most practical applications (this error can be reduced further if required). The same comment applies to other cases considered below.

The effect of the choice of the number of iterations on the prediction of the integral solution is illustrated in Figs. 4.3 and 4.4 for the same case as shown in Figs. 4.1 and 4.2. This effect is shown only for the times when the deviation between the results predicted by the linear and integral solutions is maximal. For the first iteration, the time evolution of droplet radius is the same as predicted by the conventional model. The deviation of the corresponding droplet temperatures predicted by the integral and linear solutions appears to be quite noticeable. For the fifth iteration, the droplet surface temperatures predicted by the integral and linear solutions practically coincide up to $t \approx 0.45$ ms. The corresponding plots of $R_d(t)$, predicted by the integral solution, turned out to be closer to those predicted by the linear solution than those predicted by the conventional model. The closeness between the plots predicted by the linear and integral solutions improved as the number of iterations increased. However, even for the 15th iteration, the deviation between the results remains visible, although not important for practical applications (cf. Figs. 4.1 and 4.2). For higher iterations, the results are practically indistinguishable from those predicted by the 15th iteration. Interestingly, odd iterations predicted smaller $R_d(t)$, and even iterations predicted larger $R_d(t)$ compared with those predicted by the linear solution. At the qualitative level, this could be related to the fact that the faster evaporation rate assumed for the first iteration (conventional model) leads to a lower droplet surface temperature. At the second iteration, this lower droplet surface temperature leads to a slower evaporation rate, and so on.

As to the computational efficiency of the new integral model, we note that for a PC Xeon 3,000 Hz (the calculations were processed on one kernel only) with 2 GB RAM, the conventional approach requires 3586 s to calculate 1191 steps. Once these calculations have been completed, the integral model requires 453 s to calculate 15 iterations. This makes this model potentially suitable for incorporation into computational fluid dynamics (CFD) codes.

Calculations similar to those shown in Figs. 4.1 and 4.2 were performed for droplets with initial radii 50 and 100 μm [90]. Time evolution of droplet surface temperatures and radii were shown to be largely unaffected by the initial droplet radii. This agrees with similar results reported in [88] (see Figs. 4–6 in that paper).

The effect of nonconstant initial distribution of droplet temperature on the time and space evolution of this distribution is illustrated in Fig. 4.5. Two cases are compared in this figure. In both cases, the initial droplet radii are assumed to be equal to 5 μm, and gas temperature is assumed to be constant and equal to $T_g = 1{,}000$ K. In the first case, the initial distribution of temperature was assumed to be independent of R (or $r = R/R_d$) and equal to 300 K. In this case, the analysis was based on (4.104). In the second case, the initial distribution of droplet temperature was approximated as

$$T_{d0}(R) = 300 + 10(R/R_{d0})^2 = 300 + 10(r)^2, \qquad (4.117)$$

and the analysis was based on (4.116).

Comparing the plots referring to both cases, shown in Fig. 4.5, one can see that these plots visibly converge with time. This can be related to the fact that increased droplet surface temperature in the general case leads to decreased convective heating of droplets. Hence the droplet surface temperature increases more slowly in the general case than in the case of constant initial temperature inside the droplets. When the initial temperatures inside the droplets are constant, the predictions of (4.104) and (4.116) coincide [90]. This shows the correctness of both approaches to the problem.

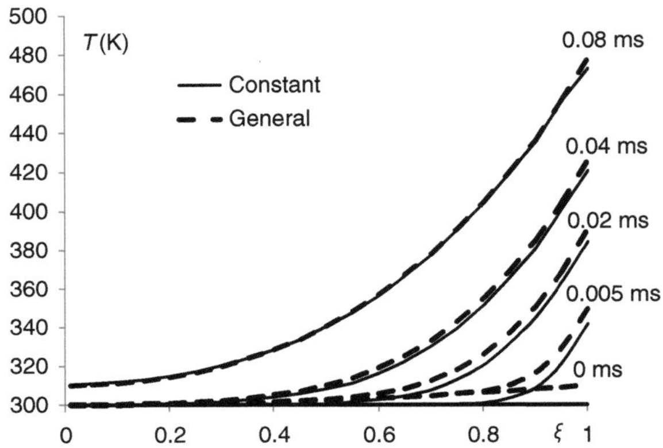

Fig. 4.5 The plots of T versus $r = R/R_d$ for a stationary n-dodecane ($M_f = 170$ kg/kmol) droplet with initial radius 5 μm, evaporating in ambient air at a pressure of $p = 30$ bar $= 3{,}000$ kPa, and temperature 1,000 K. The instants of time are indicated near the curves. The calculations were performed based on (4.104) (derived for constant initial distribution of temperatures inside droplets) and (4.116) (derived for the general case of the arbitrary distribution of the initial temperature inside the droplet). Equation (4.116) applied to the case in which the initial distribution of temperature inside the droplet is given by (4.117)). Reprinted (with minor modification) from [90], Copyright Elsevier (2011)

The above results for the linear model were validated by the model based on direct numerical solution to (4.79) [67]. The authors of that paper used the boundary immobilization method coupled to a Keller Box discretisation scheme of the one-phase one-dimensional time-dependent governing equations. This algorithm is implicit, therefore not having any limitation on the time step size, and was in addition shown to be second-order accurate in the time and space variables.

We appreciate that the errors associated with the conventional assumption that the droplet radii remain constant during the time step can be comparable with or even larger than those associated with other effects, including uncertainties in gas temperature measurements, convection heat transfer coefficient approximations, and the effect of interactions between droplets in realistic sprays. The importance of the latter effect is discussed in recent papers [51, 87], but its analysis lies beyond the scope of this monograph.

4.5 Modelling versus Experimental Data

The experimental validation of the models discussed earlier in this chapter is not easy. As mentioned in the previous chapter, perhaps the first experimental evidence of the need to take into account temperature gradients in Diesel fuel droplets was provided by Bertoli and Migliaccio [9], who compared pressure traces in a Diesel engine predicted by the models taking and not taking into account the effects of

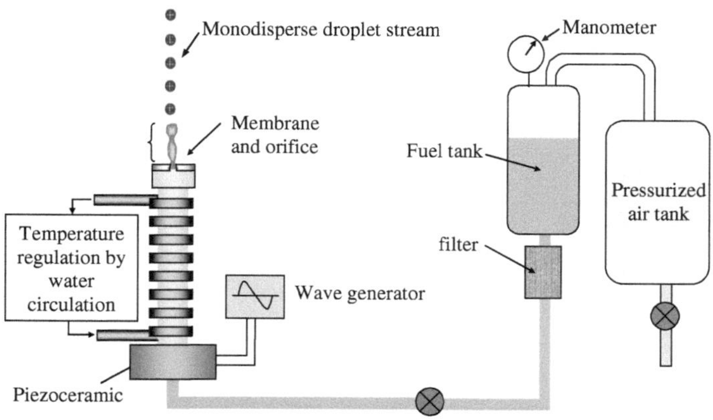

Fig. 4.6 Generation of a monodispersed stream of droplets Reprinted from [62], Copyright Elsevier (2008)

the above-mentioned temperature gradients and showed that the prediction of the former model agrees better with experimental data than that of the latter one. This result, however, can provide only indirect support for the model, and not its direct experimental validation.

The first opportunity to provide a direct experimental validation of the models described in this chapter, taking into account the effects of temperature gradient inside droplets, seems to have been provided by a series of experiments performed at the University of Nancy by the group of Professor F. Lemoine [27, 55, 60–62]. In these experiments, linear monodisperse droplet streams were generated by Rayleigh disintegration of a liquid jet undergoing vibrations generated in a piezoelectric ceramic (see Fig. 4.6). The voltage applied to the piezoceramic was a square wave, whose the amplitude determined the position of the breakup zone for a given fuel at a given temperature. The fuel was preheated in the injector by means of externally heated circulating water. The temperature of the fuel was measured exactly at the injection point with a K-type thermocouple. For specific frequencies of forced mechanical vibration, the liquid jet broke up into almost equally spaced and monosized droplets [34]. By adjusting the liquid flow rate and the piezoceramic frequency, it was possible to increase the droplet spacing up to about six times the droplet diameter. This, however, was accompanied by a modification of droplet sizes.

Downstream distance from the injector x was converted into time t with the help of the space evolution of the droplet velocity

$$t = \int_0^x \frac{\mathrm{d}x}{U_{\mathrm{drop}}(x)}.$$

Droplets were injected into a quiescent atmosphere at room temperature. Droplet temperatures were measured using the technique described in [61, 62]. The input

Table 4.2 Approximations of acetone and ethanol droplet velocities

Substance	Approximation of U_{drop} in m/s (t is in ms)
100 % acetone	$12.81 - 0.316t$
100 % ethanol	$12.30 - 0.344t$

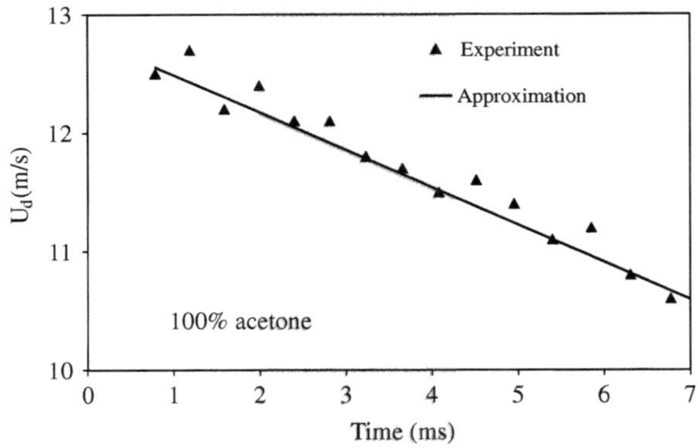

Fig. 4.7 Experimentally observed velocities for pure acetone droplets (*triangles*). These were approximated by the straight line $U_d = 12.81 - 0.316t$, where t is measured from the moment of injection. Reprinted from [87], Copyright Elsevier (2010)

parameters for the models were the initial droplet temperature (assuming that this temperature is homogeneous), ambient gas temperature (assuming that this temperature remains constant during the experiment, although this temperature could vary depending on the distance from the nozzle), the distance parameter (ratio of the distances between droplets and their diameters), and the droplet velocities (assuming that these velocities remain constant during the experiment, although they could vary depending on the distance from the nozzle). The analysis of this section is focused on pure acetone and ethanol droplets, although a number of other substances and their mixtures were studied (see [51, 62]).

The measured time evolution of the droplet velocities in the axial direction was shown to be close to a linear function. The relevant approximations of the experimental results are summarised in Table 4.2 .

Figure 4.7 shows a comparison between experimentally observed velocities for pure acetone droplets and their approximation given in Table 4.2. As one can see from this figure, the linear approximation for the experimental data presented in Table 4.2 is reasonably accurate for practical applications. The same conclusion was obtained for ethanol droplets.

The measured initial values of droplet temperature, diameter, ambient gas temperature, and initial distance parameter C (ratio of the distance between droplets to their diameters) for the same cases as in Table 4.2 are shown in Table 4.3.

The changes in C from the previous to the current time step were taken into account based on the following equation:

Table 4.3 The measured initial values of droplet temperature, droplet diameter, ambient gas temperature and distance parameter for the same cases as in Table 4.2

Substance	Droplet temperature (°C)	Diameter (μm)	Gas temperature (°C)	Distance parameter
100 % acetone	35.1	143.4	21.5	7.7
100 % ethanol	38.0	140.8	22.0	7.1

$$C_{\text{new}} = C_{\text{old}} \frac{U_{\text{drop, new}}}{U_{\text{drop, old}}} \frac{R_{d,\,\text{old}}}{R_{d,\,\text{new}}}, \tag{4.118}$$

where subscripts $_{\text{new}}$ and $_{\text{old}}$ refer to the values of variables at the previous time step and one time step behind respectively. In this case, the values of $R_{d,\,\text{old}}$ and $R_{d,\,\text{new}}$ are known at the current time step. Note that the changes in C during the experiments were noticeable in all cases and cannot be ignored, as was done in [87].

The parameter C was used for quantifying the effect of interaction between droplets on the values of Nu and Sh [87]:

$$\eta(C) = \frac{\text{Sh}}{\text{Sh}_{\text{iso}}} = \frac{\text{Nu}}{\text{Nu}_{\text{iso}}} = 1 - 0.57\left(1 - \frac{1 - \exp[-0.13(C-6)]}{1 + \exp[-0.13(C-6)]}\right). \tag{4.119}$$

Note that there is a mistake in the corresponding formula given in [87]. This was corrected in [27]. Other approximations for $\eta(C)$ are presented and discussed in [21, 31, 32].

As shown in [23], the above ratios of Sherwood and Nusselt numbers are controlled not only by the distance parameter but also by the nature of the evaporation process of a given substance. To take into account the latter effect, the following dimensionless time t^* was introduced:

$$t^* = \frac{\delta f}{v_r}, \tag{4.120}$$

where f is the frequency of droplet production (in Hertz) (set up for each particular experiment and directly linked with the distance parameter), δ is the film thickness, which is different for mass and thermal boundary layers (δ_M and δ_T), v_r is the radial velocity of the vapour released at the droplet surface, estimated as

$$v_r = \frac{\dot{m}_d}{4\pi \rho_v R_d^2}, \tag{4.121}$$

ρ_v is the density of vapour. Thus the parameter t^* takes into account the contributions of both the distance parameter (via f) and the volatility of the substance (via v_r).

The values for δ_M and δ_T for monocomponent droplets were estimated based on (4.32). The definitions of these thicknesses and v_r were generalised to the case of multicomponent droplets via introduction of the average density and mass averaged values of transport coefficients. Having introduced the concept of t^*, Deprédurand et al. [23] suggested the following correlations for $\frac{\mathrm{Nu}}{\mathrm{Nu}_{\mathrm{iso}}}$ for acetone, ethanol, 3-pentanone, n-decane, n-dodecane, and n-heptane, and $\frac{\mathrm{Sh}}{\mathrm{Sh}_{\mathrm{iso}}}$ for ethanol, 3-pentanone, and n-heptane ($\frac{\mathrm{Sh}}{\mathrm{Sh}_{\mathrm{iso}}}$ for acetone was approximated by the same correlation as for $\frac{\mathrm{Nu}}{\mathrm{Nu}_{\mathrm{iso}}}$):

$$\frac{\mathrm{Nu}}{\mathrm{Nu}_{\mathrm{iso}}} = \frac{1 - A_T}{(G_T t^* + 1)^{C_T}} + A_T, \tag{4.122}$$

$$\frac{\mathrm{Sh}}{\mathrm{Sh}_{\mathrm{iso}}} = \frac{1 - A_M}{(G_M t^* + 1)^{C_M}} + A_M, \tag{4.123}$$

where the coefficients used in these equations are given in Table 4.1 of [89]; $\delta = \delta_T$ is used in correlation (4.122), while $\delta = \delta_M$ is used in correlation (4.123).

Although correlations (4.122) and (4.123) were derived for a limited number of substances, they can be used for a wider range of substances and their mixtures (e.g., [89]).

Droplet temperatures were measured using a technique based on two-colour laser-induced fluorescence (LIF). The possibility of measuring the droplet temperature using LIF was described in earlier papers [16, 55]. The liquid in the injector was initially seeded with a very low concentration of a fluorescent dye, and a laser beam, having a wavelength tuned to the absorption band of the dye, illuminated the droplets downstream of the injector. The fluorescent emissions from two different spectral bands with different temperature sensitivities were utilised. The ratio of the intensities of these two bands allowed the authors to eliminate the effects of parameters that are unknown or difficult to control, such as variations in laser intensity, tracer concentration, and measurement volume during the acquisitions [55]. At the same time, this ratio depends on the temperature, making possible the temperature measurement. Although fluorescent tracers like Rhodamine B and Kitton red are well known for the temperature sensitivity of their fluorescent signal and their high quantum yield, the fluorescence spectrum of these molecules can vary significantly with the composition of the liquid mixture in which they are dissolved. To account for this effect, a special method was developed, based on the inclusion of a third spectral band [60]. The probe volume (intersection between the laser beams and the detection field of view) was about $150 \times 150 \times 1{,}200\ \mu\mathrm{m}^3$. It was larger than the droplet in order to provide a global excitation of the whole droplet volume. The signal was averaged over the total time of droplet transit in the probe volume.

The measurements can be affected by errors, as described in [55, 60]. In particular, the measurement method requires a calibration and a reference temperature, which can lead to systematic errors. However, these errors are not expected to be significant. Calibration was performed over an extended range of temperatures. The reference value used to convert the fluorescence ratio into temperature was obtained from a thermocouple placed at the exit of the injector, which limits the risk of bias. The uncertainty related to the statistical convergence of the data was small (typically the RMS did not exceed 0.5 °C). Given the periodicity of the droplet streams, the fluorescence signal can be averaged over thousands of droplets; this is achieved easily through acquisition times of a few seconds. Finally, the overall accuracy of the technique was estimated to be about ±1.2 °C [55, 60].

In the presence of thermal gradients inside droplets, the temperatures obtained based on the fluorescence signal could differ from the volume-averaged droplet temperatures. This is related to the fact that the laser intensity (and thus the fluorescence field) was not distributed uniformly inside the droplet, mainly due to light refraction at the droplet surface. Also, the receiving optical equipment collected the fluorescent signal from the droplet interior with a spatial distortion. The optical system was designed to obtain a measurement volume much larger than the droplet, which reduces this adverse effect (this refers to the size of the laser beam and focal lengths of the lenses). Nevertheless, even if the measurements allowed the authors to estimate the volume-averaged droplet temperature, there was systematically more weighting in the zones near the centre of the droplet than at its edge. A complete modelling of the interactions between the laser beams and the droplet would be required to account for this effect. The evaluation of the differences between the volume-averaged droplet temperature and the measured temperature would also be related to the preconceived temperature distribution within the droplet (i.e., the local temperature gradients). Examples of such calculations are presented in [16, 18]. Geometrical optics and ray tracing were used in combination with a two-dimensional description of heat transfer inside the droplet. One-dimensional models were used. These models may be good at predicting the droplet heating in terms of volume-averaged temperature, but they can be rather inappropriate for estimating the local gradients. It will be assumed that the values of the measured temperatures are between the temperatures at the centre of the droplets and the average droplet temperatures, being closer to the droplet average temperatures than to the temperatures at the centre of the droplets.

In what follows, the focus is on the comparison of the results of calculations based on the analytical solution to the heat transfer Eq. (3.1) (Eq. (3.37)) inside droplets (solution A), those based on the numerical solutions to (3.1) (solution B), and experimental data. The code for the numerical solution of (3.1) was developed by Castanet et al. [17]. The effects of the movement of the droplet surface due to evaporation and thermal swelling/contraction during individual time steps are ignored in both solutions. To capture these effects in the validation process, much more refined experiments compared with those available at the moment would be required. Only one-way solutions were considered, so that the effects of droplets on gas were ignored. The latter effects would modify the results slightly, but would not change the trends of the predicted results [89]. The effects of interaction between droplets were taken into account based on (4.119).

The plots of the time-dependence of the surface, average, and centre temperatures (T_s, T_{av}, and T_c) for pure acetone, obtained based on solutions A and B, and experimentally observed average droplet temperatures are shown in Fig. 4.8. The calculations started with the first observed droplets approximately 1 ms after the start of injection. At earlier times, the liquid fuel formed an unstable jet whose temperature evolution cannot be interpreted using our model. Also, we assumed that there

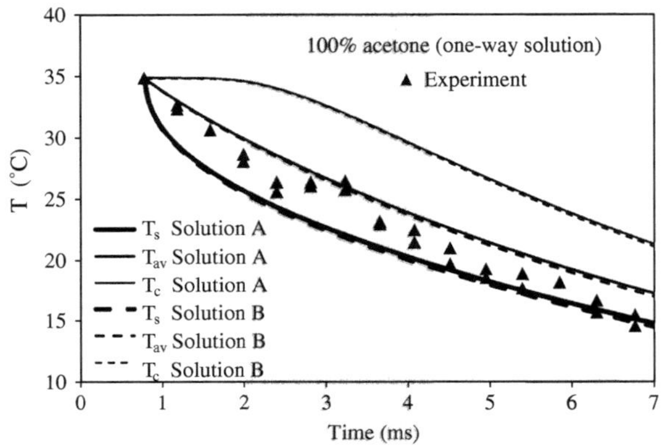

Fig. 4.8 The time evolution of droplet surface, average, and centre temperatures (T_s, T_{av}, and T_c), predicted by one-way solutions A and B and experimentally observed temperatures for pure acetone droplets with the values of the initial parameters, droplet velocity, and gas temperature given in Tables 4.2 and 4.3. Reproduced from Fig. 3 of [27] with permission of Begell House

is no temperature gradient inside the droplets at the initial moment of time used in calculations.

As follows from Fig. 4.8, all three temperatures are well separated. Hence, the difference between them needs to be taken into account in the analysis of experimental data. The results predicted by solutions A and B coincide within the accuracy of plotting, which gives us confidence in the results predicted by both solutions. The observed temperature values lie close to the average temperatures. In the case of ethanol (figure is not presented), the observed temperature values are close to or below the surface temperature of the droplets. The reason for this rather poor quantitative agreement between the predicted and observed temperatures for ethanol is not clear to us.

Detailed experimental measurements of temperature distribution inside droplets, using the setup shown in Fig. 4.6, was recently reported by Castanet et al. [18]. The temperature field within the droplets was measured with the help of the two-colour laser-induced fluorescence technique. Experiments were undertaken on droplets made of different substances including acetone, ethanol, 3-pentanone, n-heptane, n-decane, and n-dodecane, which have different physical properties such as their volatility and their viscosity. In some cases, the isotherms appeared circular and concentric, suggesting that thermal conduction is the dominant mechanism of heat transfer in the droplet. In other cases, measurements showed rather significant temperature differences between the leading and the trailing edges of the moving droplets. In the latter case, the experimental results cannot be interpreted using simplified models described in this section, and the effects of recirculation inside droplets (Hill vortex) would need to be taken into account. This was done in the numerical solution presented in [18].

As mentioned in [82], the models described in this chapter are applicable only to spherical droplets, while the shapes of most of the droplets observed in internal combustion engines are far from spherical [22]. Heating and evaporation of aspherical droplets was investigated by Tonini and Cossali [95]. In that paper, the effect on the steady-state evaporation rate of a liquid spheroidal drop immersed in stagnant gas at constant density was investigated, showing the analogy between heat and mass transfer. The vapour distribution around a spheroidal droplet and the local surface mass and heat fluxes under steady-state conditions were predicted for oblate and prolate droplets. A comparison between isovolumetric spherical and spheroidal droplets suggested that the deformed droplets are characterised by larger evaporation rate, with the oblate shape yielding a lower evaporation rate than the prolate one with the same surface area. The local vapour mass flux at drop surface was found to be directly dependent on the nondimensional Gaussian curvature.

The results of further investigation of this problem are presented in [96]. In this paper, the evaporation process of a liquid spheroidal drop floating in a gaseous atmosphere was considered, accounting for the oscillation between oblate and prolate states. A previously developed exact solution for the heat and mass transfer equations [95] was extended to investigate the effect of oscillation on drop evaporation under the assumption of quasi-steady-state conditions. The results are compared with the prediction of the approximate model [63]. The validity of the quasi-steady-state assumption, on which the analysis of [96] was based, was discussed.

References

1. Abou. Al-Sood, M., & Birouk, M. (2008). Droplet heat and mass transfer in a turbulent hot airstream. *International Journal of Heat and Mass Transfer*, *51*, 1313–1324.
2. Abramzon, B., & Sazhin, S. (2005). Droplet vaporization model in the presence of thermal radiation. *International Journal of Heat and Mass Transfer*, *48*, 1868–1873.
3. Abramzon, B., & Sazhin, S. (2006). Convective vaporization of fuel droplets with thermal radiation absorption. *Fuel*, *85*, 32–46.
4. Abramzon, B., & Sirignano, W. A. (1989). Droplet vaporization model for spray combustion calculations. *International Journal of Heat and Mass Transfer*, *32*, 1605–1618.
5. Aggarwal, S. K., & Chitre, S. (1991). Computations of turbulent evaporating sprays. *Journal of Propulsion*, *7*, 213–220.
6. Archambault, M. R. (2010). A maximum entropy approach to modeling the dynamics of a vaporizing spray. *Atomization and Sprays*, *20*, 1017–1031.
7. Atkins, P., & de Paula, J. (2002). *Atkins' physical chemistry* (7th ed.). Oxford: Oxford University Press.
8. Awasthi, I., Gogos, G., & Sundararajan, T. (2013). Effects of size on combustion of isolated methanol droplets. *Combustion and Flame*, *160*, 1789–1802.
9. Bertoli, C., & Migliaccio, M. (1999). A finite conductivity model for diesel spray evaporation computations. *International Journal of Heat and Fluid Flow*, *20*, 552–561.
10. Bird, R. B., Stewart, W. E., & Lightfoot, E. N. (2002). *Transport phenomena*. Chichester: John Wiley & Sons.
11. Birouk, M., & Gökalp, I. (2002). A new correlation for turbulent mass transfer from liquid droplets. *International Journal of Heat and Mass Transfer*, *45*, 37–45.
12. Brenn, G. (2005). Concentration fields in evaporating droplets. *International Journal of Heat and Mass Transfer*, *48*, 395–402.

13. Bykov, V., Goldfarb, I., Gol'dshtein, V., & Greenberg, J. B. (2002). Thermal explosion in a hot gas mixture with fuel droplets: a two reactants model. *Combustion Theory and Modelling*, *6*, 1–21.
14. Caldwell, J., & Kwan, Y. Y. (2003). On the perturbation method for the Stefan problem with time-dependent boundary conditions. *International Journal of Heat and Mass Transfer*, *46*, 1497–1501.
15. Caldwell, J., Savović, S., & Kwan, Y. Y. (2003). Nodal integral and finite difference solution of one-dimensional Stefan problem. *ASME Journal of Heat Transfer*, *125*, 523–527.
16. Castanet, G., Lavieille, P., Lebouché, M., & Lemoine, F. (2003). Measurement of the temperature distribution within monodisperse combusting droplets in linear stream using two color laser-induced fluorescence. *Experiments in Fluids*, *35*, 563–571.
17. Castanet, G., Maqua, C., Orain, M., Grisch, F., & Lemoine, F. (2007). Investigation of heat and mass transfer between the two phases of an evaporating droplet stream using laser-induced fluorescence techniques: comparison with modelling. *International Journal of Heat and Mass Transfer*, *50*, 3670–3683.
18. Castanet, G., Labergue, A., & Lemoine, F. (2011). Internal temperature distributions of interacting and vaporizing droplets. *International Journal of Thermal Science*, *50*, 1181–1190.
19. Chauveau, C., Birouk, M., & Gökalp, I. (2011). An analysis of the d^2-law departure during droplet evaporation in microgravity. *International Journal of Multiphase Flow37*, 252–259.
20. Chiang, C. H., Raju, M. S., & Sirignano, W. A. (1992). Numerical analysis of convecting, vaporizing fuel droplet with variable properties. *International Journal of Heat and Mass Transfer*, *35*, 1307–1324.
21. Chiang, C. H., & Sirignano, W. A. (1993). Interacting convecting vaporising fuel droplets with variable properties. *International Journal of Heat and Mass Transfer*, *36*, 875–886.
22. Crua, C., Shoba, T., Heikal, M., Gold, M., & Higham, C. (2010). High-speed microscopic imaging of the initial stage of diesel spray formation and primary breakup, SAE, Report 2010–01-2247.
23. Deprédurand, V., Castanet, G., & Lemoine, F. (2010). Heat and mass transfer in evaporating droplets in interaction: influence of the fuel. *International Journal of Heat and Mass Transfer*, *53*, 3495–3502.
24. Desantes, J. M., López, J. J., García, J. M., & Pastor, J. M. (2007). Evaporative diesel spray modelling. *Atomization and Sprays*, *17*, 193–231.
25. Dombrovsky, L. A., & Sazhin, S. S. (2003). A simplified non-isothermal model for droplet heating and evaporation. *International Communications in Heat Mass Transfer*, *30*, 787–796.
26. Ebrahimian, V., & Habchi, C. (2011). Towards a predictive evaporation model for multi-component hydrocarbon droplets at all pressure conditions. *International Journal of Heat and Mass Transfer*, *54*, 3552–3565.
27. Elwardany, A. E., Gusev, I. G., Castanet, G., Lemoine, F., & Sazhin, S. S. (2011). Mono- and multi-component droplet cooling/heating and evaporation: comparative analysis of numerical models. *Atomization and Sprays*, *21*, 907–931.
28. El Wakil, M. M., Uyehara, O. A., & Myers, P. S. (1954). A theoretical investigation of the heating-up period of injected fuel droplets vaporizing in air. NACA Technical Note 3179.
29. Faeth, G. M. (1977). Current status of droplet and liquid combustion. *Progress in Energy and Combustion Science*, *3*, 191–224.
30. Faeth, G. M. (1983). Evaporation and combustion of sprays. *Progress in Energy and Combustion Science*, *9*, 1–76.
31. Fieberg, C., Reichelt, L., Martin, D., Renz, U., & Kneer, R. (2009). Experimental and numerical investigation of droplet evaporation under diesel engine conditions. *International Journal of Heat and Mass Transfer*, *52*, 3738–3746.
32. Frackowiak, B., Lavergne, G., Tropea, C., & Strzelecki, A. (2010). Numerical analysis of the interactions between evaporating droplets in a monodisperse stream. *International Journal of Heat and Mass Transfer*, *53*, 1392–1401.
33. Frank-Kamenetskii, D. A. (1969). *Diffusion and heat transfer in chemical kinetics*. New York, London: Plenum Press.

34. Frohn, A., & Roth, N. (2000). *Dynamics of droplets*. Berlin and Heidelberg GmbH: Springer-Verlag.
35. Fuchs, N. A. (1959). *Evaporation and droplet growth in gaseous media*. London: Pergamon Press.
36. Goldfarb, I., Gol'dshtein, V., Kuzmenko, G., & Greenberg, J. B. (1998). On thermal explosion of a cool spray in a hot gas. *Proceedings of the 27th International Symposium on Combustion (Vol. 2, pp. 2367–2374)*. Colorado, USA.
37. Goldfarb, I., Gol'dshtein, V., Kuzmenko, G., & Sazhin, S. S. (1999). Thermal radiation effect on thermal explosion in gas containing fuel droplets. *Combustion Theory and Modelling*, *3*, 769–787.
38. Goldfarb, I., Sazhin, S., & Zinoviev, A. (2004). Delayed thermal explosion in flammable gas containing fuel droplets: asymptotic analysis. *International Journal of Engineering Mathematics*, *50*, 399–414.
39. Gupta, S. C. (2003). *The classical Stefan problem: basic concepts, modelling and analysis*. Amsterdam: Elsevier.
40. Haywood, R. J., Nafziger, R., & Renksizbulut, M. (1989). A detailed examination of gas and liquid transient processes in convection and evaporation. *ASME Journal of Heat Transfer*, *111*, 495–502.
41. Hohmann, S., & Renz, U. (2003). Numerical simulation of fuel sprays at high ambient pressure: the influence of real gas effects and gas solubility on droplet vaporisation. *International Journal of Heat and Mass Transfer*, *46*, 3017–3028.
42. Incropera, F. P., & DeWitt, D. P. (1996). *Fundamentals of heat and mass transfer*. New York: John Wiley & Sons.
43. Javed, I., Baek, S. W., Waheed, K., Ali, G., & Cho, S. O. (2013). Evaporation characteristics of kerosene droplets with dilute concentrations of ligand-protected aluminum nanoparticles at elevated temperatures. *Combustion and Flame*, *160*, 2955–2963.
44. Johnson, M. V., Zhu, G. S., Aggarwal, S. K., & Goldsborough, S. S. (2010). Droplet evaporation characteristics due to wet compression under RCM conditions. *International Journal of Heat and Mass Transfer*, *53*, 1100–1111.
45. Kartashov, E. M. (2001). *Analytical methods in the heat transfer theory in solids*. Moscow: Vysshaya Shkola (in Russian).
46. Kim, H., & Sung, N. (2003). The effect of ambient pressure on the evaporation of a single droplet and a spray. *Combustion and Flame*, *135*, 261–270.
47. Kim, H., Baek, S. W., & Chang, D. (2014). A single n-heptane droplet behavior in rapid compression machine. *International Journal of Heat and Mass Transfer*, *69*, 247–255.
48. Klingsporn, M., & Renz, U. (1994). Vaporization of a binary unsteady spray at high temperature and high pressure. *International Journal of Heat and Mass Transfer*, *37*(Suppl. 1), 265–272.
49. Kolaitis, D. I., & Founti, M. A. (2006). A comparative study of numerical models for Eulerian-Lagrangian simulations of turbulent evaporating sprays. *International Journal of Heat and Mass Transfer*, *27*, 424–435.
50. Kontogeorgis, G. M., & Folas, G. K. (2010). *Thermodynamic models for industrial applications*. Chippenham (UK): John Wiley & Sons.
51. Kristyadi, T., Deprédurand, V., Castanet, G., Lemoine, F., Sazhin, S. S., Elwardany, A., et al. (2010). Monodisperse monocomponent fuel droplet heating and evaporation. *Fuel*, *89*, 3995–4001.
52. Kulmala, M., Vesala, T., Schwarz, J., & Smolik, J. (1995). Mass transfer from a drop-II. Theoretical analysis of temperature dependent mass flux correlation. *International Journal of Heat and Mass Transfer*, *38*, 1705–1708.
53. Kuo, K.-K. (1986). *Principles of combustion*. Chichester: John Wiley & Sons.
54. Labowsky, M., Rosner, D. E., & Arias-Zugasti, M. (2011). Turbulence effects on evaporation rate-controlled spray combustor performance. *International Journal of Heat and Mass Transfer*, *54*, 2683–2695.

55. Lavieille, P., Lemoine, F., Lebouché, M., & Lavergne, G. (2001). Evaporating and combusting droplet temperature measurement using two color laser induced fluorescence. *Experiments in Fluids*, *31*, 45–55.
56. Lefebvre, A. H. (1989). *Atomization and sprays*. Bristol, PA: Taylor & Francis.
57. Luijten, C. C. M., & Kurvers, C. (2010). Real gas effects in mixing-limited diesel spray vaporization models. *Atomization and Sprays*, *20*, 595–609.
58. Mandal, D. K., & Bakshi, S. (2012). Internal circulation in a single droplet evaporating in a closed chamber. *International Journal of Multiphase Flow*, *42*, 42–51.
59. Mao, C.-P., Szekely, G. A., & Faeth, G. M. (1980). Evaluation of a locally homogeneous flow model of spray combustion. *Journal of Energy*, *4*, 78–87.
60. Maqua, C., Castanet, G., Lemoine, F., Doué, N., & Lavergne, G. (2006). Temperature measurements of binary droplets using three-color laser-induced fluorescence. *Experiments in Fluids*, *40*, 786–797.
61. Maqua, C., Castanet, G., & Lemoine, F. (2008). Bi-component droplets evaporation: temperature measurements and modelling. *Fuel*, *87*, 2932–2942.
62. Maqua, C., Castanet, G., Grisch, F., Lemoine, F., Kristyadi, T., & Sazhin, S. S. (2008). Monodisperse droplet heating and evaporation: experimental study and modelling. *International Journal of Heat and Mass Transfer*, *51*, 3932–3945.
63. Mashayek, F. (2001). Dynamics of evaporating drops. Part 1: Formulation and evaporation model. *International Journal of Heat and Mass Transfer*, *44*, 1517–1526.
64. Maxwell, J. B. (1950). *Data book on hydrocarbons: application to process engineering*. New York: D. van Nostrand Company, INC.
65. McIntosh, A. C., Gol'dshtein, V., Goldfarb, I., & Zinoviev, A. (1998). Thermal explosion in a combustible gas containing fuel droplets. *Combustion Theory and Modelling*, *2*, 153–165.
66. Miliauskas, G., & Garmus, V. (2009). The peculiarities of hot liquid droplets heating and evaporation. *International Journal of Heat and Mass Transfer*, *52*, 3726–3737.
67. Mitchell, S. L., Vynnycky, M., Gusev, I. G., & Sazhin, S. S. (2011). An accurate numerical solution for the transient heating of an evaporating spherical droplet. *Applied Mathematics and Computation*, *217*, 9219–9233.
68. Mukhopadhyay, A., & Sanyal, D. (2005). A semi-analytical model for evaporating fuel droplets. *ASME Journal of Heat Transfer*, *127*, 199–203.
69. Myers, T. G. (2010). Optimal exponent heat balance and refined integral methods applied to Stefan problems. *International Journal of Heat and Mass Transfer*, *53*, 1119–1127.
70. Naaktgeboren, C. (2007). The zero-phase Stefan problem. *International Journal of Heat and Mass Transfer*, *50*, 4614–4622.
71. Nafziger, R. (1988). Convective droplet transport phenomena in high temperature air streams. *M.A.Sc. Thesis*, University of Waterloo, Ontario, Canada.
72. Park, T. W., Aggarwal, S. K., & Katta, V. R. (1995). Gravity effects on the dynamics of evaporating droplets in a heated jet. *Journal of Propulsion and Power*, *11*, 519–528.
73. Paschedag, A. R., Piarah, W. H., & Kraume, M. (2005). Sensitivity study for the mass transfer at a single droplet. *International Journal of Heat and Mass Transfer*, *48*, 3402–3410.
74. Peng, D.-Y., & Robinson, D. B. (1976). A new two-constant equation of state. *Industrial and Engineering Chemistry Fundamentals*, *15*, 59–64.
75. Rayapati, N. P., Panchagnula, M. V., Peddieson, J., Short, J., & Smith, S. (2011). Eulerian multiphase population balance model of atomizing, swirling flows. *International Journal of Spray and Combustion Dynamics*, *3*, 111–136.
76. Rayapati, N. P., Panchagnula, M. V., & Peddieson, J. (2013). Application of population balance model to combined atomization and evaporation processes in dense sprays. *Atomization and Sprays*, *23*, 505–523.
77. Reid, R. C., Prausnitz, J. M., & Sherwood, T. K. (1977). *The properties of gases and liquids* (3rd ed.). New York: McGraw-Hill.
78. Renksizbulut, M., & Yuen, M. C. (1983). Experimental study of droplet evaporation in a high-temperature air stream. *ASME Journal of Heat Transfer*, *105*, 384–388.

79. Renksizbulut, M., & Yuen, M. C. (1983). Numerical modeling of droplet evaporation in a high-temperature air stream. *ASME Journal of Heat Transfer, 105*, 389–397.
80. Savović, S., & Caldwell, J. (2003). Finite difference solution of one-dimensional Stefan problem with periodic boundary conditions. *International Journal of Heat and Mass Transfer, 46*, 2911–2916.
81. Sazhin, S. S. (2006). Advanced models of fuel droplet heating and evaporation. *Progress in Energy and Combustion Science, 32*(2), 162–214.
82. Sazhin, S. S., & Heikal, M. R. (2012). Droplet heating and evaporation—recent results and unsolved problems. *Computational Thermal Sciences, 4*(6), 485–496.
83. Sazhin, S. S., Feng, G., Heikal, M. R., Goldfarb, I., Goldshtein, V., & Kuzmenko, G. (2001). Thermal ignition analysis of a monodisperse spray with radiation. *Combustion and Flame, 124*, 684–701.
84. Sazhin, S. S., Krutitskii, P. A., Abdelghaffar, W. A., Mikhalovsky, S. V., Meikle, S. T., & Heikal, M. R. (2004). Transient heating of diesel fuel droplets. *International Journal of Heat and Mass Transfer, 47*, 3327–3340.
85. Sazhin, S. S., Abdelghaffar, W. A., Krutitskii, P. A., Sazhina, E. M., & Heikal, M. R. (2005). New approaches to numerical modelling of droplet transient heating and evaporation. *International Journal of Heat and Mass Transfer, 48*, 4215–4228.
86. Sazhin, S. S., Abdelghaffar, W. A., Sazhina, E. M., & Heikal, M. R. (2005). Models for droplet transient heating: effects on droplet evaporation, ignition, and break-up. *International Journal of Thermal Sciences, 44*, 610–622.
87. Sazhin, S. S., Elwardany, A., Krutitskii, P. A., Castanet, G., Lemoine, F., Sazhina, E. M., et al. (2010). A simplified model for bi-component droplet heating and evaporation. *International Journal of Heat and Mass Transfer, 53*, 4495–4505.
88. Sazhin, S. S., Krutitskii, P. A., Gusev, I. G., & Heikal, M. R. (2010). Transient heating of an evaporating droplet. *International Journal of Heat and Mass Transfer, 53*, 2826–2836.
89. Sazhin, S. S., Elwardany, A., Krutitskii, P. A., Deprédurand, V., Castanet, G., Lemoine, F., et al. (2011). Multi-component droplet heating and evaporation: numerical simulation versus experimental data. *International Journal of Thermal Sciences, 50*, 1164–1180.
90. Sazhin, S. S., Krutitskii, P. A., Gusev, I. G., & Heikal, M. (2011). Transient heating of an evaporating droplet with presumed time evolution of its radius. *International Journal of Heat and Mass Transfer, 54*, 1278–1288.
91. Shanthanu, S., Raghuram, S., & Raghavan, V. (2013). Transient evaporation of moving water droplets in steam-hydrogen-air environment. *International Journal of Heat and Mass Transfer, 64*, 536–546.
92. Sirignano, W. A., & Law, C. K. (1978). Transient heating and liquid phase mass diffusion in droplet vaporization. In J. T. Zung (Ed.), *Evaporation-combustion of fuels. Advances in chemistry series* (Vol. 166, pp. 1–26). Washington, DC: American Chemical Society.
93. Tikhonov, A. N., & Samarsky, A. A. (1972). *Equations of mathematical physics*. Moscow: Nauka Publishing House (in Russian).
94. Tonini, S., & Cossali, G. E. (2012). An analytical model of liquid drop evaporation in gaseous environment. *International Journal of Thermal Sciences, 57*, 45–53.
95. Tonini, S., & Cossali, G. E. (2013). An exact solution of the mass transport equations for spheroidal evaporating drops. *International Journal of Heat and Mass Transfer, 60*, 236–240.
96. Tonini, S., & Cossali, G. E. (2014). An evaporation model for oscillating spheroidal drops. *International Communications in Heat and Mass Transfer, 51*, 18–24.
97. Voller, V. R., & Falcini, F. (2013). Two exact solutions of a Stefan problem with varying diffusivity. *International Journal of Heat and Mass Transfer, 58*, 80–85.
98. Wu, J.-S., Hsu, K.-H., Kuo, P.-M., & Sheen, H.-J. (2003). Evaporation model of a single hydrocarbon fuel droplet due to ambient turbulence at intermediate Reynolds numbers. *International Journal of Heat and Mass Transfer, 46*, 4741–4745.
99. Wu, J.-S., Liu, Y.-J., & Sheen, H.-J. (2001). Effects of ambient turbulence and fuel properties on the evaporation rate of single droplets. *International Journal of Heat and Mass Transfer, 44*, 4593–4603.

100. Xu, G., Ikegami, M., Honma, S., Ikeda, K., Ma, X., Nagaishi, H., et al. (2003). Inverse influence of the initial diameter on droplet burning rate in cold and hot ambiences: a thermal action of flame in balance with heat loss. *International Journal of Heat and Mass Transfer*, *46*, 1155–1169.
101. Yang, K., Hong, F., & Cheng, P. (2014). A fully coupled numerical simulation of sessile droplet evaporation using Arbitrary Lagrangian-Eulerian formulation. *International Journal of Heat and Mass Transfer*, *70*, 409–420.
102. Yang, J.-R., & Wong, S.-C. (2001). On the discrepancies between theoretical and experimental results for microgravity droplet evaporation. *International Journal of Heat and Mass Transfer*, *44*, 4433–4443.
103. Yang, J.-R., & Wong, S.-C. (2002). An experimental and theoretical study of the effects of heat conduction through the support fiber on the evaporation of a droplet in a weakly convective flow. *International Journal of Heat and Mass Transfer*, *45*, 4589–4598.
104. Yao, G. F., Abdel-Khalik, S. I., & Ghiaasiaan, S. M. (2003). An investigation of simple evaporation models used in spray simulations. *ASME Journal of Heat Transfer*, *125*, 179–182.
105. Zhou, Y., Wang, Y., & Bu, W. (2014). Exact solution for a Stefan problem with latent heat a power function of position. *International Journal of Heat and Mass Transfer*, *69*, 451–454.

Chapter 5
Heating and Evaporation of Multicomponent Droplets

Abstract The chapter begins with an overview of early developments of the models for multicomponent droplet heating and evaporation. Then recently developed models for bicomponent droplet heating and evaporation, based on the analytical solution to the equation for species diffusion inside liquid droplets, are discussed. These models are generalised to the case that more than two components are present in liquid droplets (biodiesel fuel droplets). The original analytical solution to the species diffusion equation was obtained under the assumption that droplet radius is constant; it was generalised to the case that the droplet radius is a linear function of time during each individual time step. A recently developed quasidiscrete model, suitable for modelling heating and evaporation of droplets consisting of a large number of components, is described and applied to the analysis of Diesel and gasoline fuel droplets.

The droplet heating and evaporation models described so far have been based on the assumption that liquid consists of one component only. This assumption is not valid for most practically important fuels, including Diesel and gasoline fuels. The application of the single-component assumption in this case was justified not by the physical nature of the problem, but by the fact that it led to considerable simplification of the modelling of the processes involved.

An overview of early developments of the models for multicomponent droplet heating and evaporation is presented in Sect. 5.1. Section 5.2 is focused on a recently developed model for bicomponent droplet heating and evaporation, based on the analytical solution to the species diffusion equation. This model can be generalised to the case of a relatively small number of components, and this generalisation is also briefly discussed in Sect. 5.2. A recently developed quasidiscrete model, suitable for modelling of heating and evaporation of droplets consisting of a large number of components, is described and analysed in Sect. 5.3.

S. Sazhin, *Droplets and Sprays*, DOI: 10.1007/978-1-4471-6386-2_5,

5.1 Background

In the case of multicomponent droplets, different components evaporate at different rates, creating concentration gradients in the liquid phase. The latter leads to the liquid phase mass diffusion described by the diffusion equation for the mass fractions of each component. The simplest form of this equation, when only the radial diffusion is accounted for, can be represented as [36, 57]

$$\frac{\partial Y_{l,i}}{\partial t} = D_l \left(\frac{\partial^2 Y_{l,i}}{\partial R^2} + \frac{2}{R} \frac{\partial Y_{l,i}}{\partial R} \right), \tag{5.1}$$

where subscripts $_l$ and $_i$ indicate liquid phase and the particular type of species respectively. It was assumed that the diffusion coefficient D_l is the same for all liquid components. An analysis of the diffusion equation for $Y_{l,i}$ without the spherical symmetry assumption was discussed in [9, 47, 58].

In most cases, including in this book, the value of D_l has been estimated based on the laminar model. The effects of turbulence on D_l are considered in [2].

If we further assume that the diffusion coefficients referring to all species in the gas phase are the same (this assumption was relaxed in [22]), then the evaporation rate of the ith species can be represented as

$$\dot{m}_i = \varepsilon_i \dot{m}_d = -4\pi \varepsilon_i R_d D_v \rho_{\text{total}} \ln(1 + B_M), \tag{5.2}$$

where ε_i describes the species evaporation rate. Assuming that species concentrations in the ambient gas are equal to zero, the values of ε_i can be found from the following relation [13, 19, 52, 62]:

$$\varepsilon_i = \frac{Y_{vis}}{\sum_i Y_{vis}}, \tag{5.3}$$

where the subscript $_v$ indicates the vapour phase.

In deriving (5.2), we took into account (4.16). Note that the latent heat of evaporation of multicomponent droplets can be found from the expression

$$L = \sum_i \varepsilon_i L_i, \tag{5.4}$$

where L_i is the latent heat of evaporation of the ith species.

If there were no supply of the ith species to the surface from the interior of the droplet, then we would have

$$Y_{l,is} = \varepsilon_i, \tag{5.5}$$

where the additional subscript $_s$ indicates 'surface'. In the general case, this supply of the ith species can be described by the following mass diffusion flow rate:

$$\dot{m}_{i\,(\mathrm{suppl})} = -4\pi R_d^2 D_l \frac{\partial Y_{l,i}}{\partial R}\bigg|_{R=R_d}. \tag{5.6}$$

In [23], Eq. (5.6) was generalised to take into account the effect of change of droplet radius during the time step. This effect is considered later in this chapter, using an approach different from that used in [23].

Note that the analysis of evaporation of multicomponent droplets by some authors (e.g., [24]) led to the following expression for $\dot{m}_d$:

$$\dot{m}_d = -2\pi R_d D_{iv} \rho_{\mathrm{total}} B_{Mi} \mathrm{Sh}_{\mathrm{iso}\,(i)}, \tag{5.7}$$

where B_{Mi} is the species Spalding mass transfer number, defined as

$$B_{Mi} = \frac{Y_{\mathrm{vis}} - Y_{vi\infty}}{\varepsilon_i - Y_{\mathrm{vis}}}, \tag{5.8}$$

D_{iv} is the diffusion coefficient of species i in air, $\mathrm{Sh}_{\mathrm{iso}\,(i)}$ was defined by (4.28), replacing B_M with B_{Mi}. Recalling (5.3), one can see that $B_M = B_{Mi}$. Hence, for stationary droplets, this leads to the paradox that the same value of $\dot{m}_d$ is predicted by (5.7) for different D_{iv}. This paradox is resolved by the fact that although (5.7) is correct, the value of $\mathrm{Sh}_{\mathrm{iso}\,(i)}$ cannot be approximated by the analogue of the corresponding equation presented in Chap. 4 (see Eq. (4.28)), which is implicitly based on the assumption that the evaporating species do not affect each other. In our analysis, it is implicitly assumed that D_{iv} are the same for all species (see [15, 59] for an in-depth discussion of this issue).

In the steady-state case, $\dot{m}_{i\,(\mathrm{suppl})}$ is equal to the difference between the actual evaporation flux and the evaporation flux which would take place in the absence of supply. This allows us to write the mass balance equation at the surface of the droplet in the following form:

$$\frac{\partial Y_{l,i}}{\partial R}\bigg|_{R=R_d-0} = \frac{D_v \rho_{\mathrm{total}} \ln(1+B_M)}{D_l \rho_l R_d} \left(Y_{l,i} - \varepsilon_i\right). \tag{5.9}$$

This equation was derived in [57] following a different procedure. It can be considered a boundary condition for (5.1) at the droplet surface. This needs to be supplemented by the conditions

$$\frac{\partial Y_{l,i}}{\partial R}\bigg|_{R=0} = 0 \tag{5.10}$$

at the droplet centre and the relevant initial conditions. Note that as in the case of the equation for temperature inside droplets, condition (5.10) can be replaced by a more general requirement that $Y_{l,i}(R,t)$ be continuously twice differentiable functions for $R \le R_d$.

In some papers (e.g., [49]), the effects of species diffusion and heat transfer inside droplets were taken into account based on a two-layer model: droplet core was assumed to be homogeneous and described by temperature T_d and species mass fractions Y_{il}, which are different from temperature and mass fractions at the surface of the droplets. This model had to be based on the introduction of semiempirical constants describing heat and species transfer processes inside droplets. It was expected to predict correct trends in the time evolution of temperature and species mass fractions inside droplets, but it cannot be used for accurate quantitative analysis of the processes. This model will not be used in our analysis.

In the equilibrium state, the partial pressure of the ith vapour species at the surface of the droplet can be found from the equation

$$p_{v,i} = \gamma_i X_{l,i} p^*_{v,i}, \tag{5.11}$$

where $X_{l,i}$ is the molar fraction of the ith species in the liquid near the droplet surface, $p^*_{v,i}$ is the partial vapour pressure of the ith species in the case $X_{l,i} = 1$, γ_i is the activity coefficient. In some applications, the latter coefficient can be assumed equal to 1. In this case, (5.11) leads to the Raoult law:

$$p_{v,i} = X_{l,i} p^*_{v,i}. \tag{5.12}$$

A molecular interpretation of (5.12) can be found in [5] (pp. 169–170). Equation (5.12) is an approximate one, and deviations from the Raoult law are possible [6, 57].

Using the Clausius–Clapeyron equation for a single species, we can write [33]

$$\ln\left(\frac{p^*_{v,i}}{p}\right) = \frac{L_i}{R_u M_i}\left(\frac{1}{T_{b,i}} - \frac{1}{T_s}\right), \tag{5.13}$$

where M_i is the molar mass, $T_{b,i}$ is the boiling temperature of the ith species, p is the ambient pressure. In deriving (5.13), we took into account that $p^*_{v,i}$ is equal to the ambient pressure when $T_s = T_{b,i}$. In view of (5.12), Eq. (5.13) can be rewritten as [57]

$$\frac{X_{v,i}}{X_{l,i}} = \tilde{p}_v^{-1} \exp\left[\frac{L_i}{R_u M_i}\left(\frac{1}{T_{b,i}} - \frac{1}{T_s}\right)\right], \tag{5.14}$$

where $\tilde{p}_v = p_v/p$.

As in the case of one-component droplets, the numerical solution of (5.1) might be too CPU intensive for practical implementation into engineering CFD codes when more than two components are involved. In practice, this approach has been applied mainly to the case of binary droplets [9, 36, 43, 44].

Two limiting cases were suggested in which the solution of this equation becomes no longer necessary. These are (a) the rapid regression or zero diffusivity limit, and (b) the uniform concentration or infinite-diffusivity limit [20, 57]. In the first limit, the composition of droplets remains constant during the whole evaporation process. The analysis of this case would be similar to the analysis of one-component droplets. In the second limit, the droplet composition changes with time in such a way that the mass fraction of its less-volatile components increases. Although the second limit is strictly applicable to the description of slow evaporation processes only [57], in practice it is used in a much wider range of applications (see, e.g., [20, 31]).

The analysis of this section so far has been based on the assumption that the droplets are stationary and there is no contribution from convection processes. In the general case of moving droplets, the situation becomes much more complicated, and recirculation inside the droplets would need to be taken into account [57]. This, however, would make the whole model far too complicated for most engineering applications, as in the case of single-component droplets. The analysis of moving droplets can be greatly simplified when the actual recirculation inside them is accounted for by an increase of diffusivity in them by a factor similar to χ_T (see (3.73)) used for the analysis of heat transfer (with Pr replaced by Sc) [14]. This increase of diffusivity inside droplets could provide further support to the infinite diffusivity limit discussed earlier. Analysis of (5.1) together with the heat transfer equation, in which the contribution of thermal radiation is taken into account, was discussed in [35], where a simplified model, based on the assumption that the absorption of thermal radiation is homogeneous, was used.

The model based on (5.1) is known as the Discrete Component Model (DCM). It is typically applicable only when the number of components in the droplets is small (e.g., [32, 63]). In realistic cases, such as Diesel or gasoline fuels, when the number of components in a droplet is measured in hundreds, this approach ceases to be practical. An alternative approach is based on the probabilistic analysis of a large number of components (e.g., Continuous Thermodynamics approach [1, 4, 26, 40, 45, 48, 61, 67] and the Distillation Curve Model [10]). In the second family of models, a number of additional simplifying assumptions were used, including the assumption that species inside droplets mix infinitely quickly.

The Continuous Thermodynamics approach is based on the introduction of the distribution function $f_m(I)$ such that

$$\int_{I_1}^{I_2} f_m(I)\mathrm{d}I = 1, \tag{5.15}$$

where I is a property of the component (usually taken as the molar mass), f_m characterises the relative contribution of the components having this property in the vicinity of I, I_1 and I_2 are limiting values of this property. For most practically important fuels, $f_m(I)$ can be approximated by relatively simple functions. For example, the following function was considered in [4, 40, 45, 67]:

$$f_m(I) = C_m(I_1, I_2)\frac{(I-\gamma)^{\alpha-1}}{\beta^{\alpha}\Gamma(\alpha)}\exp\left[-\left(\frac{I-\gamma}{\beta}\right)\right], \tag{5.16}$$

where $\Gamma(\alpha)$ is the gamma function, α and β are parameters that determine the shape of the distribution, γ determines the original shift, $C_m(I_1, I_2)$ is the normalization constant, $C_m(I_1, I_2) = 1$ for $I_1 = \gamma$ and $I_2 = \infty$. The authors of [29] considered a more general function presented as the weighted sum of the functions (5.16) (double-gamma PDF).

The main attractive feature of this approach to modelling multicomponent droplet evaporation is that the composition of fuel can be characterised by just a few key parameters (two parameters α and β for given γ when the function (5.16) is taken), instead of hundreds of values of mole fractions of individual components, required in the conventional approach (based on the solution of (5.1)). As in the conventional approach, it is assumed that the gas diffusion coefficients for all the components are the same. See [4, 27–29, 40, 45, 66, 67] for details of the implementation of this approach.

The Distillation Curve Model (DC model) is capable of taking into account the distillation curve of the actual multicomponent fuels. The most important feature of this model is the description of fractional boiling during the droplet evaporation process as a function of a single variable: the actual mean molar mass of fuel inside the droplet. An additional advantage is that the model is based on algebraic equations, which brings clear advantages from the point of view of CPU efficiency. The influence of diffusion resistance in droplets on evaporation is quantified by the liquid Peclet number [10]

$$\mathrm{Pe}_l = \frac{|\dot{m}_d|}{4\pi R_d D_l \rho_l}. \tag{5.17}$$

A more detailed analysis of this model is given in [10].

A model containing features of both discrete and continuous models was suggested in [37–39, 65]. In [65], the mixtures of multicomponent (Diesel fuel, gasoline, biodiesel) and monocomponent substances were studied based on a combination of Discrete Components and Continuous Thermodynamics models. As in the case of the classical Continuous Thermodynamics approach, it was assumed that the mixing processes inside droplets are infinitely fast for both species and temperature. The analysis of [37–39] was based on the application of the Quadrature Method of Moments (QMoM), originally developed in [34]. This method allows one to use two or three pseudocomponents for each group of components instead of dozens of real components for the whole mixture. The normal boiling point of each pseudocomponent was allowed to change during the evaporation process. This approach was shown to be particularly useful if the condensation process needs to be modelled alongside the evaporation. As in the case of the conventional Continuous Thermodynamics approach, it was assumed that the droplets are well mixed.

A model called the discrete/continuous multicomponent (DCMC) model, which focused primarily on gasoline fuel, was suggested in [64]. In this approach, gasoline was assumed to consist of five discrete families of hydrocarbons: n-paraffins, i-paraffins, naphthenes, aromatics, and olefins. Each family of hydrocarbons was composed of an infinite number of components, which are modeled using a probability density function (PDF), and the mass fraction of each family of hydrocarbons (PDF), the mean and variance of each PDF were tracked. Compared with the discrete multicomponent (DMC) model, the DCMC model saves computer time. Compared with the continuous multicomponent (CMC) model, the DCMC model has much higher accuracy. As in the case of the conventional Continuous Thermodynamics and combined approaches mentioned earlier, it was assumed that the droplets are well mixed.

The authors of [46] applied the Discrete Component Model for studying mixtures of substances containing many components (Diesel and gasoline fuels) by approximating these fuels with relatively small numbers of physical species (six species for Diesel fuel and seven species for gasoline). They took into account the effects of temperature gradient inside droplets, but assumed that the mass transfer processes inside droplets are infinitely fast. As already mentioned, this assumption is far from obvious.

One of the main limitations of the Discrete Component Models, suggested so far, is that in most cases, they are based on the numerical solution of (5.1), which can be a serious limitation when the models are implemented in a CFD code. The main limitation of the Continuous Thermodynamics and Distillation Curve approaches is that both models are based on the assumption of infinitely fast diffusivity of species inside droplets. These limitations were overcome in the recently developed version of the DCM, based on the analytical solution to (5.1), and the quasidiscrete model. These models are described in Sects. 5.2, following [17, 52, 53], and 5.3, following [16, 54], respectively.

5.2 Bicomponent Droplets

The model for bicomponent droplet heating and evaporation developed in [52, 53] is essentially the DCM described in the previous section, with the only difference that the numerical solution to the species diffusion equation (5.1) is replaced with its analytical solution. The temperature gradient and recirculation inside droplets are taken into account based on the Effective Thermal Conductivity (ETC) model (see Eq. (3.73)), taking or not taking into account the effect of the moving interface during the time step (see Sect. 4.4). To take into account the effect of the moving interface, a model based on the assumption that the droplet radius is a linear function of time was used (see Sect. 4.4).

The analytical solutions to this equation, ignoring and taking into account the effects of the moving interface during the time step, are presented and discussed in Sect. 5.2.1. These solutions are analysed and validated against experimental data, where possible, in Sect. 5.2.2.

5.2.1 Analytical Solutions to the Species Equation

Assuming that R_d = const, the analytical solution to (5.1) for a short time step, subject to boundary condition (5.9) and the initial condition $Y_{li}(t=0) = Y_{li0}(R)$, was obtained in the form [52, 53] (see Appendix H)

$$Y_{li} = \varepsilon_i + \frac{1}{R}\left\{\left[\exp\left[D_l\left(\frac{\lambda_0}{R_d}\right)^2 t\right][q_{Yi0} - Q_{Y0}\varepsilon_i]\sinh\left(\lambda_0\frac{R}{R_d}\right)\right.\right.$$
$$\left.\left.+\sum_{n=1}^{\infty}\left[\exp\left[-D_l\left(\frac{\lambda_n}{R_d}\right)^2 t\right][q_{Yin} - Q_{Yn}\varepsilon_i]\sin\left(\lambda_n\frac{R}{R_d}\right)\right]\right]\right\}, \quad (5.18)$$

where λ_0 and λ_n ($n \geq 1$) are solutions to the equations

$$\tanh\lambda = -\frac{\lambda}{h_{Y0}} \quad \text{and} \quad \tan\lambda = -\frac{\lambda}{h_{Y0}}$$

respectively, $h_{Y0} = -\left(1 + \frac{\alpha_m R_d}{D_l}\right)$,

$$Q_{Yn} = \begin{cases} -\frac{1}{||v_{Y0}||^2}\left(\frac{R_d}{\lambda_0}\right)^2(1+h_{Y0})\sinh\lambda_0 & \text{when} \quad n=0 \\ \frac{1}{||v_{Yn}||^2}\left(\frac{R_d}{\lambda_n}\right)^2(1+h_{Y0})\sin\lambda_n & \text{when} \quad n\geq 1 \end{cases} \quad (5.19)$$

$$q_{Yin} = \frac{1}{||v_n||^2}\int_0^{R_d} R Y_{li0}(R) v_{Yn}(R)\mathrm{d}R, \quad (5.20)$$

$n \geq 0$,

$$v_{Y0}(R) = \sinh\left(\lambda_0\frac{R}{R_d}\right), \quad v_{Yn}(R) = \sin\left(\lambda_n\frac{R}{R_d}\right), \quad n\geq 1,$$

$$||v_{Y0}||^2 = \int_0^{R_d} v_{Y0}^2(R)\mathrm{d}R = -\frac{R_d}{2}\left[1 + \frac{h_{Y0}}{h_{Y0}^2 - \lambda_n^2}\right], \quad (5.21)$$

$$||v_{Yn}||^2 = \int_0^{R_d} v_{Yn}^2(R)\mathrm{d}R = \frac{R_d}{2}\left[1 + \frac{h_{Y0}}{h_{Y0}^2 + \lambda_n^2}\right], \quad n\geq 1, \quad (5.22)$$

$Y_{lis} = Y_{lis}(t)$ are the liquid components' mass fractions at the droplet's surface,

$$\alpha_m = \frac{|\dot{m}_d|}{4\pi \rho_l R_d^2}, \tag{5.23}$$

$\dot{m}_d$ is the droplet evaporation rate.

Note that the norms $||v_{Y0}||^2$ and $||v_{Yn}||^2$ defined by (5.21) and (5.22) contain the term R_d (cf. the corresponding expression for $||v_n||^2$ used in expression (3.37)).

Since we are interested only in a solution which is continuously twice differentiable in the whole domain, Y_{li} should be bounded for $0 \leq R < R_d$. Moreover, the physical meaning of Y_{li}, as the mass fraction, implies that $0 \leq Y_{li} \leq 1$.

Note that solutions (5.18) and (3.37) look rather different, although Eq. (5.1) with boundary conditions (5.9) have essentially the same structure as (3.1) with boundary condition (3.36). The former equations can be obtained from the latter if we replace T with Y_{li}, κ with D_l, T_g with ε_i, T_s with Y_{lis}, and k_l/h with $-D_l/\alpha_m$. The reason for this is that the solution (3.37) is valid only for $h_0 > -1$. At the same time, the boundary condition for (5.1) at $R = R_d$ (see (5.9)) leads to the situation in which $h_{Y0} < -1$.

The condition α_m = const can always be guaranteed for sufficiently small time steps.

The solution (5.18) is based essentially on the assumption that R_d = const. To take into account the effect of the moving interface due to evaporation, we assume that $R_d(t)$ is a linear function of time and is defined by expression (4.83). In this case, the solution to Eq. (5.1) for a short time step, subject to the above boundary and initial conditions, can be represented in the form [25] (see Appendix I)

$$\begin{aligned} Y_{li}(R) = & \frac{\alpha_m \varepsilon_i \exp\left[\frac{\alpha R_{d0}}{4D_l}\left(\frac{R_{d0}R_d(t)-R^2}{R_d(t)}\right)\right]}{\alpha_m + \frac{\alpha R_{d0}}{2}} \frac{R_{d0}^{5/2}}{R_d^{5/2}(t)} + \frac{1}{R\sqrt{R_d(t)}} \exp\left[-\frac{\alpha R_{d0} R^2}{4D_l R_d(t)}\right] \\ & \times \left[\sum_{n=1}^{\infty} [q_{Yn} + f_{Yn}\mu_{Y0}(0)] \exp\left[-\frac{D_l \lambda_n^2 t}{R_{d0} R_d(t)}\right] \sin\left(\lambda_n \frac{R}{R_d(t)}\right)\right. \\ & \left. + [q_{Y0} + f_{Y0}\mu_{Y0}(0)] \exp\left[\frac{D_l \lambda_0^2 t}{R_{d0} R_d(t)}\right] \sinh\left(\lambda_0 \frac{R}{R_d(t)}\right)\right], \end{aligned} \tag{5.24}$$

where

$$f_{Yn} = \frac{R_d}{||v_{Yn}||^2} \int_0^1 f_Y(r) v_{Yn}(r) \mathrm{d}r = \begin{cases} \frac{R_d}{||v_{Y0}||^2 \lambda_0^2} \sinh \lambda_0 & \text{when} \quad n = 0 \\ -\frac{R_d}{||v_{Yn}||^2 \lambda_n^2} \sin \lambda_n & \text{when} \quad n \geq 1 \end{cases},$$

$$q_{Yn} = \frac{R_d}{||v_{Yn}||^2} \int_0^1 W_{Y0}(r) v_{Yn}(r) \mathrm{d}r,$$

$$f_Y(r) \equiv -r/(1+h_{Y0}) = \sum_{n=0}^{\infty} f_{Yn} v_{Yn}(r), \tag{5.25}$$

$$W_{Y0}(r) = \sum_{n=0}^{\infty} q_{Yn} v_{Yn}(r), \tag{5.26}$$

$$\mu_{Y0}(t) \equiv -\frac{\alpha_m \varepsilon_i \,(R_d(t))^{5/2}}{D_l} \exp\left[\frac{R_d^{'}(t) R_d(t)}{4D_l}\right], \tag{5.27}$$

$$H_{Y0}(t) \equiv h_{Y0} = -\frac{\alpha_m}{D_l} R_d(t) - 1 - \frac{R_d^{'}(t) R_d(t)}{2D_l} = \text{const} < -1, \tag{5.28}$$

$$v_{Y0}(r) = \sinh\left(\lambda_0 r\right),$$

$r = R/R_d$, λ_0 is the solution to the equation $\tanh \lambda = -\lambda/h_{Y0}$,

$$v_{Yn}(r) = \sin\left(\lambda_n r\right)$$

for $n \geq 1$, λ_n are the solutions to the equation $\tan \lambda = -\lambda/h_{Y0}$. A countable set of positive solutions to this equation (positive eigenvalues) λ_n is arranged in ascending order, as in the case of the solutions to Eq. (3.38).

The norms $||v_{Y0}||^2$ and $||v_{Yn}||^2$ are defined by (5.21) and (5.22).

The solution (5.24) was obtained under the assumption that the parameter h_{Y0} is a constant and less than -1 during the time step (it can vary from one time step to another). In the general case of a time-dependent h_{Y0}, the solution of the differential heat conduction equation was reduced to the solution of the Volterra integral equation of the second kind (see [25] and Appendix I for details).

It is essential for us to retain both α and α_m in (5.24) to be able to compare the prediction of this equation with the prediction of the conventional approach to modelling species diffusion when $\alpha = 0$ but $\alpha_m \neq 0$ during the time step. In the latter case, since we have that $R_d'(t) = 0$, Eq. (5.24) reduces to (5.18).

5.2.2 Analysis of the Results

This section is focused on the analysis of the bicomponent droplet heating and evaporation predicted by the models based on the analytical solutions to the species diffusion equation described in Sect. 5.2.1. This analysis is performed separately for the model based on the assumptions that the droplet radius is constant during the time step, radius is a linear function of time but the droplet temperature is frozen (the process of species diffusion is separated from the process of droplet heating), radius is a linear function of time and both processes of species diffusion and droplet heating are taken into account. The modelling results will be validated against experimental data where appropriate.

5.2.2.1 Heating and Evaporation When R_d = const

The above model, ignoring the effects of the moving interface, was validated against the experimental data obtained with help of the setup shown in Fig. 4.6, but for a mixture of acetone and ethanol. These results were reported in [17, 52, 53]. In what follows, they are briefly reproduced, mainly following [17].

The analysis was based on (5.11) (the results will be compared with the predictions of the Raoult law) with the activity coefficients described by the following equation [11, 41]:

$$\ln \gamma_i = \frac{1}{R_u T} \frac{\partial \left(n_{l,\,\mathrm{total}} G^E\right)}{\partial n_{li}}, \tag{5.29}$$

where R_u is the universal gas constant, T is temperature in Kelvin, $n_{l,\,\mathrm{total}} = \sum_i n_{li}$, n_{li} is the molar concentration of the ith component in the liquid phase.

The following approximation for the ethanol ($i = 1$)–acetone ($i = 2$) mixture, based on fitting experimental data, was suggested in [11]:

$$n_{l,\,\mathrm{total}} G^E = R_u T \frac{n_{l1} n_{l2}}{n_{l1} + n_{l2}} \left[\frac{\alpha n_{l2}}{n_{l1} + n_{l2}} + \frac{\beta n_{l1}}{n_{l1} + n_{l2}} - \frac{\delta n_{l1} n_{l2}}{(n_{l1} + n_{l2})^2} \right], \tag{5.30}$$

where

$$\alpha = \frac{546.3}{T} - 0.9897,$$

$$\beta = \frac{543.3}{T} - 0.9483,$$

$$\delta = \frac{15.63}{T} + 0.0759.$$

Substituting (5.30) into (5.29) yields the following expressions for $\gamma_1 = \gamma_{\mathrm{eth}}$ and $\gamma_2 = \gamma_{\mathrm{ac}}$:

$$\left. \begin{aligned} \gamma_1 &= \exp\left[X_{l2}^2 \left[\alpha + 2\left(\beta - \alpha - \delta\right) X_{l1} + 3\delta X_{l1}^2\right]\right] \\ \gamma_2 &= \exp\left[X_{l1}^2 \left[\beta + 2\left(\alpha - \beta - \delta\right) X_{l2} + 3\delta X_{l2}^2\right]\right] \end{aligned} \right\}. \tag{5.31}$$

The expressions (5.31) were used in (5.11).

As in the case of pure acetone and ethanol discussed in Sect. 4.5, the approximation for the measured time evolution of the droplet velocities for various mixtures of acetone and ethanol are summarised in Table 5.1.

Table 5.1 Approximations of the droplet velocities for acetone and ethanol mixtures

Substance	Approximation of U_{drop} in m/s (t is in ms)
25 % ethanol + 75 % acetone	$12.75 - 0.370\,t$
50 % ethanol + 50 % acetone	$12.71 - 0.448\,t$
75 % ethanol + 25 % acetone	$12.28 - 0.306\,t$

Table 5.2 The measured initial values of droplet temperature, droplet diameter, ambient gas temperature, and distance parameter for the same cases as in Table 5.1

Substance	Droplet temperature (°C)	Diameter (μm)	Gas temp. (°C)	Distance parameter
25 % ethanol + 75 % acetone	32.5	133.8	21.1	8.7
50 % ethanol + 50 % acetone	37.5	142.7	20.8	7.53
75 % ethanol + 25 % acetone	38.6	137.1	21.6	7.53

The measured initial values of droplet temperature, diameter, ambient gas temperature, and distance parameter C for the same cases as in Table 5.1 are shown in Table 5.2.

In what follows, we focus on a comparison of the results of calculations based on the analytical solutions to the equation for heat transfer (3.1) (Eq. (3.37)) and species transfer equation (5.1) (Eq. (5.24)) inside droplets (solution A), those based on the numerical solutions to (3.1) and (5.1) (solution B), and experimental data. The numerical codes for the solutions of (3.1) and (5.1) were developed by Castanet et al. [12]. The effects of the movement of the droplet interface due to evaporation and thermal swelling/contraction during individual time steps were ignored in both solutions. Only the one-way solutions were considered, so that the effects of droplets on gas were ignored. The latter effects would modify the results slightly, but would not change the trends of the predicted results [53].

Plots similar to those shown in Fig. 4.8, but for mixtures of ethanol and acetone, are presented in Figs. 5.1–5.4. The calculations were performed for cases of the ideal mixture ($\gamma_i = 1$ in (5.11)) (see (5.12)), and the nonideal mixture (γ_i in (5.11) was calculated based on (5.29)). As can be seen from these figures, in all cases the predictions of the temperatures by the ideal and nonideal models are noticeably different (by up to several degrees), especially at later times. However, both these models predict about the same trends in the evolution of temperature with time. The ideal model can be used if prediction errors of several degrees can be tolerated. This seems to be our case, where the random errors of the estimates of droplet temperatures appear to be about 2–3°. As in the cases shown in Fig. 4.8, the results predicted by solutions A and B coincide within the accuracy of plotting, which gives us confidence in the results predicted by both solutions. This closeness of the results predicted by solutions A and B is demonstrated for the 25 % ethanol – 75 % acetone mixture droplets in Fig. 5.2. For other mixtures, the closeness of the curves is similar to what was demonstrated in Fig. 5.2 (the actual plots are presented in [17]).

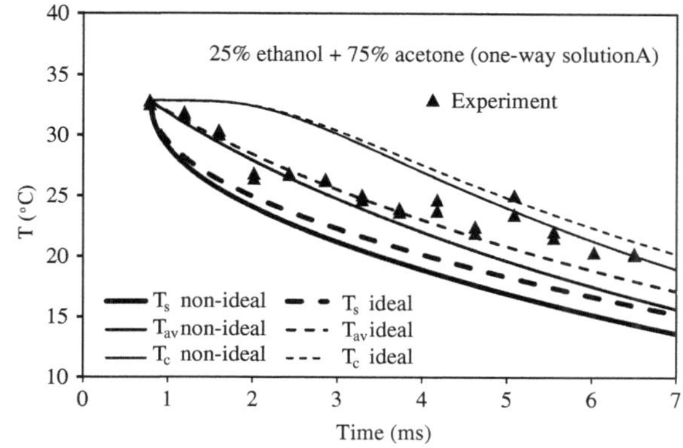

Fig. 5.1 The time evolution of droplet surface, average, and centre temperatures (T_s, T_{av}, and T_c), predicted by one-way solution A for ideal and nonideal models and experimentally observed temperatures for the 25 % ethanol – 75 % acetone mixture droplets with the values of the initial parameters, droplet velocity, and gas temperature given in Tables 5.1 and 5.2. Reproduced (with minor modification) from Fig. 5a of [17] with permission of Begell House

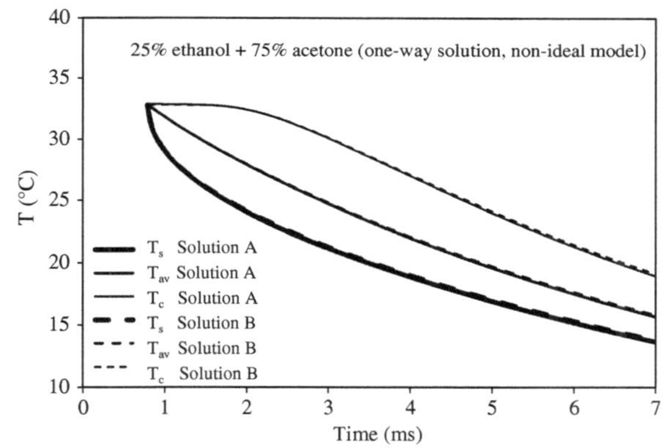

Fig. 5.2 The time evolution of droplet surface, average, and centre temperatures (T_s, T_{av}, and T_c), predicted by solutions A and B for the nonideal models for the 25 % ethanol – 75 % acetone mixture droplets with the values of the initial parameters, droplet velocity, and gas temperature given in Tables 5.1 and 5.2. Reproduced (with minor modification) from Fig. 5b of [17] with permission of Begell House

In the case of an acetone-dominated mixture (25 % ethanol – 75 % acetone; see Fig. 5.1), the agreement between the observed and predicted average droplet temperatures, for both ideal and nonideal models, is reasonably good. Most of the observed temperatures lie between the average and centre temperatures, although the scatter of experimental data in this case is more noticeable than for pure acetone (see Fig. 4.8). In the case of the 50 % ethanol – 50 % acetone mixture (see Fig. 5.3), the

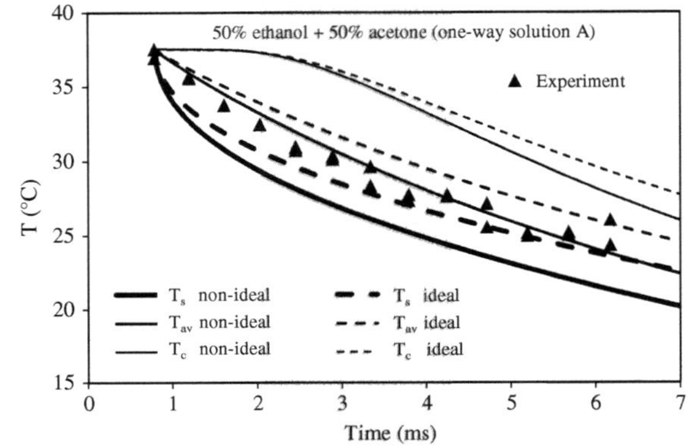

Fig. 5.3 The same as Fig. 5.1 but for the 50 % ethanol – 50 % acetone mixture droplets. Reproduced (with minor modification) from Fig. 6a of [17] with permission of Begell House

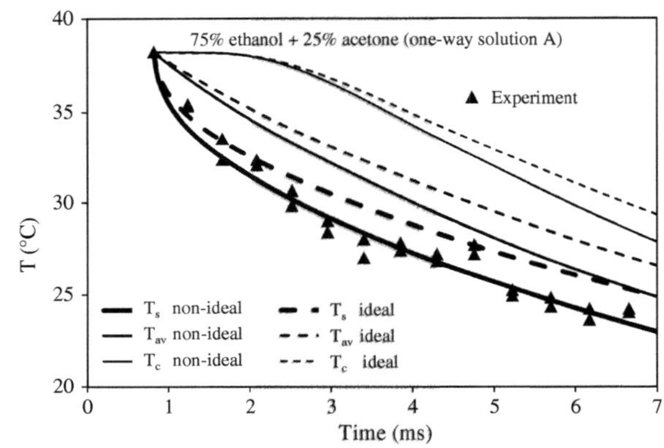

Fig. 5.4 The same as Figs. 5.1 and 5.3 but for the 75 % ethanol – 25 % acetone mixture droplets. Reproduced (with minor modification) from Fig. 7a of [17] with permission of Begell House

experimentally observed temperatures lie close to the average temperatures predicted by the nonideal model. For the 75 % ethanol – 25 % acetone mixture (see Fig. 5.4), the experimentally observed temperatures are close to the surface temperatures predicted by the nonideal model. As with pure acetone and ethanol, the reason for this deviation between the measured and predicted temperatures is not clear to us.

To take into account the effect of coupling between droplets and ambient gas, in [17, 53] it was assumed that the droplet exchanges heat and mass with a certain volume V_g surrounding the droplet, which is called the region of influence. The shape of this region can be either spherical, in the case of isolated droplets, or cylindrical, for droplet streams (as shown in Fig. 4.6). The initial value of the ratio $r_V = V_g/V_d$, where $V_d = \frac{4}{3}\pi R_d^3$, was chosen to be 3,500 in calculations performed in [17, 53].

At this value of r_V, the difference between the one-way and coupled solutions was visible, but the vapour pressure did not reach the saturation level during the droplet lifetime. The values of V_g were recalculated at each time step depending on the volume of the droplet. Since the changes in V_d were small for the cases under consideration, the values of V_g remained close to their initial values. Although this approach is not suitable for rigorous quantitative analysis of heat/mass exchange between droplets and surrounding gas, it can be used for the qualitative analysis of this effect. Only solution A was used for the comparison between the results of coupled and one-way solutions, since the predictions of solutions A and B practically coincide.

For the 25 % ethanol – 75 % acetone mixture droplets, the coupled solution showed a slightly faster drop in temperature at the beginning of evaporation and a slower reduction at later times. For the 50 % ethanol – 50 % acetone mixture droplets, the effects of the coupled solution on the results were similar to those for the 25 % ethanol – 75 % acetone mixture droplets. The experimentally observed values of droplet temperature in both cases lie close to the average droplet temperatures predicted by both one-way and coupled solutions. For the 75 % ethanol – 25 % acetone mixture droplets, the effects of the coupled solution were similar to those for the previous two cases. In the latter case, however, the deviation between the predicted and observed temperature values was larger for the coupled than for the one-way solution.

An alternative approach to modelling evaporation of nonideal liquid mixtures, using the COSMO-RS-based CFD model, was described in [30]. In contrast to the approach described above, the liquid mixture was assumed to be well mixed.

5.2.2.2 Heating and Evaporation When $R_d \neq$ const

As shown in Sect. 4.4, the assumption that the droplet radius is fixed during the time step can lead to noticeable deviations from the results predicted by the models that take into account the changes in this radius during the time steps. The cases tested in that section refer to droplet heating and evaporation in a hot gas, and the moving interface was linked only with droplet evaporation. The effects of thermal swelling/contraction were ignored. Also, only monocomponent droplets were considered, in which case the moving interface had an effect only on the heat conduction equation inside droplets.

In this section, the general case of multicomponent droplets is considered, and the effects of the moving interface on both heat transfer and species diffusion equations are taken into account based on (4.83) and (5.24), mainly following [17]. The model is first applied to the case of the acetone–ethanol mixture droplets considered in the previous section. Then other related cases are considered. The analysis of this section is restricted to the one-way solution A.

The plots of time evolutions of the temperatures at the centre and at the surface of the droplets and the average droplet temperatures predicted by the models not taking into account the effect of the moving interface and taking into account this effect for

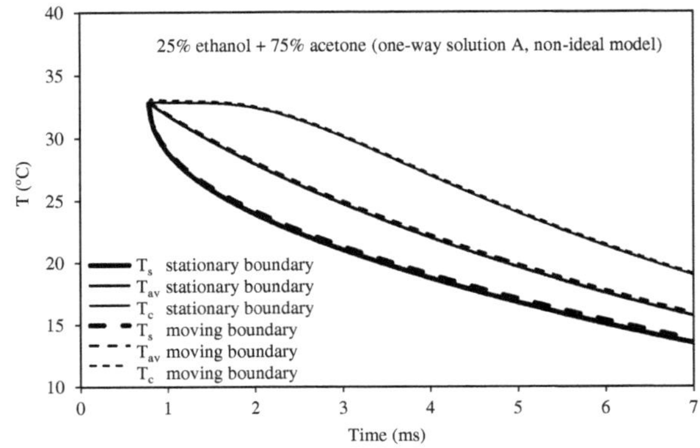

Fig. 5.5 The time evolution of droplet surface, average, and centre temperatures (T_s, T_{av}, and T_c), predicted by the one-way solution A for the nonideal model, taking and not taking into account the effects of the moving interface during individual time steps (moving and stationary boundaries) on the solutions to both heat transfer and species diffusion equations for the 25 % ethanol – 75 % acetone mixture droplets with the values of the initial parameters, droplet velocity, and gas temperature given in Tables 5.1 and 5.2. Reproduced (with minor modifications) from Fig. 11a of [17] with permission of Begell House

both temperature and species diffusion for the 25 % ethanol – 75 % acetone mixture droplets are shown in Fig. 5.5. As can be seen from this figure, the effect of the moving interface on the predicted temperatures can be safely ignored in the analysis of experimental data described in the previous section. The same conclusion can be drawn for the cases of the 50 % ethanol – 50 % acetone and 75 % ethanol – 25 % acetone mixture droplets (figures are not shown).

In Fig. 5.6, a hypothetical case is shown in which the 50 % ethanol – 50 % acetone mixture droplets are cooled or heated and evaporated until complete evaporation takes place. This example allows us to achieve good visualization of the effect of the moving interface on droplet heating and evaporation. Both plots for the droplet surface temperature and droplet radius are shown. The same values as shown in Table 5.2 for the initial droplet temperature, diameter, distance parameter, and gas temperature were used, but in contrast to the case shown in Table 5.1, it was assumed that the droplet velocity remains constant and equal to 12.71 m/s. This is the initial velocity of the droplet in the experiment described earlier. The cases of the stationary interface during individual time steps, the cases in which the effects of the moving interface are taken into account for the heat transfer and species diffusion equations separately during individual time steps, and the case in which these effects are taken simultaneously for heat transfer and species diffusion are shown.

As can be seen from this figure, the plots taking into account the effects of the moving interface on the heat transfer equation only and ignoring this effect altogether practically coincide. That means that this effect can be safely ignored for this case. Also, the plots taking into account the effects of the moving interface on the solution to the species diffusion equation and taking it into account for both

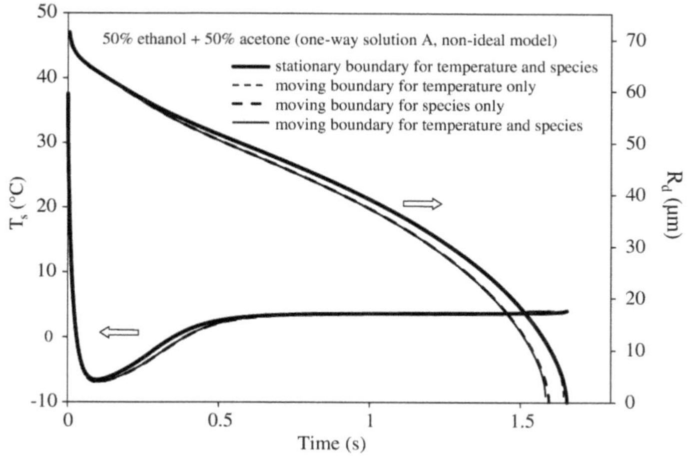

Fig. 5.6 The time evolution of droplet surface temperatures (T_s) and radius (R_d), predicted by the one-way solution A for the nonideal model, taking and not taking into account the effects of the moving interface during individual time steps on the solutions to the heat transfer equation only, species diffusion equation only, and both heat transfer and species diffusion equations for the 50 % ethanol – 50 % acetone mixture droplets with the values of the initial parameters and gas temperature given in Table 5.2, assuming that the droplet velocity is constant and equal to 12.71 m/s. Reproduced (with minor modifications) from Fig. 12 of [17] with permission of Begell House

solutions to the heat transfer and species diffusion equations practically coincide, but the difference between both these curves and those ignoring this effect altogether can be clearly seen after about 0.1 s. The effect of the moving interface is a reduction of the predicted droplet surface temperature at times between about 0.1 and 0.6 s. During this period, the droplet surface temperature is below the ambient gas temperature. Hence the reduction of the droplet surface temperature is expected to increase the heat flux from the ambient gas to the droplets, leading to the acceleration of droplet evaporation. This agrees with the predicted time evolution of the droplet radius, taking and not taking into account the effect of the moving interface, shown in Fig. 5.6.

In Fig. 5.7, the case similar to that shown in Fig. 5.6, but for gas temperature equal to 1000 K, is shown. In this case, droplet surface temperature increases during the whole period of droplet heating and evaporation, in contrast to the case shown in Fig. 5.6. As one can see from Fig. 5.7, the plots taking into account the effects of the moving interface on the solution to the heat transfer equation and ignoring this effect altogether practically coincide, as in the case shown in Fig. 5.6. Also, similarly to the case shown in Fig. 5.6, the plots taking into account the effects of the moving interface on the solution to the species diffusion equations and taking it into account for both heat transfer and species diffusion equations practically coincide, but the difference between both these curves and those ignoring this effect altogether can be clearly seen after about 5 ms. This difference between the plots is much more visible

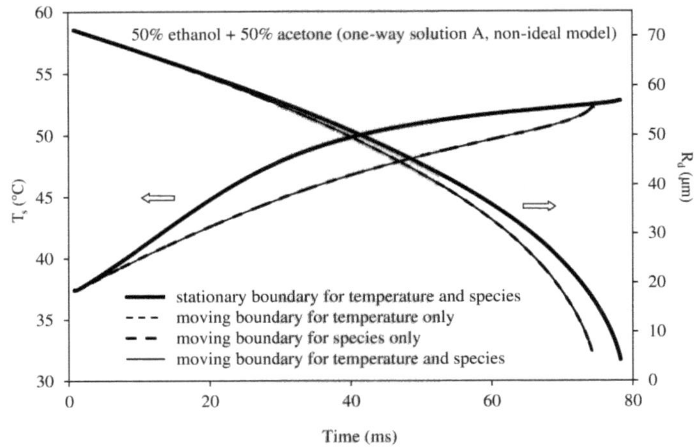

Fig. 5.7 The plots of T_s and R_d versus time for same parameters as in Fig. 5.6 but for the gas temperature equal 1,000 K. Reproduced (with minor modifications) from Fig. 13 of [17] with permission of Begell House

than in the case shown in Fig. 5.6. As in the case shown in Fig. 5.6, the effect of the moving interface is to reduce the predicted droplet surface temperature, leading to an increase in the heat flux from the ambient gas to the droplets and acceleration of droplet evaporation. This agrees with the predicted time evolution of droplet radius, taking and not taking into account the effect of the moving interface, shown in Fig. 5.6.

The plots of time evolution of the surface mass fraction of ethanol $Y_{\mathrm{l,s,eth}}$ for the same case as shown in Fig. 5.7 are shown in Fig. 5.8. Similarly to the case shown in Fig. 5.7, the main effect of the moving interface on the solution to the species diffusion equation is its influence on the values of $Y_{\mathrm{l,s,eth}}$. This effect leads to visible reductions of the values of $Y_{\mathrm{l,s,eth}}$ until the complete evaporation of the droplet takes place.

Note that the analysis of this section can be generalised to the case of more than two components in droplets. In this case, however, we need to make a number of additional assumptions, such as the assumption that the diffusion coefficients of all liquid (D_l) and vapour (D_v) components are the same. The limitations of this assumption for the vapour phase were discussed in [15]. The authors of that chapter offered a new model to deal with this problem, but they had to impose a rather stringent assumption that the thermal conductivity and species diffusivity in the liquid phase are infinitely large. As shown in a number of papers, including [53], this assumption is almost never satisfied in the case of heating and evaporation of multicomponent droplets, especially at the early stages of the process.

The model described above was applied to the analysis of biodiesel fuel droplet heating and evaporation in realistic Diesel engine-like conditions [55]. The analysis was focused on five types of biodiesel fuels: Palm Methyl Ester (PME) produced from

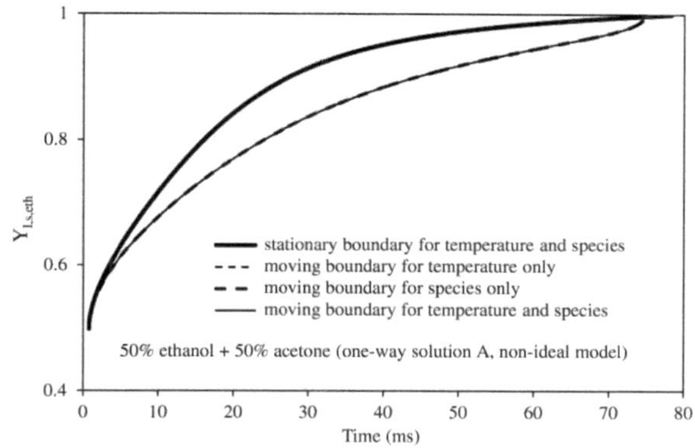

Fig. 5.8 The same as Fig. 5.7, but for the mass fraction of ethanol at the surface of the droplet. Reproduced (with minor modifications) from Fig. 14 of [17] with permission of Begell House

palm oil, Hemp Methyl Esters, produced from hemp seed oil in Ukraine (HME1) and the European Union (HME2), Rapeseed oil Methyl Ester (RME), produced from rapeseed oil in the Ukraine, and Soybean oil Methyl Ester (SME) produced from soybean oil. These fuels contain up to 15 methyl esters and possibly small amounts of unspecified additives, which were treated as methyl esters with average characteristics. Calculations were performed (1) taking into account the contribution of all components of biodiesel fuel and (2) assuming that this fuel can be treated as a one-component fuel with averaged transport and thermodynamic coefficients (as a widely used approximation for the analysis of realistic multicomponent fuels, including biodiesel fuels). It was pointed out that for HME2, the droplet surface temperatures and radii predicted by the multicomponent and single-component models almost coincide, which justifies the application of the single-component model for the analysis of droplet heating and evaporation. For PME, HME1, and SME, the predictions of the multicomponent and single-component models were rather close (the droplet evaporation times predicted by these models differed by less than about 2 %). The maximal deviation between the predictions of the multicomponent and single-component models was observed for RME. Even in the latter case, however, the difference between the evaporation times predicted by these models was less than about 5.5 %. This difference is much smaller than that observed in the case of Diesel and gasoline fuel droplets (see the next section). The difference in the predictions of biodiesel, Diesel, and gasoline droplet heating and evaporation was related to the fact that in the case of Diesel and gasoline fuel droplets, the contribution of components in a wide range of molar masses, boiling temperatures, and enthalpies of evaporation needs to be taken into account, while in the case of biodiesel droplets, the main contribution comes from the components in a narrow range of these parameters. For biodiesel droplets, the multicomponent model predicted higher droplet surface temperatures and longer evaporation times than the single-component model. This is related to the fact that at the final stages of droplet evaporation, the mass fraction of heavier species increases at the expense of lighter species. The heavier species evaporate more slowly than the lighter species and have higher wet-bulb temperatures.

A more general model of bicomponent droplet heating and evaporation, taking into account the deformation of droplet surfaces and formation of vortices inside droplets, was discussed in [7, 60].

5.3 Quasidiscrete Model

Although the DCM, described in the previous section, could potentially be generalised to any number of components, its practical application is limited to cases in which the number of these components is relatively small. As mentioned in Sect. 5.1, in the case of a large number of components, a different approach based on the Continuous Thermodynamics or Distillation Curve models was suggested. The main limitation of these models is that they are incapable in principle of taking into account the diffusion of liquid species inside droplets. As follows from our analysis in Sect. 5.2, this process cannot be ignored in practical engineering applications.

A new approach to modelling heating and evaporation of multicomponent droplets suitable for the case in which a large number of components is present in the droplets was suggested in [16, 54]. In contrast to the previously suggested models, designed for large numbers of components, the new model takes into account the diffusion of liquid species and thermal diffusion as in the classical DCM. This model was called the 'quasidiscrete model'. In what follows, it will be described following mainly [16].

It is possible to draw a parallel between our approach to modelling the heating and evaporation of multicomponent droplets and the modelling of absorption of thermal radiation in molecular gases, using the *weighted-sum-of-grey-gases* method [42, 50]. In this method, the medium is assumed to consist of several fractions of grey gases with different (but grey) absorption coefficients. The accuracy of this method turned out to be sufficient for most practical engineering applications. At the same time, its application is much more CPU efficient than the rigorous approach in which all or most of the molecular absorption bands are accounted for.

5.3.1 Description of the Model

As in the case of the Continuous Thermodynamics approach, the quasidiscrete model is based on the distribution function $f_m(I)$, introduced in (5.15). An obvious limitation of this approach is that it is applicable only in the case that all properties of the components depend on only one parameter I. Although molar mass M is almost universally used to describe the property I, this choice is certainly far from being a unique one. Since most practically important hydrocarbon fuels consist mainly of molecules of the type C_nH_{2n+2} (alkanes), where $n \geq 1$ in the general case and $n \geq 5$ for liquid fuels, it is more practical to write the distribution function f_m as a function of the carbon number n rather than M. These two parameters are linked by the following equation:

$$M = 14n + 2, \tag{5.32}$$

where M is measured in kg/kmol.

We assume that $f_m(n)$ can be approximated by (5.16), which can be rearranged to

$$f_m(n) = C_m(n_0, n_f)\frac{(M(n)-\gamma)^{\alpha-1}}{\beta^\alpha \Gamma(\alpha)} \exp\left[-\left(\frac{M(n)-\gamma}{\beta}\right)\right], \tag{5.33}$$

where $n_0 \le n \le n_f$, and the subscripts $_0$ and $_f$ stand for initial and final (the smallest and the largest values of n),

$$C_m(n_0, n_f) = \left\{\int_{n_0}^{n_f} \frac{(M(n)-\gamma)^{\alpha-1}}{\beta^\alpha \Gamma(\alpha)} \exp\left[-\left(\frac{M(n)-\gamma}{\beta}\right)\right] \mathrm{d}n\right\}^{-1}. \tag{5.34}$$

This choice of C_m ensures that

$$\int_{n_0}^{n_f} f_m(n)\mathrm{d}n = 1.$$

The meaning of other parameters in expression (5.33) is the same as in expression (5.16).

Note that real-life automotive fuels, including Diesel and gasoline fuels, apart from alkanes, contain significant amounts of alkenes, alkynes, naphthenes, and aromatics. The contribution of these elements is not taken into account in [16, 54], and this is a serious limitation of the model described in those papers.

The following approximation for the dependence of the saturation vapour pressure of alkanes (in MPa) on n was used [4]:

$$p^{\mathrm{sat}}(n) = \exp\left(A(n) - \frac{B(n)}{T - C(n)}\right), \tag{5.35}$$

where

$$A(n) = 6.318\, n^{0.05091}, \quad B(n) = 1178\, n^{0.4652}, \quad C(n) = 9.467\, n^{0.9143},$$

T is in Kelvins. The above approximations for $A(n)$, $B(n)$, $C(n)$ were originally derived for $4 < n < 17$, but it was assumed that they can be applied for $n \ge 17$ as well if the contribution of hydrocarbon fuels with these n is relatively small.

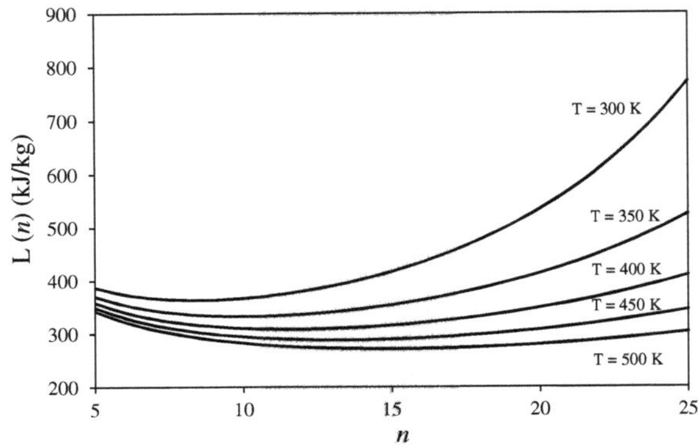

Fig. 5.9 A plot of $L(n)$ versus n as predicted by formula (5.37). Reprinted from Ref. [16]. Copyright Elsevier (2011)

From the Clausius–Clapeyron equation, it follows that [4]

$$L = -\frac{R_u}{M(n)} \frac{\mathrm{d} \ln p^{\mathrm{sat}}(n)}{\mathrm{d}(1/T)}, \tag{5.36}$$

where R_u is the universal gas constant. Using (5.35), formula (5.36) can be rewritten as

$$L = \frac{R_u B(n) T^2}{M(n)(T - C(n))^2}. \tag{5.37}$$

The plots of $L(n)$ for temperatures in the range from 300 to 500 K, as predicted by formula (5.37), are shown in Fig. 5.9. As one can see from that figure, $L(n)$ increases with decreasing temperature, as expected. For small n ($n < 10$), $L(n)$ slowly decreases with increasing n, while at larger n, it increases with increasing n, and this increase is particularly strong for low temperatures. This is consistent with the expectation that heavier components are generally less volatile than lighter ones. The values of $L(n)$ for $n = 10$ and $n = 12$ (n-decane and n-dodecane), predicted by (5.37), are close to those used in [3].

The approximations of the dependence of liquid density, viscosity, specific heat capacity, and thermal conductivity on n are presented and discussed in Appendix K.

As follows from (5.35), (5.37), and the results presented in Appendix K, the transport and thermodynamic properties of the fuels under consideration are relatively weak functions of n. In this case, it would be sensible to assume that the properties of hydrocarbons in a certain narrow range of n are close, and replace the continuous distribution (5.33) with a discrete one, consisting of N_f quasicomponents with carbon numbers

$$\overline{n}_j = \frac{\int_{n_{j-1}}^{n_j} n f_m(n)\mathrm{d}n}{\int_{n_{j-1}}^{n_j} f_m(n)\mathrm{d}n}, \tag{5.38}$$

the corresponding molar fractions

$$X_j = \int_{n_{j-1}}^{n_j} f_m(n)\mathrm{d}n, \tag{5.39}$$

and mass fractions

$$Y_j = \frac{M(\overline{n}_j)X_j}{\sum_{j=1}^{j=N_f}\left[M(\overline{n}_j)X_j\right]}, \tag{5.40}$$

where j is an integer in the range $1 \le j \le N_f$. Note that

$$\sum_{j=1}^{j=N_f} X_j = \sum_{j=1}^{j=N_f} Y_j = 1. \tag{5.41}$$

The choice of n_j can be arbitrary. In our model we assume that all $n_j - n_{j-1}$ are equal, i.e., all quasicomponents have the same range of values of n. For the case $N_f = 1$, this approach reduces the analysis of multicomponent droplets to monocomponent ones.

These new quasicomponents are not the actual physical hydrocarbon components ($\overline{n}_j$ are not integers in the general case). Hence, this model is called a quasidiscrete model. These quasicomponents are treated as actual components in the conventional DCM, including taking into account diffusion of liquid species in droplets. This model is expected to be particularly useful when N_f is much less than the number of actual species in the hydrocarbon mixture.

The mixtures are treated as ideal (Raoult's law is assumed to be valid). In this case, partial pressures of individual quasicomponents can be estimated as

$$p_v(\overline{n}_j) = X_{lsi}(\overline{n}_j)p^{\mathrm{sat}}(\overline{n}_j), \tag{5.42}$$

where X_{lsi} is the molar fraction of liquid quasicomponents at the surface of the droplet, $p^{\mathrm{sat}}(\overline{n}_j)$ is determined by (5.35).

Having replaced n in (5.37) with $\overline{n}_j$, we obtain the required values of L for all quasicomponents.

Table 5.3 The values of α, β, γ, n_0 and n_f for Diesel and gasoline fuels

Fuel	α	β (kg/kmol)	γ (kg/kmol)	n_0	n_f
Diesel	18.5	10	0	5	25
Gasoline	5.7	15	0	5	18

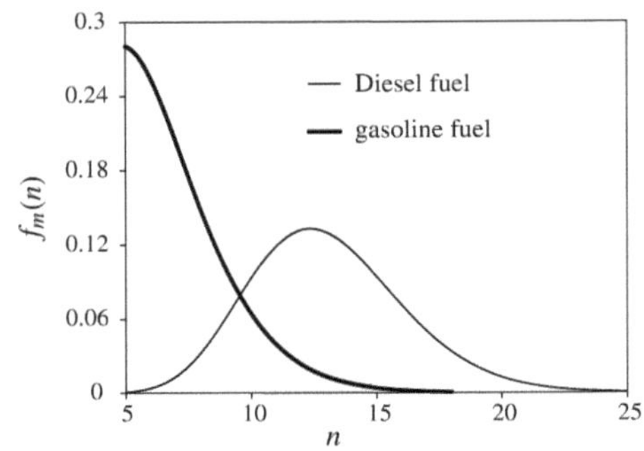

Fig. 5.10 Plots of $f_m(n)$ versus n as predicted by (5.33) for Diesel (*thin solid*) and gasoline (*thick solid*) fuels with the values of parameters given in Table 5.3. Reprinted from Ref. [16]. Copyright Elsevier (2012)

5.3.2 *Application to Diesel and Gasoline Fuel Droplets*

Following [4], we assume the values of parameters for the distribution function (5.33) for Diesel and gasoline fuels shown in Table 5.3.

Plots of f_m versus n for Diesel and gasoline fuels for the values of parameters given in Table 5.3 are shown in Fig. 5.10. As follows from this figure, the forms of the plots of f_m versus n for Diesel and gasoline fuels appear to be rather different. The values n for which f_m is maximal are equal to 12.4 and 5 for Diesel and gasoline fuels respectively. The average values of n ($\overline{n}$) for these fuels are 12.56 and 7.05 respectively. The analysis of droplet heating and evaporation for both fuels is performed separately below.

5.3.2.1 Diesel Fuel

We assume that the initial droplet temperature is equal to 300 K and is homogeneous throughout its volume. Gas temperature and pressure are assumed to be equal to 880 K and 3 MPa = 30 bar respectively. The initial composition of droplets is described by the distribution function (5.33) with the values of parameters for Diesel fuel given in Table 5.3, as shown in Fig. 5.10.

Plots of droplet surface temperatures T_s and droplet radii R_d versus time for the initial droplet radius equal to 10 μm and velocity 1 m/s are shown in Fig. 5.11. The droplet velocity is assumed to be constant during the whole process. The calculations were performed for the cases $N_f = 1$ (one-quasicomponent droplet, $\overline{n} = 12.56$), and $N_f = 20$ (twenty-quasicomponent droplet), using the ETC/ED and ITC/ID models.

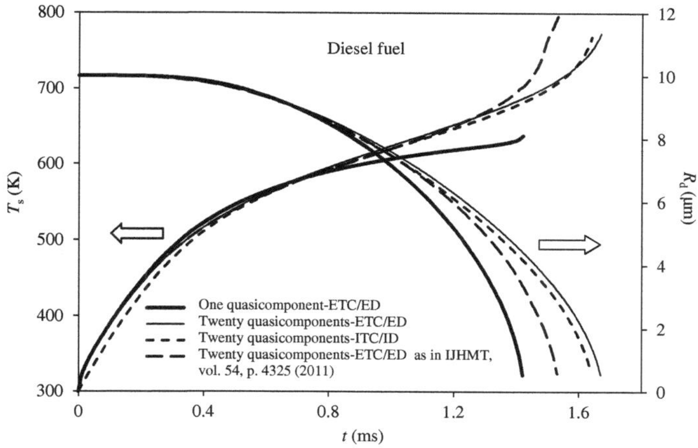

Fig. 5.11 Plots of T_s and R_d predicted by four models, versus time. The initial droplet radius and temperature are assumed to be equal to 10 μm and 300 K respectively; the droplet velocity is assumed to be equal to 1 m/s, and its changes during the heating and evaporation process are ignored; gas temperature is assumed equal to 880 K. These are the models used for calculations: the Effective Thermal Conductivity (ETC)/Effective Diffusivity (ED) model using one quasicomponent; the ETC/ED model using twenty quasicomponents; and the Infinite Thermal Conductivity (ITC)/Infinite Diffusivity (ID) model using twenty quasicomponents. The approximations for density, viscosity, heat capacity, and thermal conductivity of liquid components for these three models are given in Appendix K. The results are shown by *thick solid, thin solid, and thin dashed curves*. The fourth model is the ETC/ED model using twenty quasicomponents with the density, viscosity, heat capacity, and thermal conductivity of the liquid components assumed to be equal to those on n-dodecane (*thick dashed*), as used in [54]. Reprinted from Ref. [16]. Copyright Elsevier (2012)

In the same figure, the plots of T_s and R_d versus time for $N_f = 20$, using the ETC/ED model but assuming that the density, viscosity, heat capacity, and thermal conductivity of all components are the same and equal to those of n-dodecane, are presented.

As one can see from this figure, the droplet radii R_d and surface temperatures T_s predicted by the ETC/ED model using one and twenty quasicomponents are noticeably different, especially at the final stages of droplet heating and evaporation. The model using twenty quasicomponents predicts higher surface temperatures and longer evaporation times than the model using one quasicomponent. This can be related to the fact that at the final stages of droplet evaporation, the species with large n become dominant, as will be demonstrated later. These species evaporate more slowly than the species with lower n and have higher wet-bulb temperatures. One can see from this figure that the results predicted by a simplified model used in [54] (density, viscosity, heat capacity, and thermal conductivity of the liquid components assumed to be equal to those on n-dodecane) are noticeably different from those predicted by the more rigorous model used in [16] (the approximations for density, viscosity, heat capacity, and thermal conductivity of liquid components are given in

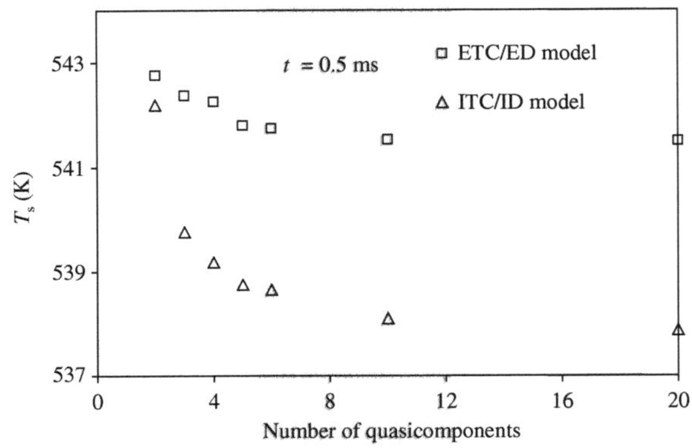

Fig. 5.12 Plots of T_s versus the number of quasicomponents N_f for the same conditions as in Fig. 5.11 at time 0.5 ms as predicted by the ETC/ED (*squares*) and ITC/ID (*triangles*) models. Reprinted from Ref. [16]. Copyright Elsevier (2012)

Appendix K). This shows the limitations of the simplified model used earlier for the density, viscosity, heat capacity, and thermal conductivity of the liquid components.

Also, there are noticeable differences in predictions of the ETC/ED and ITC/ID models using twenty quasicomponents, especially in the case of the surface temperature at the initial stage of droplet heating and evaporation. The accurate prediction of this temperature is particularly important for the prediction of the autoignition timing in Diesel engines [51]. This questions the reliability of the models for heating and evaporation of multicomponent droplets based on the ITC/ID approximations. As mentioned in Sect. 5.1, these models are almost universally used for modelling these processes, especially when large numbers of components are involved in the analysis.

Plots of T_s and R_d at time equal to 0.5 ms versus the number of quasicomponents N_f predicted by the ETC/ED and ITC/ID models are shown in Figs. 5.12 and 5.13 respectively for the same conditions as in Fig. 5.11. Symbols refer to those N_f for which calculations were performed. The approximations for density, viscosity, heat capacity, and thermal conductivity of liquid components given in Appendix K were used. As follows from these figures, for $N_f \geq 10$ the predicted T_s and R_d no longer depend on N_f. In fact, the difference between the values for surface temperature and radius predicted for $N_f = 5$ and $N_f = 20$ can be considered negligible compared with the difference between the values for temperature predicted by the ETC/ED and ITC/ID models. Hence, heating and evaporation of Diesel fuel droplets can be safely modelled using just five quasicomponents, in agreement with our earlier results [54] obtained for time equal to 0.25 ms using a simplified version of the quasidiscrete model. The errors due to the ITC/ID approximation for $N_f \geq 3$ are significantly larger than those due to the choice of a small number of quasicomponents, especially for the surface temperature. These errors cannot be ignored in most engineering

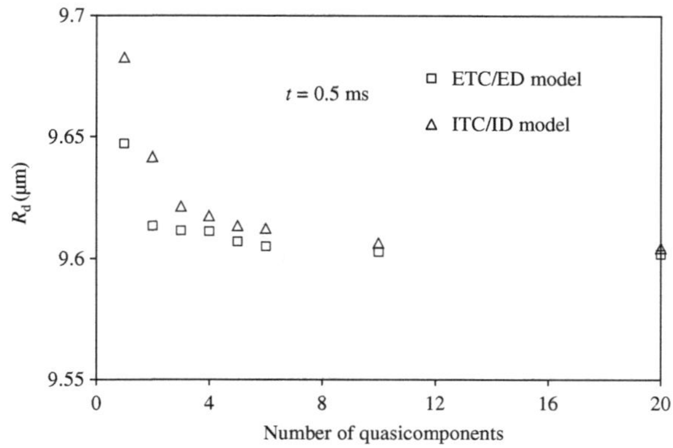

Fig. 5.13 The same as Fig. 5.12 but for droplet radii R_d. Reprinted from Ref. [16]. Copyright Elsevier (2012)

applications, and this questions the applicability of the models using the ITC/ID approximation, including the widely used Continuous Thermodynamics models.

The results shown in Figs. 5.12 and 5.13 are consistent with those obtained for other instants of time, including time equal to 1 ms [16]. The closeness of the temperatures predicted by the ETC/ED and ITC/ID models at the later stages of droplet heating and evaporation can be related to the fact that at this stage, the droplet temperature becomes almost homogeneous (see Fig. 9 of [54]), and the effects of the temperature gradient inside droplets can be ignored. In agreement with [54], smaller droplet radii are predicted by the ITC/ID model compared with the ETC/ED model at the final stages of droplet heating and evaporation.

Note that at early stages of droplet heating and evaporation ($t = 0.5$ ms), the predicted droplet radius is slightly reduced with the increase in the number of quasicomponents used, while at a later stage ($t = 1$ ms), the opposite effect is observed, in agreement with the results reported in [54]. This could be related to the fact that at the early stages, droplet evaporation is controlled by the most-volatile quasicomponents, while at the later stages it is controlled by less-volatile quasicomponents. When the number of quasicomponents increases, the volatility of the most-volatile component increases, and that of the least-volatile decreases.

5.3.2.2 Gasoline Fuel

Plots similar to those shown in Fig. 5.11, but for gasoline fuel, are presented in Fig. 5.14. The maximal number of quasicomponents for gasoline fuel is 13. The initial conditions are assumed to be the same as in the case of Diesel fuel droplets to enable us to perform a direct comparison between heating and evaporation of Diesel and gasoline fuel droplets in identical conditions. As in the case shown in Fig. 5.11, the droplet velocity is assumed to be constant and equal to 1 m/s during

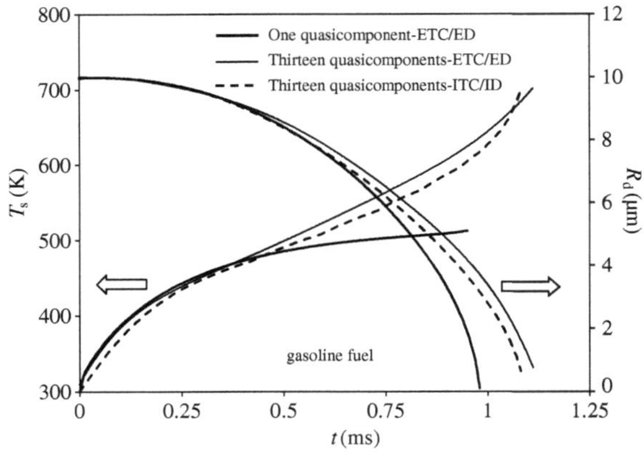

Fig. 5.14 The same as Fig. 5.11 for the first three curves, but for the gasoline fuel with the maximal number of quasicomponents $N_f = 13$. All plots are based on the approximations for density, viscosity, heat capacity, and thermal conductivity of liquid components given in Appendix K. Reprinted from Ref. [16]. Copyright Elsevier (2012)

the whole process. The calculations were performed for the case of $N_f = 1$ (one-quasicomponent droplet, $\overline{n} = 7.05$) and $N_f = 13$ (thirteen-quasicomponent droplet), using the ETC/ED and ITC/ID models. The density, viscosity, heat capacity, and thermal conductivity of all liquid components are described in Appendix K.

As in the case of Diesel fuel droplets, the droplet radii R_d and surface temperatures T_s predicted by the ETC/ED models using one and thirteen quasicomponents are noticeably different, especially at the final stages of droplet heating and evaporation. The model using thirteen quasicomponents predicts higher surface temperatures and longer evaporation time than the model using one quasicomponent. As in the case of Diesel fuel droplets, this can be related to the fact that at the final stages of droplet evaporation, the species with large n become dominant, as demonstrated later. These species evaporate more slowly than the species with lower n and have higher wet-bulb temperatures.

The differences in predictions of the ETC/ED and ITC/ID models using thirteen quasicomponents are more noticeable than in the case of Diesel fuel droplets. This difference can be seen not only at the initial stage of droplet heating and evaporation, but also at the later stages of these processes. This provides additional support for our questioning the reliability of the models for heating and evaporation of multi-component droplets based on the ITC/ID approximations.

Plots of T_s and R_d at time equal to 0.75 ms versus the number of quasicomponents N_f predicted by the ETC/ED and ITC/ID models for gasoline fuel droplets are shown in Figs. 5.15 and 5.16 for the same conditions as in Fig. 5.14. As follows from these figures, for $N_f \geq 6$ the predicted T_s and R_d no longer depend on N_f. In fact, the difference between the values for temperatures and radii predicted for $N_f = 3$ and $N_f = 13$ can be considered negligible compared with the difference between the

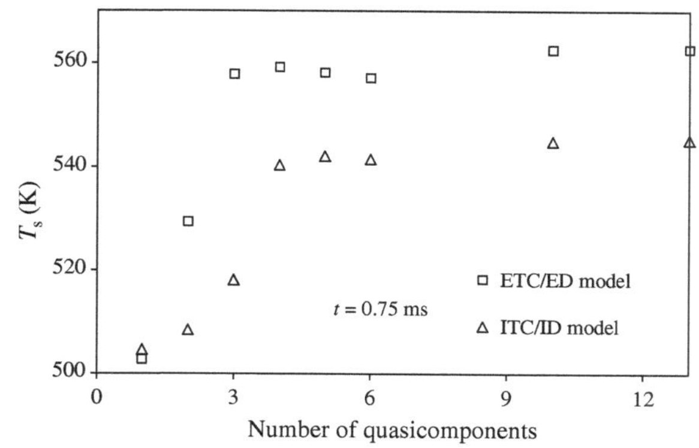

Fig. 5.15 Plots of T_s versus the number of quasicomponents N_f for the same conditions as in Fig. 5.14 at time 0.75 ms as predicted by the ETC/ED (*squares*) and ITC/ID (*triangles*) models. Reprinted from Ref. [16]. Copyright Elsevier (2012)

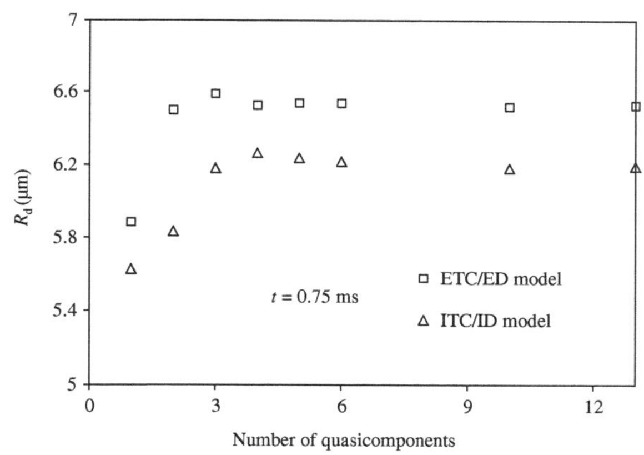

Fig. 5.16 The same as Fig. 5.15 but for droplet radii R_d. Reprinted from Ref. [16]. Copyright Elsevier (2012)

values for temperatures and radii predicted by the ETC/ED and ITC/ID models. Hence, heating and evaporation of gasoline fuel droplets can be safely modelled using just three quasicomponents. As in the case of Diesel fuel droplets, the errors due to the ITC/ID approximation for $N_f \geq 3$ are significantly larger than those due to the choice of a small number of quasicomponents, especially for the surface temperature.

Note that at the early stages of droplet heating and evaporation ($t = 0.2$ ms), the predicted droplet radii reduce slightly with the increase in the number of quasicomponents used, while at a later stage ($t = 0.75$ ms), the opposite effect is observed, in agreement with the results reported in [54] and those discussed earlier for Diesel fuel.

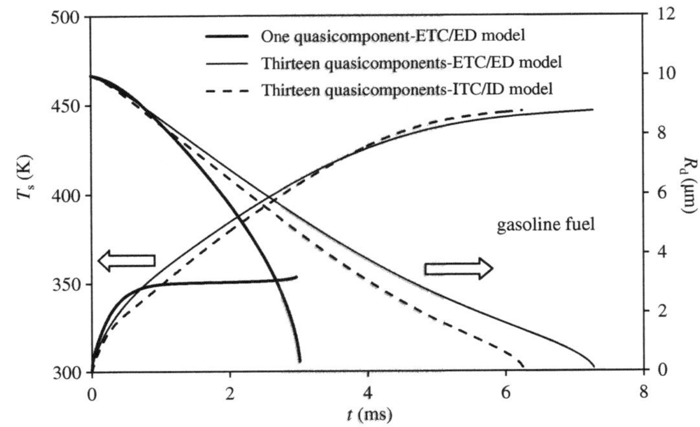

Fig. 5.17 The same as Fig. 5.14 but for droplet velocity equal to 10 m/s, gas temperature equal to 450 K, and pressure equal to 0.3 MPa. Reprinted from Ref. [16]. Copyright Elsevier (2012)

Plots similar to those shown in Fig. 5.14, but for more realistic conditions in gasoline engines, are presented in Fig. 5.17. Following [8], we assume that gas temperature is equal to 450 K, gas pressure is equal to 0.3 MPa, and droplet velocity is equal to 10 m/s. As in the case shown in Fig. 5.14, we assume that the initial droplet temperature is equal to 300 K and is homogeneous throughout its volume, while the droplet initial radius is equal to 10 μm.

Comparing Figs. 5.14 and 5.17, one can see that in the latter case, the difference between the predicted temperatures and droplet radii for one and thirteen quasicomponents is much more visible than in the former. The same conclusion applies to the predictions of the ETC/ED and ITC/ID models. This can be attributed to much slower evaporation for the case shown in Fig. 5.17 than in the case shown in Fig. 5.14. The general trends of the curves shown in Fig. 5.17 are similar to those shown in Fig. 5.14. In the case in which thirteen quasicomponents are considered, it is mainly the heavier components in the droplets that remain at the end of the evaporation process. These can reach higher temperatures and evaporate more slowly compared with the light and middle-range components.

Plots of T_s and R_d at time equal to 0.5 ms versus the number of quasicomponents N_f predicted by the ETC/ED and ITC/ID models for gasoline fuel droplets are shown in Figs. 5.18 and 5.19 for the same conditions as in Fig. 5.17. As follows from these figures, for $N_f \geq 6$ the predicted T_s and R_d no longer depend on N_f (cf. the cases shown in Figs. 5.15 and 5.16). This range can be extended to $N_f \geq 3$ at least for the ETC/ED model. In contrast to the cases shown in Figs. 5.15 and 5.16, the temperatures and radii predicted by the ETC/ED and ITC/ID models appear to be very close for small numbers of quasicomponents. This can be related to the fact that in this case, the temperature reaches saturation level by 0.5 ms when one or two quasicomponents are considered.

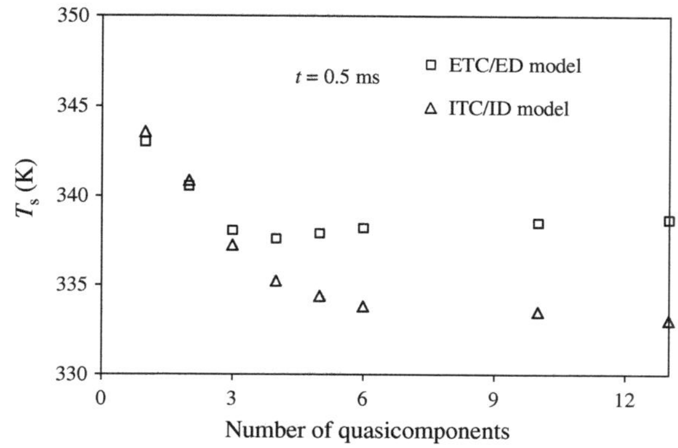

Fig. 5.18 Plots of T_s versus the number of quasicomponents N_f for the same conditions as in Fig. 5.17 at time 0.5 ms as predicted by the ETC/ED (*squares*) and ITC/ID (*triangles*) models. Reprinted from Ref. [16]. Copyright Elsevier (2012)

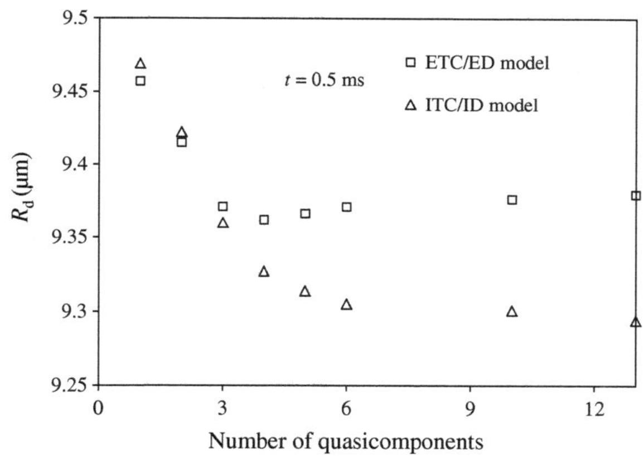

Fig. 5.19 The same as Fig. 5.18 but for droplet radii R_d. Reprinted from Ref. [16]. Copyright Elsevier (2012)

The results shown in Figs. 5.18 and 5.19 are consistent with those obtained for time equal to 2 ms. In both cases, droplet surface temperatures and radii can be well predicted by the ETC/ED model if only three quasicomponents are used. In contrast to the case shown in Figs. 5.15 and 5.16, at time equal to 2 ms, the temperatures and radii predicted by the ETC/ED and ITC/ID models are very close only for the case in which one quasicomponent is used.

Plots of Y_{si} versus time for the four quasicomponents for the same case as shown in Fig. 5.17, are presented in Fig. 5.20. The results presented in this figure are consistent with those shown in Fig. 8 of [54] for Diesel fuel using a simplistic approach to approximating density, viscosity, heat capacity, and thermal conductivity of the liquid

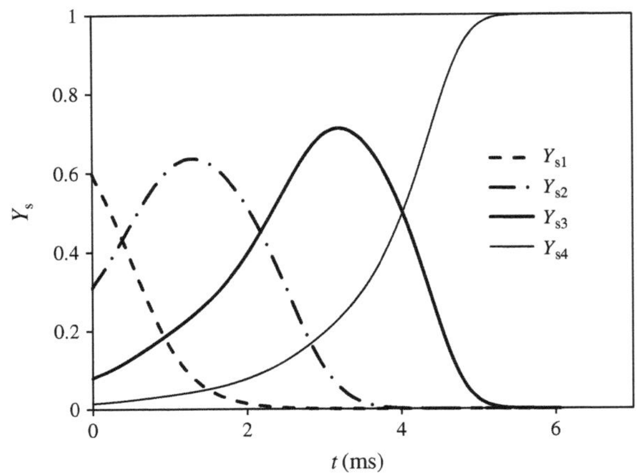

Fig. 5.20 Plots of Y_{si} versus time for four quasicomponents ($i = 1, 2, 3, 4$) for the same case as shown in Fig. 5.17. Reprinted from Ref. [16]. Copyright Elsevier (2012)

components. The values of Y_{s1} decrease monotonically with time, while those of Y_{s4} increase monotonically with time. The values of Y_{s2} and Y_{s3} initially increase with time, but at later times they rapidly decrease with time. At times close to the moment when the droplet completely evaporates, only the quasicomponent Y_{s4} remains. Since this quasicomponent is the most slowly evaporating one and has the highest wet-bulb temperature, the model based on four quasicomponents is expected to predict longer evaporation times and larger droplet surface temperatures at the final stages of droplet evaporation than the model using one quasicomponent. This result can be generalised to the case in which the number of quasicomponents is greater than four. These results are consistent with results shown in Fig. 5.17.

Effects of gasoline fuel approximations on heating and evaporation of gasoline fuel droplets were investigated in [18]. The following approximations were considered: three quasicomponents introduced in the quasidiscrete model (molar fractions 83 % n-$C_{6.26}H_{14.58}$ + 15.6 % n-$C_{10.26}H_{22.48}$ + 1.4 % n-$C_{14.42}H_{30.84}$) and their approximations: surrogate I (molar fractions: 83 % n-C_6H_{14} + 15.6 % n-$C_{10}H_{22}$ + 1.4 % n-$C_{14}H_{30}$) and surrogate II (molar fractions 83 % n-C_7H_{16} + 15.6 % n-$C_{11}H_{24}$ + 1.4 % n-$C_{15}H_{32}$)), surrogate approximations of gasoline fuel based on its ignition characteristics: surrogate A (molar fractions 56 % n-C_7H_{16} + 28 % iso-C_8H_{18} + 17 % C_7H_8) and surrogate B (molar fractions 63 % n-C_7H_{16} + 20 % iso-C_8H_{18}+ 17 % C_7H_8).

Surrogates I and II allowed the authors of [18] to investigate the sensitivity of the results to the choice of the values of the number of carbon atoms for each of three quasicomponents. Also, this rounding up or down of the values of the carbon numbers would allow one to use these approximations of quasicomponents in Computational Fluid Dynamics (CFD) codes, which do not recognise substances with noninteger n. Surrogates A and B were considered in [21].

The results were compared with the predictions of the model based on the approximation of gasoline fuel by 13 quasicomponents and 1 quasicomponent [18]. It was shown that the predictions of the quasidiscrete model based on the approximation of gasoline fuel with three quasicomponents, especially surrogate II, are much more accurate than those based on other surrogate approximations compared with the predictions of the quasidiscrete model based on 13 quasicomponents. This demonstrated the limitation of using fuel surrogates, as suggested based on fuel ignition characteristics in [21], for the analysis of liquid gasoline fuel droplet heating and evaporation.

A new multi-dimensional quasidiscrete model has been suggested and tested for the analysis of heating and evaporation of Diesel fuel droplets in [56]. In this model, as in the original quasidiscrete model suggested earlier, the components of Diesel fuel with close thermodynamic and transport properties are grouped together to form quasicomponents. In contrast to the original quasidiscrete model, the new model takes into account the contribution of not only alkanes, but also various other groups of hydrocarbons in Diesel fuels; quasicomponents are formed within individual groups. Also, in contrast to the original quasidiscrete model, the contributions of individual components are not approximated by the distribution function of carbon numbers. The formation of quasicomponents is based on taking into account the contributions of individual components without any approximations. Groups contributing small molar fractions to the composition of Diesel fuel (less than about 1.5 %) are replaced with characteristic components. The actual Diesel fuel is simplified to form six groups: alkanes, cycloalkanes, bicycloalkanes, alkylbenzenes, indanes & tetralines, and naphthalenes, and 3 components $C_{19}H_{34}$ (tricycloalkane), $C_{13}H_{12}$ (diaromatic), and $C_{14}H_{10}$ (phenanthrene). It is shown that the approximation of Diesel fuel by 15 quasicomponents and components, leads to errors in estimated temperatures and evaporation times in typical Diesel engine conditions not exceeding about 3.7 % and 2.5 % respectively, which is acceptable for most engineering applications.

References

1. Abdel-Qader, Z., & Hallett, W. L. H. (2005). The role of liquid mixing in evaporation of complex multicomponent mixtures: modelling using continuous thermodynamics. *Chemical Engineering Science*, *60*, 1629–1640.
2. Abianeh, O. S., & Chen, C. P. (2012). A discrete multicomponent fuel evaporation model with liquid turbulence effects. *International Journal of Heat and Mass Transfer*, *55*, 6897–6907.
3. Abramzon, B., & Sazhin, S. (2006). Convective vaporization of fuel droplets with thermal radiation absorption. *Fuel*, *85*, 32–46.
4. Arias-Zugasti, M., & Rosner, D. E. (2003). Multicomponent fuel droplet vaporization and combustion using spectral theory for a continuous mixture. *Combustion and Flame*, *135*, 271–284.
5. Atkins, P., & de Paula, J. (2002). *Atkins' physical chemistry* (7th ed.). Oxford: Oxford University Press.

6. Bader, A., Keller, P., & Hasse, C. (2013). The influence of nonideal vapor-liquid equilibrium on the evaporation of ethanol/iso-octane droplets. *International Journal of Heat and Mass Transfer*, *64*, 547–558.
7. Banerjee, R. (2013). Numerical investigation of evaporation of a single ethanol/iso-octane droplet. *Fuel*, *107*, 724–739.
8. Basshuysen, R. V. (2009). *Gasoline engine with direct injection: Processes, systems, development, potential* (1st ed.). Wiesbaden: GWV Fachverlage GmbH.
9. Brenn, G. (2005). Concentration fields in evaporating droplets. *International Journal of Heat and Mass Transfer*, *48*, 395–402.
10. Burger, M., Schmehl, R., Prommersberger, K., Schäfer, O., Koch, R., & Wittig, S. (2003). Droplet evaporation modelling by the distillation curve model: accounting for kerosene fuel and elevated pressures. *International Journal of Heat and Mass Transfer*, *46*, 4403–4412.
11. Campbell, S. W., Willsak, R. A., & Thodos, G. (1987). Vapor-liquid equilibrium measurements for the ethanol-acetone system at 372.7, 397.7 and 422.6 K. *Journal of Chemical & Engineering Data*, *32*, 357–362.
12. Castanet, G., Maqua, C., Orain, M., Grisch, F., & Lemoine, F. (2007). Investigation of heat and mass transfer between the two phases of an evaporating droplet stream using laser-induced fluorescence techniques: comparison with modelling. *International Journal of Heat and Mass Transfer*, *50*, 3670–3683.
13. Continillo, G., & Sirignano, W. A. (1991). Unsteady, spherically-symmetric flame propagation through multicomponent fuel spray clouds. In G. Angelino, L. De Luca & W. A. Sirignano (Eds.), *Modern research topics in aerospace propulsion* (pp. 173–198). New York: Springer.
14. Delplanque, J.-P., Rangel, R. H., & Sirignano, W. A. (1991). Liquid-waste incineration in a parallel-stream configuration: effect of auxiliary fuel. *Progress in Aeronautics Astronautics*, *132*, 164–184.
15. Ebrahimian, V., & Habchi, C. (2011). Towards a predictive evaporation model for multi-component hydrocarbon droplets at all pressure conditions. *International Journal of Heat and Mass Transfer*, *54*, 3552–3565.
16. Elwardany, A. E., & Sazhin, S. S. (2012). A quasi-discrete model for droplet heating and evaporation: application to Diesel and gasoline fuels. *Fuel*, *97*, 685–694.
17. Elwardany, A. E., Gusev, I. G., Castanet, G., Lemoine, F., & Sazhin, S. S. (2011). Mono- and multi-component droplet cooling/heating and evaporation: comparative analysis of numerical models. *Atomization and Sprays*, *21*, 907–931.
18. Elwardany, A. E., Sazhin, S. S., & Farooq, A. (2013). Modelling of heating and evaporation of gasoline fuel droplets: a comparative analysis of approximations. *Fuel*, *111*, 643–647.
19. Faeth, G. M. (1983). Evaporation and combustion of sprays. *Progress in Energy and Combustion Science*, *9*, 1–76.
20. Gauthier, J. E. D., Bardon, M. F., & Rao, V. K. (1991). Combustion characteristics of multi-component fuels under cold starting conditions in a gas turbine. *Proceedings of the American Society of Mechanical Engineers*, Orlando, Florida, paper 91-GT-109.
21. Gauthier, B. M., Davidson, D. F., & Hanson, R. K. (2004). Shock tube determination of ignition delay times in full-blend and surrogate fuel mixtures. *Combustion and Flame*, *139*, 300–311.
22. Gopalakrishnan, V., & Abraham, J. (2004). Effects of multicomponent diffusion on predicted ignition characteristics of an n-heptane diffusion flame. *Combustion and Flame*, *136*, 557–566.
23. Gopireddy, S. R., & Gutheil, E. (2013). Numerical simulation of evaporation and drying of a bi-component droplet. *International Journal of Heat and Mass Transfer*, *66*, 404–411.
24. Gu, X., Basu, S., & Kumar, R. (2012). Dispersion and vaporization of biofuels and conventional fuels in a crossflow pre-mixer. *International Journal of Heat and Mass Transfer*, *55*, 336–346.
25. Gusev, I. G., Krutitskii, P. A., Sazhin, S. S., & Elwardany, A. (2012). A study of the species diffusion equation in the presence of the moving boundary. *International Journal of Heat and Mass Transfer*, *55*, 2014–2021.
26. Hallett, W. L. H. (2000). A simple model for the vaporization of droplets with large numbers of components. *Combustion and Flame*, *121*, 334–344.

27. Hallett, W. L. H., & Beauchamp-Kiss, S. (2010). Evaporation of single droplets of ethanol-fuel oil mixtures. *Fuel, 89*, 2496–2504.
28. Hallett, W. L. H., & Legault, N. V. (2011). Modelling biodiesel droplet evaporation using continuous thermodynamics. *Fuel, 90*, 1221–1228.
29. Harstad, K., & Bellan, J. (2004). Modeling evaporation of Jet A, JP-7, and RP-1 drops at 1 to 15 bars. *Combustion and Flame, 137*, 163–177.
30. Járvás, G., Quellet, C., & Dallos, A. (2011). COSMO-RS based CFD model for flat surface evaporation of non-ideal liquid mixtures. *International Journal of Heat and Mass Transfer, 54*, 4630–4635.
31. Ju, D., Xiao, J., Geng, Z., & Huang, Z. (2014). Effect of mass fractions on evaporation of a multi-component droplet at dimethyl ether (DME)/n-heptane-fueled engine conditions. *Fuel, 118*, 227–237.
32. Kazakov, A., Conley, J., & Dryer, F. L. (2003). Detailed modeling of an isolated, ethanol droplet combustion under microgravity conditions. *Combustion and Flame, 134*, 301–314.
33. Kuo, K.-K. (1986). *Principles of combustion*. Chichester: John Wiley & Sons.
34. Lage, P. L. C. (2007). The quadrature method of moments for continuous thermodynamics. *Computers & Chemical Engineering, 31*, 782–799.
35. Lage, P. L. C., Hackenberg, C. M., & Rangel, R. H. (1995). Nonideal vaporization of dilating binary droplets with radiation absorption. *Combustion and Flame, 101*, 36–44.
36. Landis, R. B., & Mills, A. F. (1974). Effect of internal diffusional resistance on the evaporation of binary droplets. *Proceedings of the Fifth International Heat Transfer Conference*, Tokyo, Japan, paper B7.9.
37. Laurent, C. (2008). Développement et validation de modèles d'évaporation multi-composant. Thèse de l'Institut Supériour de l'Aéronautique et de l'Espace de Toulouse.
38. Laurent, C., Lavergne, G., & Villedieu, P. (2009). Continuous thermodynamics for droplet vaporization: Comparison between Gamma-PDF model and QMoM. *Comptes Rendus Mecanique, 337*, 449–457.
39. Laurent, C., Lavergne, G., & Villedieu, P. (2010). Quadrature method of moments for modeling multi-component spray vaporization. *International Journal of Multiphase Flow, 36*, 51–59.
40. Lippert, A. M., & Reitz, R. D. (1997). Modelling of multicomponent fuels using continuous distributions with application to droplet evaporation and sprays. SAE Technical Paper 972882.
41. Maqua, C., Castanet, G., & Lemoine, F. (2008). Bi-component droplets evaporation: temperature measurements and modelling. *Fuel, 87*, 2932–2942.
42. Modest, M. F. (2003). *Radiative heat transfer* (2nd ed.). Amsterdam, Boston, London: Academic Press.
43. Mukhopadhyay, A., & Sanyal, D. (2001). A spherical cell model for multi-component droplet combustion in a dilute spray. *International Journal of Energy Research, 25*, 1275–1294.
44. Mukhopadhyay, A., & Sanyal, D. (2001). A parametric study of burning of multicomponent droplets in a dilute spray. *International Journal of Energy Research, 25*, 1295–1314.
45. Pagel, S., Stiesch, G., & Merker, G. P. (2002). Modelling of evaporation of a multicomponent fuel. In J. Taine (Ed.), *Proceedings of the Twelfth International Heat Transfer Conference, V. 1*. Grenoble (August 18–23, 2002). Paris: Editions Scientifique et Medicale Elsevier SAS; CD ROM.
46. Ra, Y., & Reitz, R. D. (2009). A vaporization model for discrete multi-component fuel sprays. *International Journal of Multiphase Flow, 35*, 101–117.
47. Raghuram, S., Raghavan, V., Pope, D. N., & Gogos, G. (2013). Two-phase modeling of evaporation characteristics of blended methanol-ethanol droplets. *International Journal of Multiphase Flow, 52*, 46–59.
48. Rivard, E., & Brüggemann, D. (2010). Numerical investigation of semi-continuous mixture droplet vaporization. *Chemical Engineering Science, 65*, 5137–5145.
49. Samimi Abianeh, O., Chen, C. P., & Mahalingam, S. (2014). Numerical modeling of multi-component fuel spray evaporation process. *International Journal of Heat and Mass Transfer, 69*, 44–53.

50. Sazhin, S. S., & Sazhina, E. M. (1996). The effective emissivity approximation for the thermal radiation transfer problem. *Fuel, 75*, 1646–1654.
51. Sazhin, S. S., Abdelghaffar, W. A., Sazhina, E. M., & Heikal, M. R. (2005). Models for droplet transient heating: effects on droplet evaporation, ignition, and break-up. *International Journal of Thermal Sciences, 44*, 610–622.
52. Sazhin, S. S., Elwardany, A., Krutitskii, P. A., Castanet, G., Lemoine, F., Sazhina, E. M., et al. (2010). A simplified model for bi-component droplet heating and evaporation. *International Journal of Heat and Mass Transfer, 53*, 4495–4505.
53. Sazhin, S. S., Elwardany, A., Krutitskii, P. A., Deprédurand, V., Castanet, G., Lemoine, F., et al. (2011). Multi-component droplet heating and evaporation: numerical simulation versus experimental data. *International Journal of Thermal Sciences, 50*, 1164–1180.
54. Sazhin, S. S., Elwardany, A., Sazhina, E. M., & Heikal, M. R. (2011). A quasi-discrete model for heating and evaporation of complex multicomponent hydrocarbon fuel droplets. *International Journal of Heat and Mass Transfer, 54*, 4325–4332.
55. Sazhin, S. S., Al Qubeissi, M., Kolodnytska, R., Elwardany, A., Nasiri, R., & Heikal, M. R. (2014a). Modelling of biodiesel fuel droplet heating and evaporation. *Fuel, 115*, 559–572.
56. Sazhin, S. S., Al Qubeissi, M., Nasiri, R., Gun'ko, V. M., Elwardany, A. E., Lemoine, F., et al. (2014b). A multi-dimensional quasi-discrete model for the analysis of Diesel fuel droplet heating and evaporation. *Fuel, 129*, 238–266.
57. Sirignano, W. A. (1999). *Fluid dynamics and transport of droplets and sprays*. Cambridge: Cambridge University Press.
58. Sirignano, W. A., & Wu, G. (2008). Multicomponent-liquid-fuel vaporization with complex configuration. *International Journal of Heat and Mass Transfer, 51*, 4759–4774.
59. Spille-Kohoff, A., Preuß, E., & Böttcher, K. (2012). Numerical solution of multi-component species transport in gases at any total number of components. *International Journal of Heat and Mass Transfer, 55*, 5373–5378.
60. Strotos, G., Gavaises, M., Theodorakakos, A., & Bergeles, G. (2011). Numerical investigation of the evaporation of two-component droplets. *Fuel, 90*, 1492–1507.
61. Tamim, J., & Hallett, W. L. H. (1995). Continuous thermodynamics model for multicomponent vaporization. *Chemical Engineering Science, 50*, 2933–2942.
62. Tong, A. Y., & Sirignano, W. A. (1986). Multicomponent transient droplet vaporization with internal circulation: integral equation formulation. *Numerical Heat Transfer, 10*, 253–278.
63. Torres, D. J., O'Rourke, P. J., & Amsden, A. A. (2003). Efficient multi-component fuel algorithm. *Combustion Theory and Modelling, 7*, 67–86.
64. Yang, S., Ra, Y., Reitz, R. D., VanDerWege, B., & Yi, J. (2010). Development of a realistic multicomponent evaporation model. *Atomization and Sprays, 20*, 965–981.
65. Zhang, L., & Kong, S.-C. (2010). Vaporization modeling of petroleum-biofuel drops using a hybrid multi-component approach. *Combustion and Flame, 157*, 2165–2174.
66. Zhang, L., & Kong, S.-C. (2012). Multicomponent vaporization modeling of bio-oil and its mixtures with other fuels. *Fuel, 95*, 471–480.
67. Zhu, G.-S., & Reitz, R. D. (2002). A model for high-pressure vaporization of droplets of complex liquid mixture using continuous thermodynamics. *International Journal of Heat and Mass Transfer, 45*, 495–507.

Chapter 6
Kinetic Modelling of Droplet Heating and Evaporation

Abstract The validity of the assumption used in the previous chapters that both liquid and gas phases can be treated as a continuum is no longer obvious when the interface between liquid droplets and the ambient gas is modelled, even when the gas pressure is well above one atmosphere. The chapter begins with a review of early kinetic models of droplet evaporation. Then more rigorous models, based on numerical solutions to Boltzmann equations for vapour and air, are discussed. Two regions of gas above the surface of an evaporating droplet are considered: the kinetic and hydrodynamic regions. Vapour and air dynamics in the first region are described by the Boltzmann equations, while the conventional hydrodynamic analysis is applied in the second region. Collisions between molecules are assumed to be inelastic in the general case. The evaporation coefficient is estimated based on molecular dynamics analysis of *n*-dodecane molecules, using the united atoms model (bonding between hydrogen and carbon molecules is assumed to be much stronger than bonding between carbon molecules).

The analysis presented so far has been implicitly based on the assumption that both liquid and gas phases can be treated as a continuum. This means that the actual velocity distribution of molecules does not affect the physical properties of these phases. The latter are controlled by the number of molecules per unit volume, their average mass, and their average velocities. This assumption is generally valid for both phases, unless the ambient pressure is low enough and as long as the droplets are not too small. However, the validity of this assumption is no longer obvious when the interface between liquid droplets and the ambient gas is modelled, even when the pressure is well above one atmosphere. This means that although the liquid phase can always be treated as a continuum, the properties of gas in the vicinity of the liquid–gas interface can depend not only on the average velocities of molecules, but also on the distribution of molecules by velocities. The latter is generally described in terms of distribution functions.

An overview of early results is presented in Sect. 6.1. Section 6.2 is focused on a description of the kinetic algorithm used in the analysis. Approximations of the

S. Sazhin, *Droplets and Sprays*, DOI: 10.1007/978-1-4471-6386-2_6,

kinetic results are discussed in Sect. 6.3. A mathematical model and kinetic algorithm for taking into account the effects of inelastic collisions are described in Sect. 6.4. Section 6.5 is focused on a description of the molecular dynamics algorithm used for the estimate of the evaporation coefficient for n-dodecane droplets. The results of calculations taking into account the effects of inelastic collisions between molecules and previously obtained values of the evaporation coefficient are presented and discussed in Sect. 6.6.

6.1 Early Results

In general, the choice of the model which needs to be used for the analysis of droplet evaporation is determined by the so-called Knudsen number $\mathrm{Kn} = l_{\mathrm{coll}}/R_d$, where l_{coll} is the characteristic mean free path of molecules. It is generally believed that when $\mathrm{Kn} < 0.01$, the gas may be approximated as a continuum. In another limiting case, $\mathrm{Kn} > 3$, the mean free path of molecules is large compared with the dimensions of the system, and a free molecular regime takes place. In the transitional regime $0.01 \leq \mathrm{Kn} \leq 3$, the mean free path of molecules is comparable with the dimensions of the system. This is known as the slip regime. In this case, gas can be regarded as a continuum at several mean paths away from the droplet surface, but kinetic effects should be taken into account in the vicinity of the droplet surface [58]. In the case of fuel droplets in internal combustion engines, Kn is generally well below 0.01, and this justifies the application of the continuum medium approximation for their analysis. The validity of this approximation near the droplet surfaces, however, is not at first obvious, as was shown in [82]. Hence, the analysis of kinetic models of droplet heating and evaporation can be justified even in the case of high-pressure gases. This section is focused on a general introduction to the problem and earlier results of kinetic modelling obtained mainly before 2006. This will be essentially an extended presentation of Sect. 3.4 of [117]. The following sections will be focused on the most interesting and important results obtained by the author and his colleagues after 2006 and not included in [117].

The most general approach to the analysis of the velocity distributions of gas molecules is based on the so called multiparticle distribution functions, which take into account not only the positions and velocities of individual molecules, but also the correlations between them. This eventually leads to the infinite chain of Bogolubov–Born–Green–Kirkwood–Yvon (BBGKY) equations for multiparticle distribution functions [2]. In practice, this chain has to be truncated. In the zeroth approximation, assuming that interactions between individual molecules take place only during collision processes, this chain reduces to a single Boltzmann equation, which can be presented in the following form [2, 39]:

$$\frac{\partial f}{\partial t} + \mathbf{v} \cdot \frac{\partial f}{\partial \mathbf{x}} + \mathbf{F} \cdot \frac{\partial f}{\partial \mathbf{v}} = \left(\frac{\partial f}{\partial t} \right)_{\mathrm{coll}}, \tag{6.1}$$

where $f = f(t, \mathbf{v}, \mathbf{x})$ is the molecular distribution function, $\mathbf{v}$, $\mathbf{x}$, and $\mathbf{F}$ are velocity, position, and force acting on individual molecules; the right-hand side of this equation is the collision integral, which takes into account collisions between molecules. The physical meaning of f can be defined via the product

$$\Delta f = f(t, \mathbf{v}, \mathbf{x})\Delta\mathbf{x}\Delta\mathbf{v},$$

where $\Delta\mathbf{x}$ is a small element of volume around $\mathbf{x}$, $\Delta\mathbf{v}$ is a small range of velocities around $\mathbf{v}$; Δf indicates the number of molecules in the element of volume $\Delta\mathbf{x}$ around $\mathbf{x}$ having velocities in the range $\mathbf{v}$ to $\mathbf{v} + \Delta\mathbf{v}$.

Integration of both parts of (6.1) over velocities gives the well-known continuity equation (analogous to conservation of mass). Similarly, the integration of both parts of this equation with weights (components of momenta and energy of molecules) gives the equations of conservation of momentum (Navier–Stokes equation) and energy.

The contribution of $\mathbf{F}$ in (6.1) is generally ignored, and this can be simplified to

$$\frac{\partial f}{\partial t} + \mathbf{v} \cdot \frac{\partial f}{\partial \mathbf{x}} = \left(\frac{\partial f}{\partial t}\right)_{\text{coll}}. \tag{6.2}$$

This could not be done in the case of ionised gases even in the absence of external fields [2, 39, 116], but in the case of nonionised gases, $\mathbf{F}$ could come only from the gravitational force, which is negligible for small molecules.

The expression for the collision integral, assuming that collisions are elastic, can be presented as [18, 19]

$$\left(\frac{\partial f}{\partial t}\right)_{\text{coll}} = \frac{d_m^2}{2} \int\limits_{-\infty}^{+\infty} \mathrm{d}\mathbf{v_1} \int\limits_{0}^{\pi} \sin\theta \mathrm{d}\theta \int\limits_{0}^{2\pi} \mathrm{d}\phi \left(f' f_1' - f f_1\right) |\mathbf{v} - \mathbf{v}_1|, \tag{6.3}$$

where d_m is the diameter of colliding molecules, θ and ϕ are relative angular coordinates of molecules, the superscript $'$ indicates the velocities and the distribution functions after collisions, $f = f(\mathbf{v})$, $f_1 = f(\mathbf{v}_1)$, $f' = f(\mathbf{v}')$, $f_1' = f(\mathbf{v}_1')$. The integral on the right-hand side of (6.3) is calculated in the three-dimensional velocity space. In deriving (6.3), it was assumed that molecules are rigid elastic spheres. In the case of evaporating droplets, the collision integral represents the sum of two terms: one describing collisions between molecules of vapour and the other describing collisions between molecules of vapour with molecules of other gases. In most cases, the focus has been on the analysis of the first term. A more general analysis taking into account the contribution of both processes was reported in [9, 83, 84, 154, 155]. A more general approach to the derivation of the expression for the collision integral, taking into account the probability of elastic scattering of molecules after collisions, is described in Sect. 13 of [3].

A general analysis of (6.2) and (6.3) is possible using numerical methods (see [11, 18, 19, 158, 159]). Applications of these methods to analysis of the droplet evaporation process were reported in [9, 83, 84, 154, 155]. A detailed analysis of these methods and their applications is beyond the scope of this monograph.

The analysis of (6.2) and (6.3) is considerably simplified when the collision integral (6.3) is not calculated but modelled. Several explicit analytical expressions for $\left(\frac{\partial f}{\partial t}\right)_{\mathrm{coll}}$ have been suggested (see, e.g., [2, 16, 25, 115]). An alternative simplification of the Boltzmann equation can be based on an analysis focused on the perturbation of the velocity distribution function from the equilibrium state (linearisation of the Boltzmann equation [32]).

Further analysis of (6.2) depends largely on the relative values of the mean free path of molecules between collisions and the characteristic size of the system. In the case of droplets, this is described by the Knudsen number Kn, as mentioned earlier. In what follows, our review of the approaches to the analysis of this equation suggested so far will mainly follow [82].

When $\mathrm{Kn} \gg 1$, the contribution of collisions can be ignored altogether, and we have a free molecular flow. In this case, the right-hand side of (6.2) can be assumed equal to zero, and this equation can allow analytical solutions. Assuming that both incoming and outgoing molecular fluxes are Maxwellian with temperatures T_∞ and T_s respectively, one can derive from (6.2) the Hertz–Knudsen–Langmuir formula for mass flux of vapour from the droplet surface [59, 73, 75, 104, 134]:

$$j_{\mathrm{lg}} \equiv \dot{m}_d'' = \frac{|\dot{m}_d|}{4\pi R_d^2} = \frac{\beta_m}{\sqrt{2\pi R_v}} \left(\frac{p_{\mathrm{vs}}}{\sqrt{T_s}} - \frac{p_{v\infty}}{\sqrt{T_g}} \right), \tag{6.4}$$

where $|\dot{m}_d|$ is the rate of droplet mass loss, β_m is the evaporation coefficient (assumed to be equal to the condensation coefficient), R_v is the gas constant for vapour, p_{vs} is the saturated vapour partial pressure corresponding to T_s, $p_{v\infty}$ is the vapour partial pressure at large distances from the droplet, T_s is the droplet surface temperature, and T_g is the ambient gas temperature.

The evaporation coefficient $\beta_m \leq 1$ is defined as the ratio of the actual mass flux leaving the surface of the droplet (before the first collision) j_{es} and the maximal possible flux [124]:

$$\beta_m = \frac{j_{\mathrm{es}}}{\rho_{\mathrm{vs}} \sqrt{\frac{R_v T_s}{2\pi}}}, \tag{6.5}$$

where ρ_{vs} is the density of saturated fuel vapour corresponding to the liquid temperature $T_s = T_{\mathrm{ls}}$ at the droplet surface. The condensation coefficient, defined also by (6.5), shows what proportion of vapour molecules striking the liquid surface is absorbed by this surface. The remaining portion $(1 - \beta_m)$ shows the fraction of reflected molecules. The value of β_m depends on the contacting surfaces [130] and can be determined experimentally [61, 81, 103]. Methods of molecular dynamics and direct modelling of droplet evaporation can also be used to estimate this coefficient.

The most recent developments in this direction will be discussed later in this chapter. Among earlier investigations, the following papers can be mentioned [30, 38, 164, 171]. A comprehensive review of theoretical and experimental studies of this coefficient for water up to 2001 is presented in [98] (see also [81]).

Formula (6.4) has been widely used for the analysis of evaporation and condensation processes despite the simplistic assumptions used for its derivation (see, e.g., [36, 51, 56, 105]). A more general approach to the evaporation–condensation problem takes into account the effects of collisions. This leads to the introduction of the concept of the Knudsen layer separating the liquid surface from the bulk of the vapour, which is described using the continuum equations. The thickness of the Knudsen layer l_K is typically estimated as several molecular mean free pass lengths l_{coll} for small drift vapour velocities v_{dr}, $10\,l_{\mathrm{coll}}$ for $v_{\mathrm{dr}} = 0.5v_{\mathrm{sound}}$, and 100–200 l_{coll} for v_{dr} close to v_{sound} [148]. The value of l_{coll} is estimated based on the temperature of the surface T_s.

For practical modelling purposes, Diesel fuel can be approximated by n-dodecane $C_{12}H_{26}$ with reservations discussed in Sect. 5.3. For this substance ($M_{C_{12}H_{26}} = 170.3$ kg/kmol), the gas constant is estimated as $R_{\mathrm{n-dod}} = 48.88$ J/(kg K). Assuming that vapour temperature is equal to surface temperature $T_s = 600$ K, we obtain $v_{\mathrm{sound}} = 202.6$ m/s. For this T_s, the saturated vapour pressure of n-dodecane is equal to $p_{\mathrm{vs}} = 6.4 \times 10^5$ Pa. Using (6.4), assuming that $p_{v\infty} = 0$ and $\beta_m = 0.5$, we obtain $j_{\mathrm{lg}} = 745\ \ \mathrm{kg/(m^2 s)}$. From the gas law, we have $\rho_{\mathrm{vs}} = 21.8\ \mathrm{kg/m^3}$. Hence, $v_{\mathrm{dr}} = j_{\mathrm{lg}}/\rho_{\mathrm{sv}} = 34$ m/s. This value can be further modified if different β_m are used. Note that this result is not very sensitive to the value of T_s. For example, for $T_s = 400$ K, we obtain $v_{\mathrm{sound}} = 165$ m/s and $v_{\mathrm{dr}} = 28$ m/s. More accurate analysis of the problem (see Eq. (6.7)) would predict slightly larger values of v_{dr}, but the assumption $v_{\mathrm{dr}} \ll v_{\mathrm{sound}}$ would remain valid. Assuming that liquid evaporates into its own vapour, l_{coll} can be estimated as [75]

$$l_{\mathrm{coll}} = \left[\sqrt{2}\pi d_f^2 \rho_{\mathrm{vs}} N_{\mathrm{A}}/M\right]^{-1},$$

where M is the molar mass, N_A is Avogadro's number.

Taking the diameter of n-dodecane molecules $d_{\mathrm{nd}} = 10^{-9}$ m, $\rho_{\mathrm{vs}} = 21.8\ \mathrm{kg/m^3}$ (corresponding to $T_s = 600$ K) and $M = 170.3$ kg/kmol, we obtain $l_{\mathrm{coll}} = 2.9 \times 10^{-9}$ m. Assuming that $l_K < 5l_{\mathrm{coll}}$, we can estimate l_K for Diesel engines as $l_K < 1.5 \times 10^{-8}\ \mathrm{m} = 0.015\ \mu\mathrm{m}$. This is about two orders of magnitude less than the droplet radii. As shown later, following [82], despite the small thickness, the contribution of the Knudsen layer cannot be *a priori* ignored, even in this case.

Strictly speaking, (6.2) can be applied only when l_{coll} is much greater than d_{nd} [95]. From this point of view, the results following from this equation are expected to be reliable only for the set of parameters for which the condition $l_{\mathrm{coll}} \gg d_{\mathrm{nd}}$ is satisfied. When l_{coll} is close to d_{nd} (which is expected near the critical state), the model can show the trends of the processes rather than give reliable quantitative estimates. Also, the solutions to the Boltzmann equation in most cases do not take into account the contribution of processes inside the vapour molecules (possible approaches to modelling inelastic collisions between molecules will be discussed later in this chapter).

The velocity distribution for the molecules in the Knudsen layer is affected by collisions, and it can be obtained from the solution to (6.2). Schrage [130] drew attention to the fact that the effects of collisions lead to the formation of the shifted Maxwellian distribution of molecules near the outer boundary of the Knudsen layer. The authors of [87] assumed that this distribution function is formed in the whole Knudsen layer up to the liquid surface. Considering the case in which the vapour drift velocity is much less than the velocity of sound (weak evaporation, this assumption is consistent with the above-mentioned estimates) and using matching boundary conditions in this layer, they derived a modified expression for j_{lg} at $\mathrm{Kn} \ll 1$ and $(T_g - T_s)/T_s \ll 1$:

$$j_{\mathrm{lg}} = \frac{2\beta_m}{(2-\beta_m)\sqrt{2\pi R_v}} \left(\frac{p_{\mathrm{vs}}}{\sqrt{T_s}} - \frac{p_{v\infty}}{\sqrt{T_g}} \right). \tag{6.6}$$

Formula (6.6) takes into account the convection of vapour and collision processes. For $\beta_m = 1$, this formula predicts a value of j_{lg} twice as large as the one predicted by (6.4). A simple derivation of (6.6) is given in [58].

Further development of the kinetic theory of evaporation and condensation was reported in [88–90, 101]. The problem was solved using two forms of the collision term in the Boltzmann equation: the conventional one and the one suggested by Bhatnagar, Gross, and Krook [16]. The latter form of the collision term secures the conservation of particles during the collision process and has been widely used to model collisions in gases and plasma (see, e.g., [115]). The results have turned out to be practically independent of the form of the collision term used.

The analysis performed in [101] allowed the authors to obtain a number of equations useful for computing the dynamics of evaporation and condensation processes, including the one describing the mass flux of vapour leaving the droplet and moving into a dense medium ($\mathrm{Kn} \ll 1$):

$$j_{\mathrm{lg}} = \frac{\beta_m (p_{\mathrm{vs}} - p_{\mathrm{Rd}})}{(1-0.4\beta_m)\sqrt{2\pi R_v T_s}}, \tag{6.7}$$

where p_{Rd} is the vapour pressure just outside the Knudsen layer.

Equation (6.7) has been widely used in various applications (see, e.g., [77]). It is valid in the case of weak evaporation and condensation, which is expected to take place when $(T_{\mathrm{Rd}} - T_s)/T_s \ll 1$. Further development of the theory of weak evaporation has been reported in [27, 29, 74, 131, 146, 147, 160, 173, 175]. These developments, however, do not undermine the usefulness of (6.7) for practical applications. This equation is more accurate than (6.4) and (6.6) and has been widely used for various applications (see, e.g., [82]).

The increase in intensity of evaporation leads to further deformation of the molecular distribution function in the Knudsen layer. At a certain stage, the linearisation of this distribution function (which was used in deriving (6.7)) is no longer possible. This leads to a situation in which v_{dr} becomes comparable with the velocity of sound, and the theory of weak evaporation and condensation is no longer valid. A theory of intensive evaporation has been developed in [1, 6, 7, 26, 45, 72, 76, 91, 92, 100, 112, 139, 144, 145, 149–151, 153, 156, 162, 174, 176]. A review of computational and experimental results referring to intensive condensation, with particular emphasis on the problem of determining β_m, was presented in [111].

The authors of [92] showed that the mass flux of vapour leaving the droplet can be approximated by the following expression (valid for both weak and strong evaporation):

$$j_{\mathrm{lg}} = 0.6\sqrt{2R_v T_s}(\rho_{\mathrm{vs}} - \rho_{\mathrm{Rd}})\sqrt{\rho_{\mathrm{Rd}}/\rho_{\mathrm{vs}}}, \tag{6.8}$$

where ρ_{Rd} is the vapour density at the outer boundary of the Knudsen layer. This equation was derived under the assumption that $\beta_m = 1$.

For arbitrary β_m, the value of j_{lg} can be obtained via the replacement of ρ_{vs} in (6.8) by [76, 92]:

$$\rho_\beta = \left(1 - 2\sqrt{\pi}\frac{j_{\mathrm{lg}}}{\rho_{\mathrm{vs}}\sqrt{2R_v T_s}}\frac{1-\beta_m}{\beta_m}\right)\rho_{\mathrm{vs}}. \tag{6.9}$$

The solution of the system of Eqs. (6.8) and (6.9) with two unknowns j_{lg} and ρ_β would allow us to find the required value of j_{lg} for arbitrary $\beta_m \leq 1$. In the limit of weak evaporation $(\rho_{\mathrm{vs}} - \rho_{\mathrm{Rd}})/\rho_{\mathrm{vs}} \ll 1$, Eq. (6.8) reduces to (6.7) for $\beta_m = 1$ (see [82]).

One of the fundamental differences in predictions of kinetic and hydrodynamic models lies in the behavior of temperature, velocity, and species concentrations near the boundaries. It is well known that the hydrodynamic models predict continuous changes of all these parameters near the boundaries, while the kinetic models predict jumps in their values over distances of the order of several mean free paths between intermolecular collisions (jumps in the values of concentrations and temperature and velocity slips). The values of these jumps are usually calculated from an analysis of the Boltzmann equation or its momenta [50, 58, 70, 132, 133, 172]. Once they have been calculated, the analysis of the processes in the gas, including the processes of evaporation, can be based on hydrodynamic theory. The boundary conditions in this case take into account jumps in the values of the above-mentioned parameters near the boundaries [35, 40, 118]. The author is unaware of any detailed comparison between this approach to kinetic modelling of the evaporation processes and the approach based on calculating the mass flux from the droplet surface, as discussed earlier.

The effects of temperature and velocity slip on the values of the Nusselt number for nanoparticles was investigated in [41] using asymptotical analysis. It has been observed that the velocity slip at the interface does not affect significantly the overall Nusselt number. However, the temperature slip is a more important parameter in the heat transfer, and it affects Nu significantly. This leads to the conclusion that if a temperature slip is possible at the interface, it must be taken into account. If the temperature discontinuity becomes large, the droplets can become almost adiabatic.

The kinetic model based on (6.7) and the hydrodynamic model based on (4.16) were applied to modelling heating and evaporation processes of typical fuel droplets in Diesel engines [82]. The application of (6.7) in the kinetic model describing weak evaporation was justified by the predicted results. The values of drift vapour velocities were much less than the velocity of sound. Also, in the kinetic model it was assumed that fuel vapour leaving the Knudsen layer was removed from the outer boundary of this layer via diffusion and convection in the surrounding hydrodynamic region. Thus the mass flux of fuel vapour leaving the droplets was assumed to be equal to the diffusion and convection mass flux from the outer boundary of the Knudsen layer j_{diff}. This approach is known as the flux matching method and is widely used in kinetic modelling of droplet evaporation and condensation (see, e.g., [106]). The location of the interface between the Knudsen layer and the hydrodynamic region was chosen rather arbitrarily. Hence, it would be more appropriate to refer in this and similar models discussed later not to the Knudsen layer (the thickness of which is usually well determined; see the earlier discussion) but to the kinetic region (the region in which the analysis is based on the Boltzmann equations rather than hydrodynamic equations).

The condition $j_{\text{lg}} = j_{\text{diff}}$ allowed the authors of [82] to determine the value of the mass fraction of fuel vapour at the outer boundary of the Knudsen region ($Y_{v\text{Rd}}$). This equation was solved under the assumption that $T_{\text{Rd}} = T_s$. The value of $Y_{v\text{Rd}}$ was less than or equal to the saturated value Y_{vs}. The value of j_{diff} was determined from the conventional hydrodynamic theory (see Chap. 4). The gas pressure and the initial droplet temperature were taken to be equal to 30 bar and 300 K respectively. The effects of temperature gradients inside droplets were ignored. In the kinetic model, two values of β_m were used: $\beta_m = 0.5$ and $\beta_m = 0.04$. These are the average and minimal values of β_m measured for water [61, 103]. The value of $\beta_m = 0.04$ is close to $\beta_m = 0.06$ estimated by Shusser et al. [138] for butane.

The calculations in [82] were performed for an initial droplet radius equal to 5 μm and initial ambient gas temperature equal to 750 K. Droplets with this initial radius are likely to be observed in Diesel engines [43]. The chosen initial gas temperature is typical for the end of the compression stroke in Diesel engines [43, 122]. As follows from the calculations presented by the authors of the above-mentioned paper, kinetic effects lead to a small increase in evaporation time and droplet temperature for $\beta_m = 0.5$. This increase becomes larger for smaller values of β_m. The overall shapes of the diameter versus time plots, as predicted by hydrodynamic and kinetic models, appeared to be rather similar. A similar analysis was performed for the initial ambient gas temperature 2,000 K and for both these temperatures and the initial droplet radius equal to 20 μm. In all cases, the kinetic effects were shown to be more pronounced for smaller droplets (5 μm) than for larger ones (20 μm), and for higher initial gas temperatures. As expected, the kinetic model predicted longer evaporation times and higher droplet temperatures compared with the hydrodynamic model. The droplet evaporation time and droplet temperature increased with decreasing evaporation coefficient.

The kinetic effects appeared to be even more pronounced for droplets with an initial radius of 1 μm. In this case, however, the kinetic model was less reliable, since it did not take into account the effects of surface tension. For droplets with initial radii comparable to 1 μm, these effects are expected to be noticeable [97].

The main conclusion of the analysis reported in [82] is that the kinetic effects on droplet evaporation are always noticeable, despite the fact that this evaporation takes place at rather high pressures (up to 30 atm and even more). This showed the limitation of the applicability of the hydrodynamic approach to accurate modelling of this process. The authors of [82] recommended that kinetic effects be taken into account in modelling the evaporation process of Diesel fuel droplets in realistic internal combustion engines.

Although the results presented in [82] clearly show the importance of taking into account the kinetic effects in modelling droplet heating and evaporation even in Diesel engine-like conditions, the approach used in [82] is based on too many simplifying assumptions, which makes it difficult to recommend it for quantitative analysis of the process. The authors of [82] assumed that the evaporation takes place into the fuel's own vapour. This is obviously not true in the case of Diesel engine-like conditions when Diesel fuel evaporates into air. Among early papers taking into account the presence of two components in the kinetic region, we can mention [10, 156]. In [156], the two-surface problems of a multicomponent mixture of vapour and noncondensable gases in the continuum limit were studied based on asymptotic analysis of the Boltzmann equation. An asymptotic analysis of the linearised Boltzmann equation for the binary mixture was presented in [10].

A more rigorous approach to modelling Diesel fuel droplet evaporation used in [124, 135] is based on the rigorous calculation of the collisional term in the Boltzmann equation using numerical methods, such as those developed in [11, 135]. The algorithm developed in [124, 135] was specifically designed to model the evaporation and condensation processes in binary mixtures consisting of molecules with different radii and masses. This algorithm was applied to the same problem of evaporation of Diesel fuel into high pressure air as considered earlier in [82], but taking into account the contribution of air molecules. As in [82], two regions above the surface of the evaporating droplet were considered. These are the kinetic region, where the analysis was based on the Boltzmann equations for vapour and air, and the hydrodynamic region. Also, it was assumed that the mass fluxes leaving the kinetic region and the corresponding diffusion fluxes in the hydrodynamic region are matched. It was assumed that the vapour mass flux leaving the droplet's surface is the maximal one (evaporation coefficient is equal to 1). This value of β_m was predicted by the approximate model described in [102] (see Sect. 2.1 of [124] for a more detailed discussion of this matter). The effects predicted by this model appeared to be noticeable, and larger than those predicted by the approximate analysis, if the contribution of air in the kinetic region was taken into account. As in [82], it was recommended that the kinetic effects be taken into account when an accurate analysis of Diesel fuel droplet evaporation is essential.

One of the main limitations of the model described in [124, 135] is that the authors did not take into account the contribution of the heat flux in the kinetic region, assuming that there is no temperature gradient within it. This assumption could not be justified without solving a more general problem, taking into account this effect. This is done in [119], and the results are reproduced in the Sect. 6.2. The algorithm of the solution of the Boltzmann equation used in that paper is the same as that used in [124, 135], but the effects of the heat flux in the kinetic region were taken into account (this led to an additional nontrivial problem of formulating the boundary conditions for temperature). Also, an updated version of the hydrodynamic model was used. More accurate expressions for the binary diffusion coefficient and the convection heat transfer coefficient than in [124] were used, and the temperature dependencies of the fuel vapour specific heat capacity and thermal conductivity were taken into account.

6.2 Kinetic Algorithm: Effects of the Heat and Mass Fluxes

The results presented in this section are mainly reproduced from our previous paper [119]. As in [124, 135], two regions of gas above the surface of the evaporating droplet are considered: the kinetic and hydrodynamic regions. These are schematically shown in Fig. 6.1. As in [124], we assume that gas consists of two components, vapour and background air, both in the kinetic and hydrodynamic regions. It is assumed that the contribution of chemical reactions of vapour and oxygen can be ignored. Vapour and air dynamics in the first region are described by the Boltzmann equations, while conventional hydrodynamic analysis is applied in the second region. This means that we investigate a two-surface problem. As in [124], the effects of droplet motion, thermal radiation, and the temperature gradient inside droplets are ignored. None of these assumptions can be rigorously justified in Diesel engine-like conditions, but they allow us to separate these effects from the kinetic effects. In a comprehensive model, these effects need to be taken into account alongside the kinetic effects. A model taking into account the temperature gradient inside droplets is described later in this chapter.

6.2.1 *Boltzmann Equations for the Kinetic Region*

6.2.1.1 Formulation of the Problem

As in [124, 135], the evolution of the molecular velocity distribution functions of air $f_a \equiv f_a(\mathbf{r}, t, \mathbf{v})$ and vapour $f_v \equiv f_v(\mathbf{r}, t, \mathbf{v})$ in the kinetic region (see Fig. 6.1) is assumed to be controlled by the corresponding Boltzmann equations:

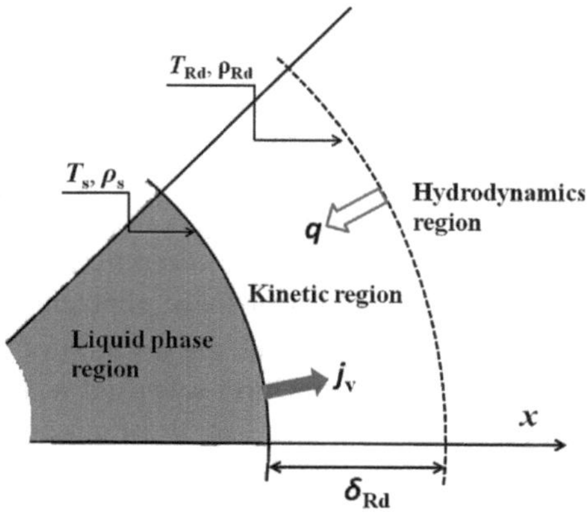

Fig. 6.1 Liquid phase, kinetic and hydrodynamic regions in the vicinity of the surface of the droplet; T_s is the droplet surface temperature, ρ_s is the vapour density in the immediate vicinity of the droplet surface, T_{Rd} and ρ_{Rd} are the temperature and density of vapour at the outer boundary of the kinetic region; δ_{Rd} indicates the thickness of the kinetic region, $\mathbf{j}_{\mathbf{v}}$ and $\mathbf{q}$ show the directions of the vapour mass flux and heat flux respectively. Reprinted from [128]. Copyright Elsevier (2013)

$$\left.\begin{aligned}\frac{\partial f_a}{\partial t} + \mathbf{v}_a \frac{\partial f_a}{\partial \mathbf{r}} &= J_{aa} + J_{av} \\ \frac{\partial f_v}{\partial t} + \mathbf{v}_v \frac{\partial f_v}{\partial \mathbf{r}} &= J_{va} + J_{vv}\end{aligned}\right\}, \tag{6.10}$$

where $J_{\alpha\beta}$ ($\alpha = a, v;\ \beta = a, v$) are collision integrals defined as (cf. (6.3))

$$J_{\alpha\beta} = \frac{\sigma_{\alpha\beta}^2}{2} \int_{-\infty}^{+\infty} \mathrm{d}\mathbf{v}_1 \int_0^{\pi} \sin\theta \mathrm{d}\theta \int_0^{2\pi} \mathrm{d}\phi \left(f'_\alpha f'_{\beta 1} - f_\alpha f_{\beta 1}\right) \left|\mathbf{v}_\alpha - \mathbf{v}_{\beta 1}\right|, \tag{6.11}$$

where $\sigma_{\alpha\beta} = (\sigma_\alpha + \sigma_\beta)/2$, σ_α, and σ_β are the corresponding diameters of molecules, θ and ϕ are angular coordinates of molecules β relative to molecules α after the collision, the superscript $'$ indicates the velocities and the molecular velocity distribution functions after collisions (in what follows, these functions will be referred to as 'molecular distribution functions'). The subscript $_1$ indicates that molecules of type β collide with molecules of type α, and as a result of this interaction, the function f_α is modified. The first integral on the right-hand side of (6.11) is calculated in the three-dimensional velocity space. In deriving (6.10) and (6.11), it was assumed that molecules are rigid spheres and that body forces acting on them are negligible. It is assumed that the evaporation coefficient β_m is equal to 1. In contrast to [124], the changes of temperature across the kinetic region are taken into account.

6.2.1.2 Numerical Algorithm

The main features of the numerical algorithm used in our calculations are described in [11, 85, 135]. In this section, these features are briefly summarised, and the new elements of the algorithm are discussed, following [124].

Time and physical space are discretised as in conventional structured computational fluid dynamics (CFD) codes [165]. The discretisation in the velocity space is performed similarly to that in the physical space, replacing continuous values of $\mathbf{v}$ by a discrete set $\{\mathbf{v}^k\}^M$, where k indicates the position of a velocity cell, M is the total number of cells. A homogeneous grid was used with the following range: $v_{i\max} - v_{i\min} = 7\sqrt{2R_v T_s}$, where $i = x, y, z$, R_v is the gas constant for fuel vapour, T_s is the droplet surface temperature. Twelve cells for each velocity component were considered, giving the total number of cells in the velocity space equal to $12^3 = 1728$. This grid was tested for calculation of the dimensionless Maxwellian distribution with normalised density $\overline{n} = 1$ and normalised temperature $\overline{T} = T_g/T_s = 1$. The results of the calculation gave us $\overline{n} = 0.9999$ and $\overline{T} = 0.9998$ [135]. Hence we can conclude that the velocity grid under consideration allows us to calculate the macroscopic variables with error much less than 1 % when the distribution function of molecules is close to the Maxwellian one. In the general case, we anticipate that these errors are, at worst, just less than 1 %, which we find acceptable for our applications.

The boundaries of the velocity domain in the v_x, v_y, v_z directions are chosen in such a way that the contribution of molecules with velocities outside this range can be ignored. For each value of $\mathbf{v}^k$, the corresponding value of f^k is specified. This allows us to present each of Eq. (6.10) in a discretised form:

$$\frac{\Delta f_\alpha^k}{\Delta t} + \mathbf{v}_\alpha^k \frac{\Delta f_\alpha^k}{\Delta \mathbf{r}} = J_{\alpha\alpha}^k + J_{\alpha\beta}^k, \tag{6.12}$$

where $k \in [1, M]$. After calculation of $J_{\alpha\beta}^k$ for each velocity cell $\mathbf{v}_\alpha^k$, the nonlinear system of integrodifferential equations (6.10) reduces to the linear system of algebraic equations (6.12).

Following [11], the numerical solution of system (6.12) for each gas component is performed in two steps. Firstly, molecular displacements are calculated ignoring the effect of collisions ($J_{\alpha\alpha}^k = J_{\alpha\beta}^k = 0$). Secondly, the collisional relaxation is calculated under the assumption of spatial homogeneity. The boundary conditions for the molecular distribution function are taken into account in the first step. The numerical solution of (6.12) is performed following the explicit approach, and using the Courant condition $\Delta t \max(|v_x|, |v_y|, |v_z|) < \min(\Delta x, \Delta y, \Delta z)$. Between 100 and 200 cells in the physical space are used, leading to errors of less than 1 %.

In the second step, the displacement of molecules stops, and they begin to collide. Using the explicit approach, the solution to each simplified equation in system (6.10) in each cell in the physical space can be written as

$$f_{\alpha}^{k,n} = \frac{\tilde{f}_{\alpha}^{k,n-1} + \Delta t \left[N_{\alpha\alpha}^{k,n-1} + N_{\alpha\beta}^{k,n-1} \right]}{1 + \Delta t \left[\nu_{\alpha\alpha}^{k,n-1} + \nu_{\alpha\beta}^{k,n-1} \right]}, \tag{6.13}$$

where

$$\left. \begin{aligned} \nu_{\alpha\beta}^{k,n-1} &= \frac{\sigma_{\alpha\beta}^2}{2} \int_{-\infty}^{+\infty} \mathrm{d}\mathbf{v_1} \int_0^{\pi} \sin\theta \mathrm{d}\theta \int_0^{2\pi} \mathrm{d}\phi \tilde{f}_{\beta 1}^{k,n-1} \left| \mathbf{v}_{\alpha}^{k} - \mathbf{v}_{\beta 1}^{k} \right| \\ N_{\alpha\beta}^{k,n-1} &= \frac{\sigma_{\alpha\beta}^2}{2} \int_{-\infty}^{+\infty} \mathrm{d}\mathbf{v_1} \int_0^{\pi} \sin\theta \mathrm{d}\theta \int_0^{2\pi} \mathrm{d}\phi \tilde{f}_{\alpha}'^{k,n-1} \tilde{f}_{\beta 1}'^{k,n-1} \left| \mathbf{v}_{\alpha}^{k} - \mathbf{v}_{\beta 1}^{k} \right| \end{aligned} \right\}, \tag{6.14}$$

$\tilde{}$ indicates the value of the molecular distribution function calculated in the first step, and the additional superscripts $^{n-1}$ and ^{n} indicate consecutive time steps.

The calculation of integrals $\nu_{\alpha\beta}^{k,n-1}$ and $N_{\alpha\beta}^{k,n-1}$ in (6.13) (see (6.14)) turned out to be a major challenge from the point of view of CPU requirements. In our algorithm, the conventional approach to calculation of these integrals is replaced by integration based on the random cubature formulas. In this case, $\nu_{\alpha\beta}^{k,n-1}$ and $N_{\alpha\beta}^{k,n-1}$ are found from the following expressions:

$$\left. \begin{aligned} \nu_{\alpha\beta}^{k,n-1} &= \frac{V}{K_0} \frac{\sigma_{\alpha\beta}^2}{2} \sum_{l=1}^{K_0} \frac{\tilde{f}_{\beta 1l}^{k,n-1} \left| \mathbf{v}_{\alpha} - \mathbf{v}_{\beta 1l} \right| \sin\theta_l}{p(\mathbf{w}_l)} \\ N_{\alpha\beta}^{k,n-1} &= \frac{V}{K_0} \frac{\sigma_{\alpha\beta}^2}{2} \sum_{l=1}^{K_0} \frac{\tilde{f}_{\alpha l}'^{k,n-1} \tilde{f}_{\beta 1l}'^{k,n-1} \left| \mathbf{v}_{\alpha} - \mathbf{v}_{\beta 1l} \right| \sin\theta_l}{p(\mathbf{w}_l)} \end{aligned} \right\}, \tag{6.15}$$

where V is the volume of five-dimensional space, $\mathbf{w}_l \equiv \mathbf{w}_l(\mathbf{v}_{\beta 1l}, \theta_l, \phi_l)$ is an arbitrary chosen point in this space, $p(\mathbf{w}_l)$ is the value of the probability density function of the distribution of these points, K_0 is the total number of these points (assumed number of collisions in a given cell in physical and velocity spaces), and the summation is performed over all these points. In this case, the relative error of calculation of $\nu_{\alpha\beta}^{k,n-1}$ and $N_{\alpha\beta}^{k,n-1}$ is proportional to $1/\sqrt{K_0}$ and does not depend on the dimension of the space [11, 12]. For practical calculations, we assumed the homogeneous distribution of $\mathbf{w}_l$, which implies that $p(\mathbf{w}_l) = 1$. The efficiency of the application of (6.15) largely depends on the choice of random nodes $\mathbf{w}_l$. One of the most widely used approaches to choosing these nodes is based on the so-called Korobov sequences [78, 85, 109, 157]. The condition $p(\mathbf{w}_l) = 1$ for these sequences is satisfied. In the case of piecewise constant functions (used in our numerical analysis), the errors of calculations using Korobov sequences are proportional to $1/K_0$. This approach was used in the algorithm with $K_0 = 200$, leading to possible errors of about 1 %. An explicit expression for V was given as

$$V = 2\pi^2 \left| v_{x(\max)} - v_{x(\min)} \right| \left| v_{y(\max)} - v_{y(\min)} \right| \left| v_{z(\max)} - v_{z(\min)} \right|. \tag{6.16}$$

The modelling of the collision processes was based on the assumption that the collisions are elastic (momentum and energy are conserved) and the directions of momenta of molecules in the coordinate system, linked with their centres of inertia, are random. The numerical implementation of this model, however, was linked with a number of difficulties. These are related to the fact that randomly chosen directions of molecular velocities after collisions are likely to lead to the values of these velocities lying between the values in the nodes of the discretised velocity space. This can eventually lead to nonconservation of momentum and energy during the collision process. In the early papers, this problem was resolved by introducing corrections to the molecular distribution function after collision [11]. Although these corrections made the system conservative, they led to additional sources of errors. In the projection method, developed later in [159], the actual molecular velocities after the collisions were replaced by pairs of velocities referring to the nearest nodes. These velocities were appropriately weighted, which ensured that the conservation of momentum and energy took place during individual collisions. This, however, led to increased complexity of the algorithm.

The approach used in [124] is different from those described above. It is based on the discretisation of the velocities, not only during the description of molecular motion, but also in the analysis of the collision processes. Two colliding molecules, with velocities $\mathbf{v}$ and $\mathbf{v}_1$, enter a certain zone of interaction. We do not know the details of the collision process, but we assume that after the collision, these molecules acquire new velocities $\mathbf{v}'$ and $\mathbf{v}_1'$ for which (1) The total momentum and energy of both molecules are conserved during the collision process (collisions are elastic); (2) vectors $\mathbf{v}'$ and $\mathbf{v}_1'$ belong to an a priori chosen set of velocities.

A conservative scheme based on a special choice of collision parameters was earlier discussed in [22, 63, 110]. In this scheme, the velocity vectors of molecules before and after collisions were taken in the nodes of the originally discretised velocity space. The approach used in [124] has some similarities with the approaches used in [22, 63, 110]. Its practical application can best be illustrated if we consider the collision process in the frame of reference linked with the centre of inertia of both molecules and describe the system dynamics in terms of momenta rather than velocities. In this frame of reference, the momenta of oncoming molecules have equal values but opposite directions. For the two-dimensional case, these are schematically shown in Fig. 6.2, where $p_{cx} = -1.5\Delta p$ and $p_{cy} = 1.5\Delta p$, the subscript $_c$ indicates that the momentum component is taken in the frame of reference linked with the centre of inertia of both molecules, Δp is the grid size in the momentum space (assumed to be the same in all directions). It can be shown that in the general case, the components of momenta in the frame of reference linked with the centre of inertia of colliding molecules are always integers of $0.5\Delta p$. The collision process leads to the rotation of momenta of both molecules in such a way that their absolute values remain the same, but the directions are opposite. All possible momenta satisfying these conditions lie on the circumferences shown in Fig. 6.2. In contrast to most previous studies of this process, we do not consider all possible values of momenta after the collision, but restrict ourselves to the cases in which the components of these momenta are integers of $0.5\Delta p$. In the two-dimensional case shown in Fig. 6.2, these

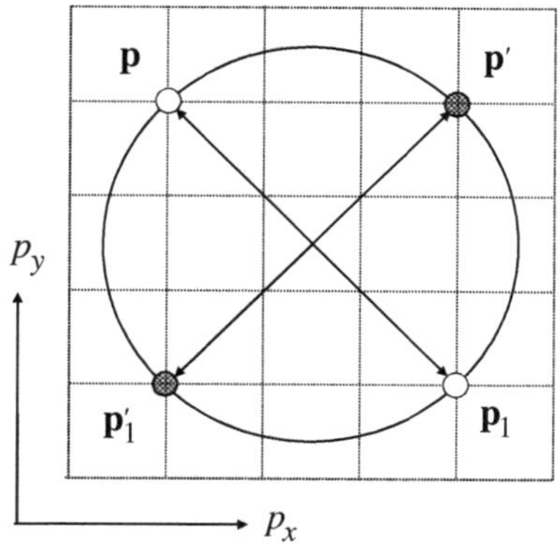

Fig. 6.2 Scheme of the collision process between two molecules in the frame of reference linked with their centre of inertia; $\mathbf{p}$ and $\mathbf{p}_1$ are momenta of molecules before the collision; $\mathbf{p}'$ and $\mathbf{p}_1'$ are their momenta after the collision. The sizes of the grid in this plane are assumed to be the same in the p_x and p_y directions and equal to Δp. Both components of all four vectors $\mathbf{p}$, $\mathbf{p}_1$, $\mathbf{p}'$, and $\mathbf{p}_1'$ are integers of $0.5\Delta p$. The absolute values of these vectors are equal to the radius of the circumference. The subscripts $_c$, indicating that the momenta are taken in the frame of reference linked with the centre of inertia of colliding molecules, are omitted. Reprinted from Ref. [124]. Copyright Elsevier (2007)

correspond to the points of intersection of the circumference with the nodes in the momentum space. In this case, there are four such points corresponding to four combinations of momenta of molecules after collision. In the three-dimensional case, the circumference shown in Fig. 6.2 turns into the surface of a sphere, and the number of possible combinations of momenta after collision increases to 12. It has been observed that the maximal number of these combinations in the three-dimensional space is 24 (although this has not been proven rigorously). In the practical implementation of this model, the calculations were performed for all possible values of θ and ϕ for each collision, and then the results were averaged over these variables. This is expected to improve the accuracy of the results compared with the random selection of θ and ϕ from the set of possible values of these variables.

It is believed that this approach provides consistency in discretisation processes used for the description of molecular dynamics and collision processes. It has been tested on numerous problems, one of which is discussed in Sect. 6.6.

6.2.2 Vapour Density and Temperature at the Boundaries

In [124], the problem of the thickness of the kinetic region was bypassed by observing that the mass flux at the outer boundary of the kinetic region is proportional to the difference between the vapour density at the surface of the droplet and the outer boundary of the kinetic region. The values of the vapour density at the outer boundary of the kinetic region, where mass fluxes in the kinetic and hydrodynamic regions coincide, gave the 'true' values of vapour density at this boundary and the corresponding values of the vapour mass flux. Unfortunately, this approach proved not to be applicable to the problem in which both the vapour density (ρ_{Rd} and temperature T_{Rd}) at the outer boundary of the kinetic region are unknown. Hence, the problem of calculating the thickness of the kinetic region had to be revisited. This is done below before the new method of calculating the values of ρ_{Rd} and T_{Rd} is described following [119]. Once these parameters have been found, the Boltzmann equations (6.2) are solved in the kinetic region, subject to the new boundary conditions at its outer boundary, following the procedure already described in Sect. 6.2.1.

Numerous coupled solutions of the Boltzmann and Navier–Stokes hydrodynamic equations for various mixtures have been reported (e.g., [86]). In these papers, it is shown that when the thickness of the kinetic region is chosen to be about 10 mean free molecular path lengths (λ_c), good matching of the kinetic and hydrodynamic solutions can be achieved. In this case, the combined kinetic/hydrodynamic solution agreed well with the kinetic solution in the whole domain. Note that the solutions of both the Navier–Stokes and Boltzmann equations were transient in the general case. The matching of these solutions was performed at each time step. The solutions of the Boltzmann equations were used as the boundary conditions for the hydrodynamic equations. In [119], it was shown that this estimate is applicable for the mixture of n-dodecane and air at temperatures and pressures typical for Diesel engines.

Once the minimal thickness of the kinetic region was estimated to be equal to $\delta_{\mathrm{Rd}} = 10\lambda_c$, the next task was to find the values of fuel vapour density and temperature at the boundaries of the kinetic region. The temperature at the surface of the droplet (T_s) was found from (4.29). The vapour density at the droplet surface (ρ_s) was found from the corresponding partial vapour pressure given by (4.7), assuming that the ideal gas law is valid. However, finding the values of these parameters at the outer boundary of the kinetic region turned out to be not trivial. In [124], it was assumed that $T_{\mathrm{Rd}} = T_s$. This, however, could not be a priori justified, and this assumption led to physical inconsistency of the formulation of the problem. It led to the conclusion that there is no heat flux in the kinetic region. In what follows, this assumption is relaxed, following [119], so that both ρ_{Rd} and T_{Rd} need to be found. They depend on the processes at the droplet surface, and on the mass and heat transfer processes in the hydrodynamic region. These values need to be found from the matching conditions of conservation of mass and heat fluxes at the interface:

$$j_{\mathrm{kin}} = j_{\mathrm{hyd}} = \dot{m}_d/(4\pi R_d^2), \tag{6.17}$$

$$q_{\mathrm{kin}} = q_{\mathrm{hyd}} = h(T_g - T_s), \tag{6.18}$$

where subscripts $_{\mathrm{kin}}$ and $_{\mathrm{hyd}}$ refer to the kinetic and hydrodynamic regions respectively.

The first step in the solution of these equations is to perform an investigation of mass and heat transfer processes in the kinetic region for a set of values of ρ_{Rd} and T_{Rd}. Given the physical background of the problem under consideration (heating

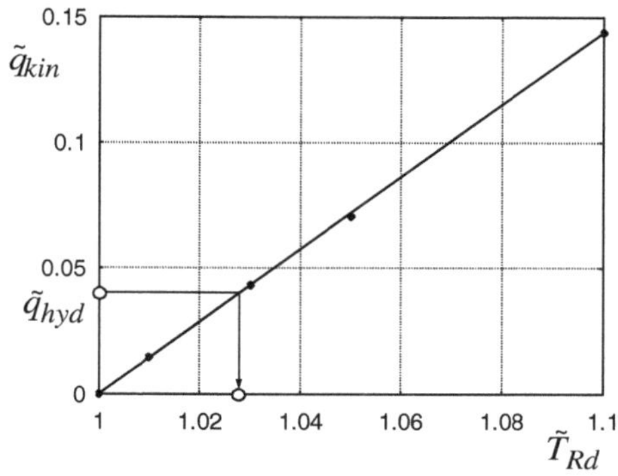

Fig. 6.3 A plot of normalised heat flux in the kinetic region $\tilde{q}_{\mathrm{kin}} = q_{\mathrm{kin}}/(p_0\sqrt{R_v T_s})$ versus normalised temperature $\tilde{T}_{\mathrm{Rd}} = T_{\mathrm{Rd}}/T_s$, assuming that $T_s = 600$ K. This figure illustrates how the value of $\tilde{T}_{\mathrm{Rd}}$ is found based on the previously calculated value of $\tilde{q}_{\mathrm{hyd}}$. Reproduced from Fig. 4b of [119] with permission of Begell House

and evaporation of droplets in a hot gas), these parameters are considered in the ranges $\rho_{\mathrm{Rd}} < \rho_s$ and $T_{\mathrm{Rd}} > T_s$. During the droplet heating process, the temperature increases away from the droplet; the evaporation process is possible when the fuel vapour density decreases away from the droplet surface. Once the values of ρ_{Rd} and T_{Rd} have been found, the solution of the Boltzmann equations (6.10) in the kinetic region allows us to calculate the mass and heat fluxes at the outer boundary of this region. The calculations are performed for the normalised values of these fluxes:

$$\tilde{j}_{\mathrm{kin}} = j_{\mathrm{kin}}/(\rho_0\sqrt{R_v T_0}), \qquad \tilde{q}_{\mathrm{kin}} = q_{\mathrm{kin}}/(p_0\sqrt{R_v T_0}),$$

where R_v is the gas constant referring to fuel vapour, T_0 is the reference temperature arbitrarily chosen as equal to 600 K, $p_0 = p_v(T_0) = p_{\mathrm{vs}}$ is the saturated vapour pressure corresponding to T_0, vapour density ρ_0 is calculated from the ideal gas law.

Firstly, it was assumed that $\rho_{\mathrm{Rd}} = 0.95\rho_s$ and $T_s = T_0$. The values of $\tilde{q}_{\mathrm{kin}}$ were calculated for a wide range of $\tilde{T}_{\mathrm{Rd}} = T_{\mathrm{Rd}}/T_0$ from 1 to 1.5. The results for the range from 1 to 1.1 are shown in Fig. 6.3. As follows from this figure, the dependence of $\tilde{q}_{\mathrm{kin}}$ on $\tilde{T}_{\mathrm{Rd}}$ is well described by a linear function. The same functional dependence of $\tilde{q}_{\mathrm{kin}}$ on $\tilde{T}_{\mathrm{Rd}}$ was observed in the whole range of $\tilde{T}_{\mathrm{Rd}}$ under consideration. The same conclusion was obtained for other values of T_s in the range 300 K (room temperature) to 659 K (critical temperature).

At the next stage, the effect of ρ_{Rd} on the values of $\tilde{q}_{\mathrm{kin}}$ was investigated. The plots of $\tilde{q}_{\mathrm{kin}}$ versus $\tilde{\rho}_{\mathrm{Rd}} = \rho_{\mathrm{Rd}}/\rho_s$ for four $\tilde{T}_{\mathrm{Rd}}$ are shown in Fig. 6.4. As follows from this figure, the dependence of $\tilde{q}_{\mathrm{kin}}$ on $\tilde{\rho}_{\mathrm{Rd}}$ is weak for all temperatures. For $\tilde{T}_{\mathrm{Rd}} \leq 1.1$ (the most important range for practical applications), this dependence can be ignored altogether. This result was very important for the analysis of [119], since it allowed the authors of that paper to decouple the effects of $\tilde{T}_{\mathrm{Rd}}$ and $\tilde{\rho}_{\mathrm{Rd}}$ on $\tilde{q}_{\mathrm{kin}}$.

At the next stage, the dependence of $\tilde{j}_{\mathrm{kin}}$ on $\tilde{\rho}_{\mathrm{Rd}}$ was investigated, assuming that $\tilde{T}_{\mathrm{Rd}} = 1.05$. The results are shown in Fig. 6.5 for $\tilde{\rho}_{\mathrm{Rd}}$ in the range 0.5 to 0.8. As follows from this figure, $\tilde{j}_{\mathrm{kin}}$ decreases linearly with increasing $\tilde{\rho}_{\mathrm{Rd}}$. Essentially, the

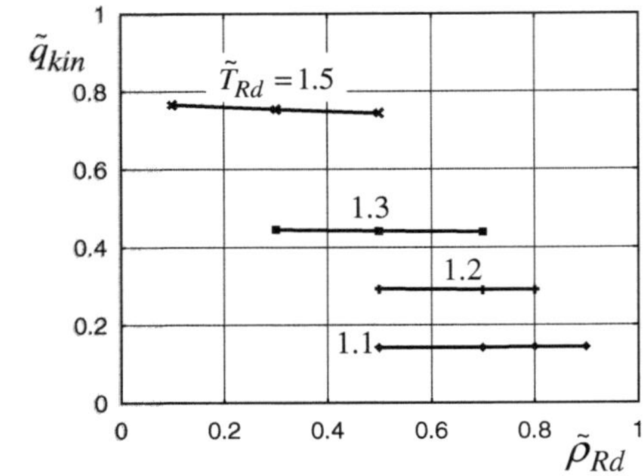

Fig. 6.4 Plots of normalised heat flux in the kinetic region $\tilde{q}_{\text{kin}} = q_{\text{kin}}/(p_{\text{vs}}\sqrt{R_v T_s})$ versus normalised fuel vapour density $\tilde{\rho}_{\text{Rd}} = \rho_{\text{Rd}}/\rho_s$ for four values of $\tilde{T}_{\text{Rd}}$, indicated near the plots. Reproduced from Fig. 5 of [119] with permission of Begell House

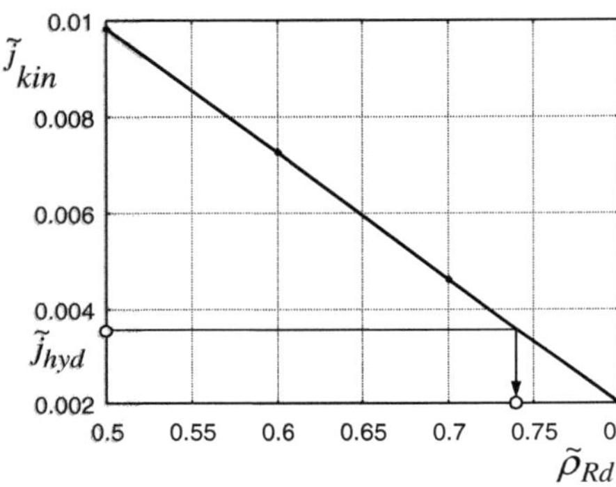

Fig. 6.5 A plot of normalised vapour mass flux in the kinetic region $\tilde{j}_{\text{kin}} = j_{\text{kin}}/(\rho_s\sqrt{R_v T_s})$ versus normalised density $\tilde{\rho}_{\text{Rd}} = \rho_{\text{Rd}}/\rho_s$, assuming that $\tilde{T}_{\text{Rd}} = 1.05$. This plot illustrates the way in which the value of $\tilde{\rho}_{\text{Rd}}$ is found, based on the previously calculated value of $\tilde{j}_{\text{hyd}}$. Reproduced from Fig. 6 of [119] with permission of Begell House

same linear dependence of $\tilde{j}_{\text{kin}}$ on $\tilde{\rho}_{\text{Rd}}$ was found for other values of $\tilde{T}_{\text{Rd}}$ in the range 1 to 1.5 and T_s in the range 300–659 K.

The above properties of $\tilde{q}_{\text{kin}}$ and $\tilde{j}_{\text{kin}}$ having been established, the following algorithm for the calculation of ρ_{Rd} and T_{Rd} was suggested [119]:

1. Calculate the value of q_{hyd} using (6.18).
2. Take a realistic value of $\tilde{\rho}_{\text{Rd}}$ (0.95 for a Diesel engine-like environment) and draw the plot of $\tilde{q}_{\text{kin}}$ versus $\tilde{T}_{\text{Rd}}$ as shown in Fig. 6.3 (as mentioned earlier, the results are expected to remain the same if a different but reasonable value of this parameter has been chosen; cf. Fig. 6.4).
3. Find the intersection of the line $\tilde{q}_{\text{hyd}} = q_{\text{hyd}}/(p_0\sqrt{R_v T_0})$ with this plot as shown in Fig. 6.3. This intersection gives the required values of $\tilde{T}_{\text{Rd}}$ and T_{Rd}.
4. For the value of T_{Rd} found in the previous step, calculate the value of j_{hyd} from (6.17).
5. Draw the plot of $\tilde{j}_{\text{kin}}$ versus $\tilde{\rho}_{\text{Rd}}$ as shown in Fig. 6.5.
6. Find the intersection of the line $\tilde{j}_{\text{hyd}} = j_{\text{hyd}}/(\rho_s\sqrt{R_v T_s})$ with this plot as shown in Fig. 6.5. This intersection gives the required values of $\tilde{\rho}_{\text{Rd}}$ and ρ_{Rd}.

The model described above was applied to calculations of heating and evaporation of fuel droplets in Diesel engine-like conditions [119]. It was pointed out that in the case of droplet heating in a relatively cool gas ($T_g = 750$ K), the effect of nonzero heat flux in the kinetic region is negligible. This effect, however, turned out to be important in the cases in which gas temperature rose to 1,000 and 1,500 K. The application of the rigorous kinetic model, taking into account the heat flux in the kinetic region, as described above, was recommended when accurate predictions of the values of droplet surface temperature and evaporation time are essential. Specific plots will be shown later in this chapter in connection with the problems of approximation of the results and comparative analysis of the models.

6.3 Approximations of the Kinetic Results

The practical application of the model described in Sect. 6.2 is ultimately based on the numerical solution of the Boltzmann equations for vapour and background gas (air) inside the kinetic region with the relevant boundary conditions. The CPU requirements of this solution, however, make it impossible to incorporate it into engineering computational fluid dynamics (CFD) codes designed to model fuel droplet heating and evaporation in realistic engineering environments, including those in Diesel engines, taking into account all complexities of the processes involved (see [43, 129]).

It seems that the only feasible way to apply the kinetic results to the analysis of Diesel fuel droplet heating and evaporation within these CFD codes would be to approximate the results of kinetic calculations using simple approximate analytical formulas. This approach to the problem is similar to what was suggested earlier in [34], where the results of complex Mie calculations describing the radiative heating of semitransparent droplets were approximated by simple analytical formulas, or to the approach suggested in [33], where a parabolic approximation was suggested for the temperature profile of a heated droplet.

The main objective of this section is to show how the best possible approximations of the results of kinetic simulation of the Diesel fuel droplet heating and evaporation process, using the approach described in Sect. 6.2, can be found. The selection of the approximations is based on finding a compromise between their accuracy and simplicity. The most important ranges of the values of fuel droplet radii and gas temperatures are considered, while the initial value of droplet temperature is assumed equal to 300 K. Although the applicability of the approximations found is restricted to a very specific problem of modelling the processes in Diesel engines, it is anticipated that the suggested approach can be generalised to a much wider range of engineering applications. The results presented in this section were earlier reported in [125–127]. The rest of this section is based mainly on [127].

6.3.1 Approximations for Chosen Gas Temperatures

6.3.1.1 Droplet Radii

Based on a direct comparison of the results of kinetic and hydrodynamic computations for gas temperatures $T_g = 750\,\text{K}$, $T_g = 1{,}000\,\text{K}$, and $T_g = 1{,}500\,\text{K}$ and initial droplet radii $R_0 = 5\,\mu\text{m}$ and $R_0 = 20\,\mu\text{m}$, it was shown that the best approximation for droplet radii predicted by the kinetic model can be achieved using the following formula:

$$\overline{\Delta R} = a_{1R} \exp\left(-\frac{\overline{R}}{t_{1R}}\right) + a_{2R} \exp\left(-\frac{\overline{R}}{t_{2R}}\right), \tag{6.19}$$

where

$$\overline{\Delta R} = \frac{R_k - R_h}{R_0}, \quad \overline{R} = \frac{R_h}{R_0},$$

R_k and R_h are droplet radii predicted by the kinetic and hydrodynamic models respectively, R_0 is the initial droplet radius, the coefficients a_{1R}, a_{2R}, t_{1R}, and t_{2R} are fitting constants.

The effects of droplets on gas temperature were ignored, and the modelling approach described in Sect. 6.2 was used.

A slightly less accurate, but simpler, approximation was found in the form

$$\overline{\Delta R} = a_{3R} \exp\left(-\frac{\overline{R}}{t_{3R}}\right), \tag{6.20}$$

where a_{3R} and t_{3R} are new fitting constants.

The fitting constants used in (6.19) and (6.20) are presented in Table 6.1 for several combinations of initial droplet radii and gas temperatures.

The plots of $\overline{\Delta R} = (R_k - R_h)/R_0$ versus $\overline{R} = R_h/R_0$, obtained using the kinetic model (calculation) and approximate formula (6.19) (approximation I) for $R_0 = 5\ \mu\text{m}$ and $T_g = 1{,}000$ K, are shown in Fig. 6.6. As follows from this figure, approximation I looks almost ideal. There are some noticeable deviations between the predictions of formula (6.20) (approximation II) and the results of calculations (figure not presented; see [127]). These deviations can be ignored in many practical applications. A similar closeness between the results of calculations and approximations was demonstrated for other combinations of R_0 and T_g considered above (the plots are not presented).

Approximations (6.19) and (6.20) with the values of coefficients given in Table 6.1 can be interpolated for the whole range of gas temperatures from 750 to 1,500 K, using the following interpolation formula for the coefficients a_{1R}, a_{2R}, t_{1R}, and t_{2R} (we restricted this interpolation to approximation (6.19) only):

$$s_R = b_{0R} + b_{1R}\tilde{T} + b_{2R}\tilde{T}^2, \tag{6.21}$$

Table 6.1 The values of fitting constants a_{1R}, t_{1R}, a_{2R}, t_{2R}, a_{3R}, and t_{3R}, used in (6.19) and (6.20), for various combinations of R_0 and T_g

R_0, μm	T_g, K	a_{1R}	t_{1R}	a_{2R}	t_{2R}	a_{3R}	t_{3R}
20	750	0.03449	0.08439	0.06401	0.25315	0.09453	0.17772
20	1,000	0.03525	0.08542	0.06424	0.25857	0.10158	0.19470
20	1,500	0.04359	0.09221	0.06605	0.26951	0.30955	0.25741
5	750	0.02096	0.08142	0.14800	0.02802	0.14532	0.25205
5	1,000	0.13070	0.29023	0.02404	0.08400	0.16415	0.26185
5	1,500	0.54256	0.50583	−0.13500	1.72373	0.44323	0.30402

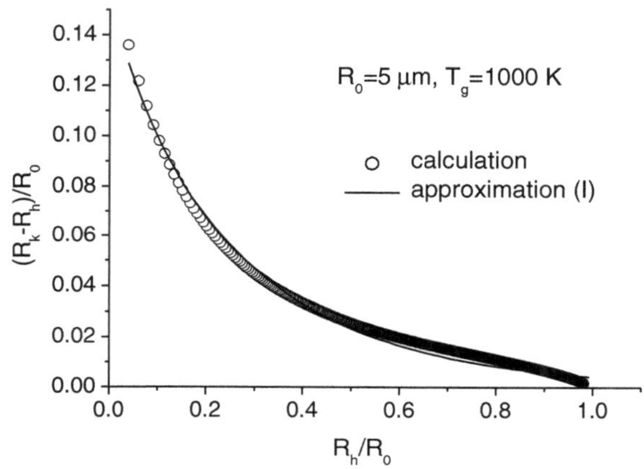

Fig. 6.6 Plots of $\overline{\Delta R} = (R_k - R_h)/R_0$ versus $\overline{R} = R_h/R_0$ for $R_0 = 5$ μm and $T_g = 1{,}000$ K, based on the kinetic model (*circles*), and approximation I, formula (6.19) (*solid curves*). Reproduced from Fig. 1 of [127] with permission of Inderscience

Table 6.2 The values of coefficients b_{iR} in approximation (6.21) for $R_0 = 20$ μm and $R_0 = 5$ μm

$R_0 =$	20 μm	20 μm	20 μm	5 μm	5 μm	5 μm
s_R	b_{0R}	b_{1R}	b_{2R}	b_{0R}	b_{1R}	b_{2R}
a_{1R}	0.04585	−0.01897	0.00790	0.07649	−0.30234	0.22279
t_{1R}	0.09076	−0.01183	0.00548	−0.94906	1.17172	−0.23396
a_{2R}	0.06602	−0.00355	0.00156	0.69765	−0.60010	0.10293
t_{2R}	0.25469	−0.00535	0.00521	2.91559	−4.55079	1.76927

where $\tilde{T} = T_g/T_c$, $T_c = 659$ K is the critical temperature of n-dodecane $C_{12}H_{26}$ (approximation of Diesel fuel), s_R stands for a_{1R}, a_{2R}, t_{1R}, or t_{2R}. The values of coefficients b_{iR}, where $i = 0, 1, 2$, for $R_{d0} = 20$ μm and $R_{d0} = 5$ μm, are given in Table 6.2.

Table 6.3 The values of fitting constants a_{1R}, t_{1R}, a_{2R}, and t_{2R} predicted by approximation (6.21)

R_0, μm	T_g, K	a_{1R}	t_{1R}	a_{2R}	t_{2R}
20	750	0.03492	0.08439	0.06400	0.25535
20	1,000	0.03555	0.08543	0.06422	0.25856
20	1,500	0.04360	0.09222	0.06602	0.26950
5	750	0.020971	0.081425	0.14800	0.02803
5	1,000	0.13071	0.29024	0.02404	0.08401
5	1,500	0.54258	0.50584	−0.13500	1.72374

The values of coefficients a_{1R}, a_{2R}, t_{1R}, and t_{2R} predicted by approximation (6.21) for $R_0 = 20$ μm and $R_0 = 5$ μm and $T_g = 750$, 1,000, and 1,500 K are shown in Table 6.3.

Comparing Tables 6.1 and 6.3, one can see that the maximal difference between the values of fitting constants does not exceed about 1 %. This justifies the application of approximation (6.21).

6.3.1.2 Droplet Temperatures

Based on a direct comparison of the results of kinetic and hydrodynamic calculations for the same gas temperatures and initial droplet radii as in the previous subsection, it was shown that a good approximation for the droplet temperatures predicted by the kinetic model can be achieved using the following formula:

$$\overline{\Delta T} = a_{1T} \exp\left(-\frac{\overline{T}}{t_{1T}}\right) + a_{2T}, \tag{6.22}$$

where

$$\overline{\Delta T} = \frac{T_k - T_h}{T_c - T_0}, \quad \overline{T} = \frac{T_h - T_0}{T_c - T_0},$$

T_k and T_h are droplet temperatures predicted by the kinetic and hydrodynamic models respectively, T_0 is the initial droplet temperature, the coefficients a_{1T}, t_{1T}, and a_{2T} are fitting constants. These constants are presented in Table 6.4 for the same combinations of initial droplet radii and gas temperatures as in the previous subsection.

The plots of $\overline{\Delta T} = (T_k - T_h)/(T_c - T_0)$ versus $\overline{T} = (T_h - T_0)/(T_c - T_0)$ obtained using the kinetic model (calculation) and approximate formula (6.22) for $R_0 = 5$ μm and $T_g = 1{,}000$ K are shown in Fig. 6.7. As follows from this figure, approximation (6.22) looks reasonably good and can be recommended for practical applications. Similar closeness between the results of calculations and approximations was demonstrated for other combinations of R_0 and T_g considered above (the plots are not presented).

Table 6.4 The values of fitting constants a_{1T}, t_{1T}, and a_{2T} used in formula (6.22) for various combinations of R_0 and T_g

R_0, μm	T_g, K	a_{1T}	t_{1T}	a_{2T}
20	750	1.25792×10^{-8}	−0.06989	1.87241×10^{-4}
20	1,000	1.28986×10^{-8}	−0.11769	1.71118×10^{-4}
20	1,500	2.48344×10^{-5}	−0.17572	4.09783×10^{-5}
5	750	1.66779×10^{-6}	−0.02012	3.62998×10^{-4}
5	1,000	4.54553×10^{-6}	−0.07213	5.78527×10^{-4}
5	1,500	2.88456×10^{-4}	−0.2292	-9.16881×10^{-4}

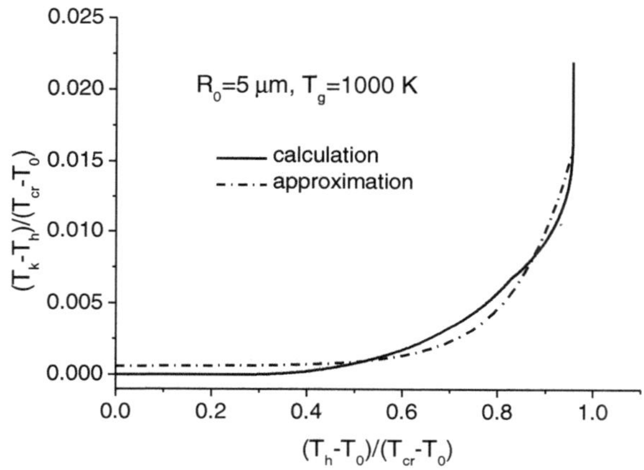

Fig. 6.7 Plots of $\overline{\Delta T} = (T_k - T_h)/(T_c - T_0)$ versus $\overline{T} = (T_h - T_0)/(T_c - T_0)$ for $R_0 = 5$ μm and $T_g = 1{,}000$ K, based on the kinetic model (solid), and approximation (6.22) (dash-dotted). Reproduced from Fig. 2 of [127] with permission of Inderscience

As in the case of droplet radii, approximation (6.22) with the values of coefficients given in Table 6.4 can be interpolated for the whole range of gas temperatures from 750 to 1,500 K, using interpolation (6.21) for the coefficients a_{1T}, a_{2T}, and t_{1T}, replacing the subscript R with T everywhere. As a result, (6.21) is replaced by

$$s_T = b_{0T} + b_{1T}\tilde{T} + b_{2T}\tilde{T}^2, \tag{6.23}$$

where s_T stands for a_{1T}, a_{2T}, or t_{1T}.

The values of these coefficients for the same initial droplet radii as in the previous subsection are shown in Table 6.5.

The values of coefficients a_{1T}, t_{1T}, and a_{2T} predicted by approximation (6.23) for $R_0 = 20$ μm and $R_0 = 5$ μm and for $T_g = 750$, 1,000, and 1,500 K are shown in Table 6.6.

Table 6.5 The values of coefficients b_{iT} in approximation (6.23) for $R_0 = 20$ μm and $R_0 = 5$ μm

s_T	R_0 μm	b_{0T}	b_{1T}	b_{2T}
a_{1T}	20	4.9653×10^{-5}	-7.6331×10^{-5}	2.87444×10^{-5}
	5	5.49341×10^{-4}	-8.47827×10^{-4}	3.22124×10^{-4}
t_{1T}	20	0.14865	−0.24154	0.04351
	5	0.02981	0.02604	−0.06143
a_{2T}	20	3.98252×10^{-5}	2.58552×10^{-4}	-1.13368×10^{-4}
	5	−0.00414	0.00649	−0.00223

Table 6.6 The values of fitting constants a_{1T}, t_{1T}, and a_{2T} predicted by approximation (6.23)

R_0, μm	T_g, K	a_{1T}	t_{1T}	a_{2T}
20	750	1.2620×10^{-8}	−0.06989	1.8724×10^{-4}
20	1,000	1.2951×10^{-8}	−0.11769	1.7112×10^{-4}
20	1,500	2.4834×10^{-5}	−0.17571	4.0978×10^{-5}
5	750	1.6685×10^{-6}	−0.02012	3.5779×10^{-4}
5	1,000	4.5463×10^{-6}	−0.07212	5.7333×10^{-4}
5	1,500	2.8845×10^{-4}	−0.22918	-9.2119×10^{-4}

Comparing Tables 6.4 and 6.6, one can see that the maximal difference between the values of fitting constants for the selected R_0 and T_g does not exceed about 1 %. This justifies the application of approximation (6.23).

6.3.2 Approximations for Chosen Initial Droplet Radii

The focus of this section is on the approximations for droplet radii and temperatures not for a given set or range of gas temperatures (as was done in the previous section), but for a range of initial droplet radii. To find these approximations, additional approximations are found for the initial droplet radius 10 μm in addition to those previously found for the initial droplet radii 5 and 20 μm. The results of the approximations are compared with rigorous kinetic calculations for $R_0 = 15$ μm. As in the previous section, mainly the results earlier presented in [127] are reproduced.

6.3.2.1 Droplet Radii

Using a trial-and-error approach, it was found that approximation (6.19) with the values of coefficients given in Table 6.1, and additional results referring to $R_0 = 10$ μm, can best be interpolated in the whole range of initial droplet radii from 5 to 20 μm using the following interpolation formula for the coefficients a_{1R}, a_{2R}, t_{1R}, and t_{2R}:

Table 6.7 The values of fitting constants c_{iR} ($i = 1, 2, 3$) in approximation (6.24) for the coefficients a_{1R}, t_{1R}, a_{2R}, and t_{2R} in approximation (6.19)

	T_g (K)	a_{1R}	t_{1R}	a_{2R}	t_{2R}
c_{0R}	750	−0.08057	0.03860	0.08288	0.25985
c_{1R}	750	0.27109	0.64352	−0.05476	−1.68683
c_{2R}	750	29.4067 μm	5.12421 μm	10.93604 μm	2.44184 μm
c_{0R}	1,000	0.04522	0.04676	0.10022	0.26271
c_{1R}	1,000	10.6517	12.7347	−3.56529	−9.64237
c_{2R}	1,000	1.03659 μm	1.26437 μm	1.21772 μm	1.31078 μm
c_{0R}	1,500	0.21792	0.24514	0.11245	0.26081
c_{1R}	1,500	0.79250	2.70877	−13.27703	−18.4742
c_{2R}	1,500	4.26859 μm	1.87024 μm	1.12864 μm	1.10535 μm

$$s_R = c_{0R} + c_{1R} \exp\left(-\frac{R_0}{c_{2R}}\right), \tag{6.24}$$

where s_R is the same as in expression (6.21), R_0 is the initial droplet radius in μm. The values of the coefficients c_{iR}, where $i = 0, 1, 2$, for three gas temperatures are shown in Table 6.7.

The main limitation of approximation (6.19) is that it can predict the values of droplet radii only up to the moment when the hydrodynamic model predicts complete droplet evaporation. Since the evaporation time predicted by the kinetic model is always longer than the evaporation time predicted by the hydrodynamic model, an additional approximation of time evolution of droplet radii at the final stage of droplet evaporation is required. To find this additional approximation, the following approach was suggested.

1. Use approximation (6.19) for the values of droplet radii predicted by the kinetic model for times until the hydrodynamic model predicts total droplet evaporation.
2. Select the time range at the final 5–10 % of the total evaporation time predicted by the hydrodynamic model.
3. Approximate the time dependence of $\tilde{R} = R_k/R_0$ in this time interval as

$$\tilde{R} = a + bt + ct^2, \tag{6.25}$$

   where t is the time (measured from the initial time) in μs, and a, b, c are fitting constants.
4. Find the evaporation time using the solution to (6.25) in the form

$$t_{1,2} = \frac{-b \pm \sqrt{b^2 - 4ac}}{2a}, \tag{6.26}$$

   where the root nearest to the time of total droplet evaporation predicted by the hydrodynamic model was selected.

As mentioned earlier, approximation (6.24), with the values of coefficients given in Table 6.7, was obtained based on an analysis of the kinetic results for initial droplet radii 5, 10, and 20 μm. To assess the effectiveness of this approximation, it was tested for the case in which the initial droplet radius is equal to 15 μm and gas temperature is equal to 1,000 K. In this case, the fitting constants used in (6.25) were found to be $a = -0.51412$, $b = 5.77135\ \mu s^{-1}$, and $c = -15.78277\ \mu s^{-2}$.

As follows from the plots of R/R_0 versus time for $R_0 = 15\ \mu$m and $T_g = 1{,}000$ K (not shown; see [127]), the agreement between the predictions of the kinetic model and the approximate model, described above, appears to be almost perfect. The difference between the evaporation times predicted by the kinetic model (6.433 μs) and the approximate model (6.505 μs) was only about 1 %.

Similar analyses were performed for other gas temperatures, and the conclusions were about the same as in the case of $T_g = 1{,}000$ K. For example, for $T_g = 750$ K, the difference between the evaporation time predicted by the kinetic model (11.535 μs) and the approximate model (11.627 μs) was only 0.8 %.

6.3.2.2 Droplet Temperatures

As shown earlier, a reasonably good approximation of $\overline{\Delta T}$ can be achieved using formula (6.22). An alternative approximation of this parameter can be given by the following expression:

$$\overline{\Delta T} = d_{1T} \exp\left(\frac{\overline{T}}{\tau_{1T}}\right), \tag{6.27}$$

where d_{1T} and τ_{1T} are new fitting constants.

Using a trial-and-error approach, it was shown that approximation (6.27) is more useful for finding the approximation for droplet temperature in the range of initial droplet radii from 5 to 20 μm than approximation (6.22). Approximation (6.27) contains fewer fitting constants than approximation (6.22) without compromising its accuracy. In this case, the coefficients d_{1T} and τ_{1T} were approximated as

$$s_T = f_{0T} + f_{1T} \exp\left(-\frac{R_0}{f_{2T}}\right), \tag{6.28}$$

where s_T stands for d_{1T} or τ_{1T}, R_0 is the initial droplet radius in μm, f_{0T}, f_{1T}, and f_{2T} are the fitting constants given in Table 6.8.

As in the case of droplet radii discussed earlier, approximation (6.27), with the values of coefficients approximated by expression (6.28) and Table 6.8, was obtained based on an analysis of the kinetic results for the droplet radii 5 μm, 10 μm, and 20 μm. To assess the effectiveness of this approximation, it was tested for the case that the initial droplet radius is equal to 15 μm and gas temperature is equal to 1,000 K.

As follows from the analysis presented in [127], this approach to the approximation of droplet temperature is less accurate than the approximation of the droplet

Table 6.8 The values of fitting constants f_{iR} ($i = 0, 1, 2$) in approximation (6.28) for the coefficients d_{1T} and τ_{1T} in approximation (6.27)

	T_g K	d_{1T}	τ_{1T}
f_{0T}	750	4.5322×10^{-10}	0.05015
f_{1T}	750	5.62537×10^{-4}	2.66871
f_{2T}	750	0.91832 μm	1.29156 μm
f_{0T}	1,000	5.6932×10^{-8}	0.07299
f_{1T}	1,000	7.6745×10^{-4}	2.70172
f_{2T}	1,000	1.07509 μm	1.27130 μm
f_{0T}	1,500	0.0	0.18478
f_{1T}	1,500	7.6745×10^{-4}	0.03506
f_{2T}	1,500	1.07509 μm	8.46655 μm

radii, although this accuracy improves at the final stage of droplet evaporation. In many practical applications, however, the difference in temperatures predicted by the hydrodynamic and kinetic models is small, and the relative error of the predicted droplet temperatures using the new approximation and kinetic calculations does not exceed about 2 %.

Unfortunately, we were not able to find a universal approximation of the kinetic results suitable in the whole range of initial droplet radii and gas temperatures relevant to Diesel engine conditions (cf. the approximation of the radiative heating of semitransparent Diesel fuel droplets suggested in [34]).

6.4 Effects of Inelastic Collisions

The most serious assumptions on which the model described in Sect. 6.2 is based is that the contribution of inelastic collisions can be ignored and the evaporation coefficient is equal to 1. The first assumption could have been justified in the case of monoatomic molecules, but appears to be highly questionable in the case of such complex molecules as $C_{12}H_{26}$ (n-dodecane) considered in Sect. 6.2. Even if the analysis of the dynamics of these molecules is simplified by considering the United Atom Model (see [23, 167] and Sect. 6.5 for details), the number of internal degrees of freedom of this molecule is expected to exceed one hundred. There seems to be no justification for ignoring their contribution. This section is focused on investigation of ways in which this assumption can be relaxed. Section 6.5 is focused on the second assumption.

Perhaps the first phenomenological model for binary collisions in a gas mixture having continuous internal energy was developed by [21]. This model was applied to Monte Carlo simulation of rarefied gas flows. Since the publication of this pioneering paper, a substantial number of papers have been published in which various models of inelastic collision have been considered. Without making any attempt to present an extensive list of these papers, we mention [13, 14, 17, 42, 44, 48, 79, 80, 93, 114, 142, 161]. This list does not include papers in which models for collisions of specific atoms and molecules were considered, such as H_2–H_2 collisions [178], N–N_2 collisions [71], and H–N_2 collisions [152]. Also, it does not include papers in which the effects of inelastic collisions on transport coefficients were considered (e.g., [113]).

The model suggested in [137] is different from those suggested earlier, although it is based on some widely used assumptions, such as the approximation of molecules by inelastic hard spheres (IHS). Although this model was tested only for some rather specific problems, its nature is rather general, and it can be applied to molecules with arbitrarily large numbers of internal degrees of freedom. It is expected to be the most effective model for the analysis of such complex molecules as n-dodecane, considered in Sects. 6.2 and 6.3. In the rest of this section, this model is described, following mainly [137].

6.4.1 Mathematical Model

6.4.1.1 A Model for Inelastic Collisions

As in Sect. 6.2, it is assumed that the molecules can be approximated as hard spheres. In contrast to that section, however, inelastic effects during the collisions between these spheres are taken into account, using a well-known inelastic hard spheres (IHS) model (see [114]). As in Sect. 6.2, only the effects of binary collisions are taken into account. This is justified by the fact that the Boltzmann equation is solved in a very thin layer, whose thickness is typically about 10 mean free paths.

Let us consider two colliding molecules. Regardless of the nature of the collision between them, their centre of mass is not affected by this collision. The state of the molecules after the collision is described in the reference system linked with this centre of mass. In this system, each of these molecules has three translational and a certain number of internal degrees of freedom, so that the total number of degrees of freedom of both molecules is equal to N. During the collisions, the energies of each molecule are redistributed between the degrees of freedom, but the total number of degrees of freedom remains the same. Also, it is assumed that none of these degrees of freedom has any preference over the others. This assumption implies that we focus our attention on systems close to thermodynamic equilibrium. The model is not applicable to systems in states far from thermodynamic equilibrium, such as gas lasers (see [120, 121]). Possible generalisations of the model to cases in which this assumption is not valid are discussed in the last paragraph of this section.

The assumption that none of the degrees of freedom has any preference over the others allows us to consider the redistribution of energy between these degrees of freedom during the collision process as random with uniform probability distribution. For each of these degrees of freedom, one dimension in the N-dimensional space, describing all degrees of freedom, is allocated. Once this has been done, we can

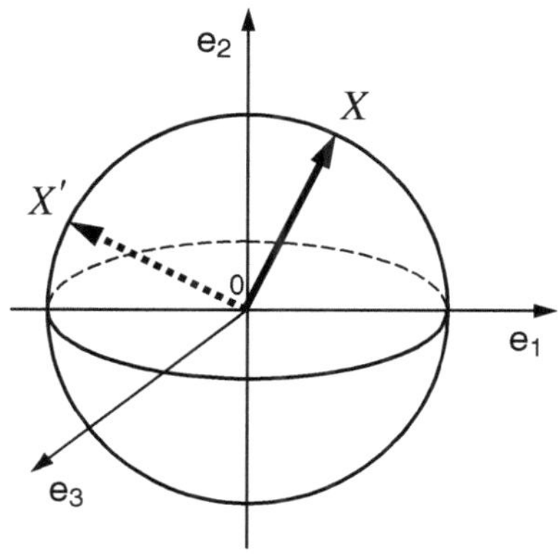

Fig. 6.8 A schematic presentation of the rotation of vector **X** in the three-dimensional space (e_1, e_2, e_3). Reprinted from [137]. Copyright Elsevier (2013)

consider a sphere in this space with its centre at the origin (where energies of all degrees of freedom are equal to zero) and radius given by the following expression:

$$r = \sqrt{\sum_{i=1}^{i=N} E_i}, \tag{6.29}$$

where E_i is the energy of the ith degree of freedom (translational or internal). Since r^2 gives the total energy of the system E_f, Eq. (6.29) can be considered an equation of the conservation of energy at the surface of the sphere.

Introducing the new coordinates $x_i = \pm\sqrt{E_i}$, we can consider an N-dimensional vector

$$\mathbf{X} = (x_1, x_2, \cdots, x_N),$$

with basis ($\mathbf{e_1}$, $\mathbf{e_2}$, $\cdots$ $\mathbf{e_N}$) and norm $||\mathbf{X}|| = r = \sqrt{E_f}$. The norms of all vectors $\mathbf{e_i}$ are equal to 1, and these vectors are mutually perpendicular.

The redistribution of energy between the degrees of freedom during the collision process can be described in terms of the rotation of vector **X** in the N-dimensional space. For $N = 3$ this is schematically shown in Fig. 6.8. If none of the degrees of freedom has any preference over the others, then this rotation of the vector **X** can be considered as random with uniform probability distribution. In the most general form it can be described by the following equation:

$$\mathbf{X}' = \mathbf{A}\mathbf{X}, \tag{6.30}$$

where $\mathbf{X}'$ is the new position of vector **X** after rotation, **A** is the rotation matrix:

$$\mathbf{A} = \begin{bmatrix} a_{11} & a_{12} & \cdots & a_{1N} \\ a_{21} & a_{22} & \cdots & a_{2N} \\ \cdots & \cdots & \cdots & \cdots \\ a_{N1} & a_{N2} & \cdots & a_{NN} \end{bmatrix}.$$

The conservation of the total energy during the collision process implies that the vector $\mathbf{X}'$ remains at the surface of the sphere of radius r. This is possible if and only if

$$\mathbf{A}^{\mathbf{T}}\mathbf{A} = \mathbf{E}, \tag{6.31}$$

where $\mathbf{A}^{\mathbf{T}}$ is the transpose of the matrix **A**, **E** is the unit matrix

$$\mathbf{E} = \begin{bmatrix} 1 & 0 & \cdots & 0 \\ 0 & 1 & \cdots & 0 \\ \cdots & \cdots & \cdots & \cdots \\ 0 & 0 & \cdots & 1 \end{bmatrix}.$$

Equation (6.31) can be represented in a more explicit form as a combination of the following systems of equations:

$$\left.\begin{array}{l} a_{11}^2 + a_{21}^2 + \cdots + a_{N1}^2 = 1 \\ a_{12}^2 + a_{22}^2 + \cdots + a_{N2}^2 = 1 \\ \cdots \quad \cdots \quad \cdots\cdots \quad \cdots \\ a_{1N}^2 + a_{2N}^2 + \cdots + a_{NN}^2 = 1 \end{array}\right\} \tag{6.32}$$

$$\left.\begin{array}{l} a_{11}a_{12} + a_{21}a_{22} + \cdots + a_{N1}a_{N2} = 0 \\ a_{11}a_{13} + a_{21}a_{23} + \cdots + a_{N1}a_{N3} = 0 \\ \cdots \quad \cdots \quad \cdots\cdots \quad \cdots \\ a_{11}a_{1N} + a_{21}a_{2N} + \cdots + a_{N1}a_{NN} = 0 \end{array}\right\} \tag{6.33}$$

$$\left.\begin{array}{l} a_{12}a_{13} + a_{22}a_{23} + \cdots + a_{N2}a_{N3} = 0 \\ a_{12}a_{14} + a_{22}a_{24} + \cdots + a_{N2}a_{N4} = 0 \\ \cdots \quad \cdots \quad \cdots \quad \cdots \\ a_{12}a_{1N} + a_{22}a_{2N} + \cdots + a_{N2}a_{NN} = 0 \end{array}\right\} \tag{6.34}$$

....

$$a_{1(N-1)}a_{1N} + a_{2(N-1)}a_{2N} + \cdots + a_{N(N-1)}a_{NN} = 0 \tag{6.35}$$

In writing (6.33)–(6.35), identical equations were excluded. The total number of Eqs. (6.33)–(6.35) is

$$(N-1)+(N-2)+(N-3)+\cdots+1=\frac{N}{2}(N-1)$$

for N^2 unknown coefficients a_{ij}. This allows us to take randomly $N^2-\frac{N}{2}(N-1)=\frac{N}{2}(N+1)$ of these coefficients with an additional restriction imposed by (6.32) (normalization condition).

The following algorithm for the construction of the matrix $\mathbf{A}$ was suggested in [137].

1. The coefficients $a_{11}, a_{21}, ..., a_{N1}$ are arbitrarily chosen but normalised based on the first equation in system (6.32).
2. The coefficients $a_{12}, a_{22}, ..., a_{(N-1)2}$ are arbitrarily chosen, while the value of the coefficient a_{N2} is found from the first equation of system (6.33):

$$a_{N2}=-\frac{1}{a_{N1}}\left(a_{11}a_{12}\ +\ a_{21}a_{22}\ +\ \cdots\ +\ a_{(N-1)1}a_{(N-1)2}\right). \tag{6.36}$$

   Then all coefficients are normalised based on the second equation in system (6.32).
3. The coefficients $a_{13}, a_{23}, ..., a_{(N-2)3}$ are arbitrarily chosen, while the values of the coefficients $a_{(N-1)3}$ and a_{N3} are found from the solution of the second equation in system (6.33) and the first equation in system (6.34). These equations can be rearranged as

$$\left.\begin{aligned} a_{(N-1)1}a_{(N-1)3}+a_{N1}a_{N3} &= b_{13} \\ a_{(N-1)2}a_{(N-1)3}+a_{N2}a_{N3} &= b_{23} \end{aligned}\right\}, \tag{6.37}$$

   where

$$b_{13}=-\left(a_{11}a_{13}\ +a_{21}a_{23}\ +\ \cdots\ +\ a_{(N-2)1}a_{(N-2)3}\right),$$

$$b_{23}=-\left(a_{12}a_{13}\ +a_{22}a_{23}\ +\ \cdots\ +\ a_{(N-2)2}a_{(N-2)3}\right).$$

   Then all coefficients are normalised based on the third equation in system (6.32).

Following the same procedure, all other components of matrix $\mathbf{A}$ are found, and this allows us to calculate $\mathbf{X}'$ based on (6.30). An example of the temporal evolution of the system with 100 degrees of freedom is shown in Fig. 6.9. The ordinate axis in this figure shows the energies E_i referring to individual degrees of freedom N_i ($i \in [1, 100]$). Initially, all degrees of freedom have energies equal to 1 ($E_i = 1$). This is shown in the top part of this figure (initial state). After the first step, energies E_i acquire random values in the range from 0 to 5, but the sum of all E_i remains the same as at the initial state ($\sum_{i=1}^{i=100} E_i = 100$). The nature of the distribution of E_i remains random after steps 2 and 3, although with different values of individual energies. In principle, one E_i can reach the maximal value of 100, provided that all other E_i are equal to zero, although this has never been observed in our simulations.

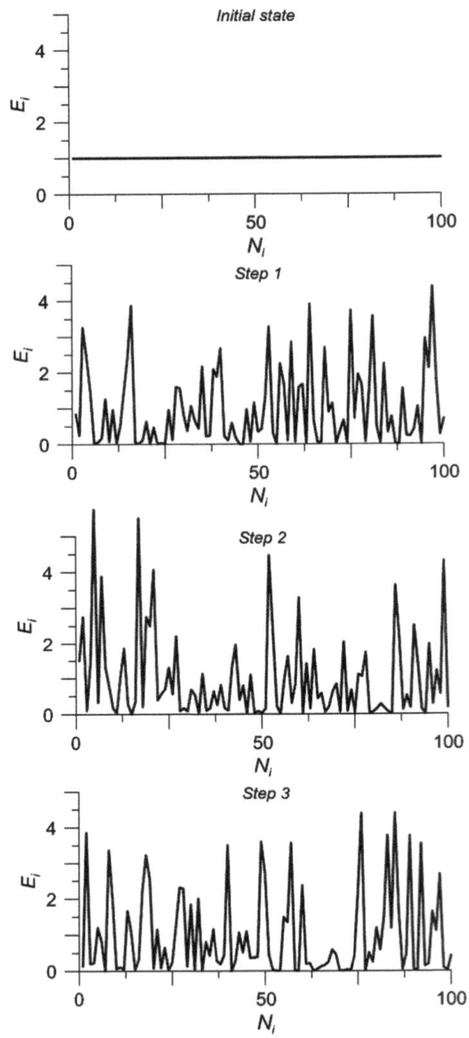

Fig. 6.9 The values of energies of individual degrees of freedom (E_i) versus N_i at the initial state and after steps 1, 2, and 3. Reprinted from [137]. Copyright Elsevier (2013)

The model described above could be generalised to the case that the probabilities of excitation of various degrees of freedom are not equal. This could be achieved by introduction of a weighting function and/or limiting the range of degrees of freedom to be activated. We cannot, however, specify this weighting function in the case of such complex molecules as *n*-dodecane, for which our model has been primarily developed, or even justify the need to introduce this function in this case.

6.4.1.2 The Solution of the Boltzmann Equation

As in Sect. 6.2, the numerical solution of the Boltzmann equation (for one or several components) is performed in two steps. Firstly, molecular displacements are calculated ignoring the effect of collisions. Secondly, the collisional relaxation is calculated under the assumption of spatial homogeneity.

Ignoring the effects of collisions, the discretised form of the Boltzmann equation describing molecular displacements for each component can be represented as

$$\frac{\Delta f}{\Delta t} + \mathbf{v}\frac{\Delta f}{\Delta \mathbf{r}} = 0, \tag{6.38}$$

where $f \equiv f(\mathbf{v}, \mathbf{r}, t)$ is the distribution function for the velocity ($\mathbf{v}$) and physical ($\mathbf{r}$) spaces.

In this case, the total internal energy in each velocity range should be conserved, which implies that

$$\frac{\Delta(E_{\text{int}} f)}{\Delta t} + \mathbf{v}\frac{\Delta(E_{\text{int}} f)}{\Delta \mathbf{r}} = 0, \tag{6.39}$$

where $E_{\text{int}} \equiv E_{\text{int}}(\mathbf{v}, \mathbf{r}, t)$ are internal energies of molecules with given $\mathbf{v}$ and $\mathbf{r}$ at time t.

The model for inelastic collisions described in the previous subsection allows us to obtain the energies of all degrees of freedom after individual collisions, as illustrated in Fig. 6.9. However, we are interested only in the net change in kinetic energy of each of two molecules during collisions in the centre of mass reference system. This change in kinetic energy can be described in terms of the change of the radius of the three-dimensional sphere that is the projection of the N-dimensional sphere described in the previous section on three-dimensional space, describing the kinetic energies of both colliding molecules in three directions. This is schematically illustrated in Fig. 6.10 for the case of the projection of the N-dimensional sphere on the two-dimensional plane. As in Sect. 6.2, this space is represented in terms of the components of momenta p_x and p_y of both colliding molecules. Points p and p_* show the positions of molecules before the collision. If the collisions were elastic, then the values of momenta after collisions would lie on the dashed circle shown in Fig. 6.10, being separated by 180°. Possible values of these momenta after the collisions are shown as empty circles.

As mentioned in Sect. 6.2, randomly chosen directions of molecular momenta after collision are likely to lead to values of these momenta lying between the values in the nodes of the discretised momenta space. This eventually can lead to nonconservation of momenta and energies during the collision process. To overcome this problem, as in Sect. 6.2 the momenta were discretised not only during the description of molecular motion but also in the analysis of the collision process. Namely, it was assumed that the momenta after the collisions belong to an *a priori* chosen set of momenta, which are nodes in the momenta space shown in Fig. 6.10. This was achieved by moving the actual point on the surface of the sphere to the nearest node. Since the nodes are

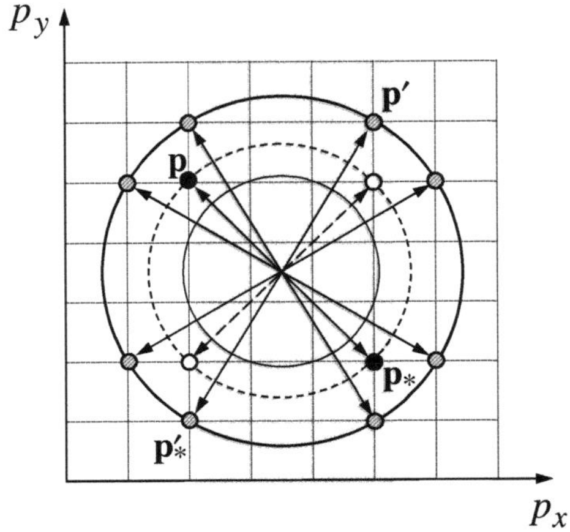

Fig. 6.10 A projection of the surface of an N-dimensional sphere, describing the energies referring to translational and internal degrees of freedom of two colliding molecules, into a two-dimensional space referring to two translational degrees of freedom; **p** and **p*** show the locations of the molecular momenta before the collision. The *dashed circle* shows possible locations of molecular momenta after the collision if the contribution of internal degrees of freedom is ignored. The *thin solid circle* shows possible locations of molecular momenta after the collision if molecular internal energy increases during the collision. The *thick solid circle* shows possible locations of molecular momenta after the collision if molecular internal energy decreases during the collision; **p**′ and **p***′ show the allowed locations of the molecular momenta after the collision if molecular internal energy decreases during the collision. Reprinted from [137]. Copyright Elsevier (2013)

not uniformly distributed on the surface of the sphere in the general case, this leads to partial loss of randomness of the distribution of momenta after collisions. This is one of the weaknesses of the model under consideration, but it seems not to lead to serious limitations of the range of its applicability, as demonstrated in Sect. 6.6.

In the case shown in Fig. 6.10, in the absence of inelastic collisions, there are four such nodes corresponding to four combinations of momenta of molecules after collision. As mentioned in Sect. 6.2, the maximal number of these combinations for the plane is eight. In the three-dimensional case, the circumference shown in Fig. 6.10 turns into the surface of a sphere, and the maximal number of possible intersection points increases to 24. This corresponds to the maximal total number of combinations of momenta after collision. This approach provides the consistency in discretisation processes used for describing the molecular dynamics and collision processes.

If during the collision the net internal energy of molecules increases, this has to be compensated by a decrease in the kinetic energies of molecules, and the radius of the corresponding circle in Fig. 6.10 is decreased. In the opposite case, when the net internal energy of molecules decreases, this has to be compensated by an increase in the kinetic energies of molecules, and the radius of the corresponding circle in Fig. 6.10 is increased. Both cases are shown in Fig. 6.10. In the case of increased net kinetic energy of molecules, possible values of the momenta of both molecules after the collision are shown as grey circles. These include points p' and p'_*. The changes in radii of the circles after collision were calculated based on Korobov's sequences ([78, 109]), enhanced by the randomization of individual points. As in the case of elastic collisions, the points p' and p'_* were chosen to coincide with the nodes of the discretised momenta space.

The model described in this section was applied to three test problems: shock wave structure in nitrogen, one-dimensional heat transfer through a mixture of n-dodecane and nitrogen, and one-dimensional evaporation of n-dodecane into nitrogen. In the first problem, the predictions of the model, taking into account the contribution of the rotational degrees of freedom, were shown to be close to experimental data and the predictions of an earlier developed model based on a different approach to taking into account the effects of inelastic collisions. This problem was generalised to a hypothetical case in which the number of internal degrees of freedom of nitrogen (N_{int}) was assumed to be in the range 0–10. It was shown that the results visibly changed when N_{int} increased from 0 to 2, but remained practically unchanged for $N_{\mathrm{int}} \geq 6$.

The predicted heat flux for the second problem was shown not to depend on the number of internal degrees of freedom of the mixture N_{int} when this number exceeds about 15. In the third problem, the predicted mass flux of n-dodecane also remained almost unchanged for $N_{\mathrm{int}} \geq 15$. These results opened the way to considering systems with arbitrarily large numbers of internal degrees of freedom by reducing the analysis of these systems to the analysis of systems with relatively small numbers of internal degrees of freedom (see [137] for further details).

6.4.2 Solution Algorithm

The idea of the solution algorithm used in the analysis is similar to the one described in Sect. 6.2 for the case in which the contribution of only elastic collisions was taken into account. As in Sect. 6.2, the first step in the solution of (6.10) is to perform an investigation of mass and heat transfer processes in the kinetic region for a set of values of ρ_{Rd} and T_{Rd}. Since the problem of heating and evaporation of droplets in a hot gas is considered, these parameters are assumed to be in the ranges $\rho_{\mathrm{Rd}} < \rho_s$ and $T_{\mathrm{Rd}} > T_s$. During the droplet heating process, the temperature increases away from the droplet; the evaporation process is possible when the fuel vapour density decreases away from the droplet surface. Once the values of ρ_{Rd} and T_{Rd} have been found, the solution to the Boltzmann equations (6.10) in the kinetic region allows us to calculate the normalised mass and heat fluxes at the outer boundary of this region:

$$\tilde{j}_{\mathrm{k}} = j_{\mathrm{k}}/(\rho_s\sqrt{R_v T_s}), \qquad \tilde{q}_{\mathrm{k}} = q_{\mathrm{k}}/(p_0\sqrt{R_v T_s}),$$

where R_v is the gas constant referring to vapour, T_0 is the reference temperature chosen equal to 600 K, p_0 and ρ_0 are the saturated fuel vapour pressure and density

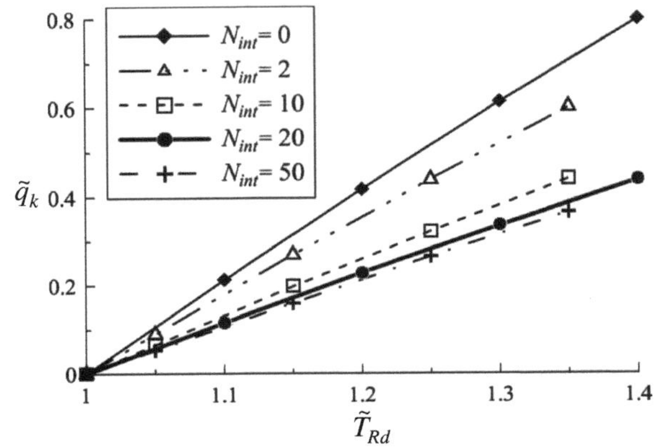

Fig. 6.11 Plots of normalised heat flux in the kinetic region $\tilde{q}_\mathrm{k} \equiv q_\mathrm{k}/(p_\mathrm{vs}\sqrt{R_v T_s})$ versus normalised temperature $\tilde{T}_\mathrm{Rd} = T_\mathrm{Rd}/T_s$ for various numbers of internal degrees of freedom N_int, assuming that $T_s = 600$ K; ρ_Rd is taken equal to $0.9\rho_s$. Reprinted from [128]. Copyright Elsevier (2013)

corresponding to T_0, ρ_0 is calculated from the ideal gas law, and the subscript $_k$ stands for kinetic.

Following [128], it was assumed that $\rho_\mathrm{Rd} = 0.9\rho_s$ and $T_s = T_0 = 600$ K. The values of $\tilde{q}_\mathrm{k}$ were calculated for the numbers of internal degrees of freedom N_int from 0 to 50 and $\tilde{T}_\mathrm{Rd} = T_\mathrm{Rd}/T_s$ from 1 to 1.4. The results are shown in Fig. 6.11. As follows from this figure, for all values of N_int, the dependence of $\tilde{q}_\mathrm{k}$ on $\tilde{T}_\mathrm{Rd}$ is well described by a linear function, in agreement with the case reported in Sect. 6.2. For any given $\tilde{T}_\mathrm{Rd}$, the values of $\tilde{q}_\mathrm{k}$ decrease with increasing N_int. The rate of this decrease, however, becomes small for $N_\mathrm{int} > 10$ and negligible for $N_\mathrm{int} > 20$. This allows us to restrict our analysis to the case of $N_\mathrm{int} = 20$, in agreement with the conclusion reached in [137].

The plots of $\tilde{q}_\mathrm{k}$ versus $\tilde{\rho}_\mathrm{Rd} \equiv \rho_\mathrm{Rd}/\rho_s$ for $\tilde{T}_\mathrm{Rd} = 1.1$ and 1.2 and $N_\mathrm{int} = 20$ are shown in Fig. 6.12. As one can see in this figure, the plots for these values of $\tilde{T}_\mathrm{Rd}$ are the lines almost parallel to the $\tilde{\rho}_\mathrm{Rd}$ axis. This allows us to ignore the dependence of $\tilde{q}_\mathrm{k}$ on $\tilde{\rho}_\mathrm{Rd}$, in agreement with the similar result obtained in Sect. 6.2 for $N_\mathrm{int} = 0$.

The plots of $\tilde{q}_\mathrm{k}$ versus $\tilde{T}_\mathrm{Rd}$ for $\tilde{\rho}_\mathrm{Rd} = 1$, $N_\mathrm{int} = 0$ and 20, and $\tilde{q}_\mathrm{h} = q_\mathrm{h}/(p_\mathrm{vs}\sqrt{R_v T_s})$ versus $\tilde{T}_\mathrm{Rd}$ (horizontal line) are shown in Fig. 6.13. For this line, it was assumed, as in Sect. 6.2, that T_Rd in (6.18) can be replaced with T_s. The following values of parameters were used: $T_g = 1{,}000$ K, $T_s = 600$ K, $R_d = 5$ μm. The intersections between the horizontal and inclined lines give the required values of $\tilde{T}_\mathrm{Rd}$. For the case of only elastic collisions ($N_\mathrm{int} = 0$), $\tilde{T}_\mathrm{Rd} = 1.014$, for the case in which the contribution of inelastic collisions with $N_\mathrm{int} = 20$ is taken into account, $\tilde{T}_\mathrm{Rd} = 1.026$. This result indicates that the contribution of internal degrees of freedom leads to an increase in $\tilde{T}_\mathrm{Rd}$.

The plots of $\tilde{j}_\mathrm{k}$ versus $\tilde{\rho}_\mathrm{Rd}$ for $\tilde{T}_\mathrm{Rd} = 1.026$, $N_\mathrm{int} = 0$ and 20, and $\tilde{j}_\mathrm{h} = j_\mathrm{h}/(p_\mathrm{vs}\sqrt{R_v T_s})$ versus $\tilde{\rho}_\mathrm{Rd}$ (horizontal line) are shown in Fig. 6.14. This figure

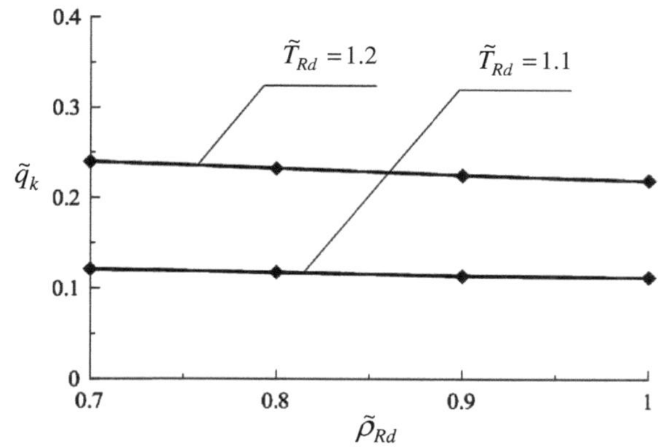

Fig. 6.12 Plots of normalised heat flux in the kinetic region $\tilde{q}_k$ versus normalised density $\tilde{\rho}_{Rd} = \rho_{Rd}/\rho_s$ for $N_{int} = 20$ and $\tilde{T}_{Rd} = 1.1$ and 1.2. Reprinted from [128]. Copyright Elsevier (2013)

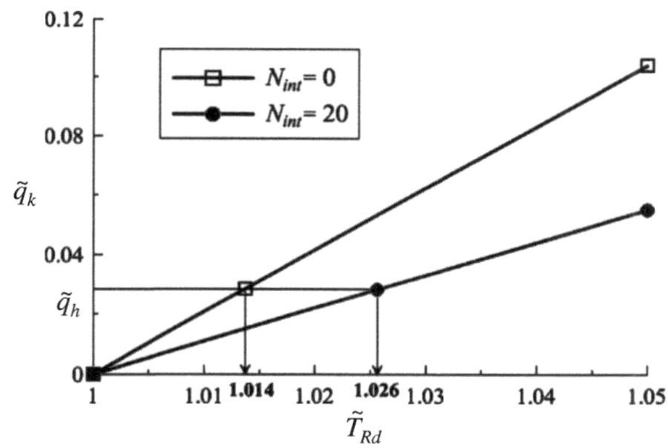

Fig. 6.13 Plots of $\tilde{q}_k$ versus $\tilde{T}_{Rd}$ for $N_{int} = 0$ and 20, and the plot of $\tilde{q}_h \equiv q_h/(p_{vs}\sqrt{R_v T_s})$ versus $\tilde{T}_{Rd}$ for $T_s = 600$ K, $T_g = 1{,}000$ K, $R_{d0} = 5$ μm, and $\tilde{\rho}_{Rd} = 1$. The intersections between the plots of $\tilde{q}_k$ and $\tilde{q}_h$ give the required values of $\tilde{T}_{Rd}$. Reprinted from [128]. Copyright Elsevier (2013)

is drawn for the same parameters as in Fig. 6.13. As in Sect. 6.2, it was assumed that ρ_{Rd} in (6.17) can be replaced with ρ_s.

The intersections between the horizontal and inclined lines give the required values of $\tilde{\rho}_{Rd}$. For the case of only elastic collisions ($N_{int} = 0$), $\tilde{\rho}_{Rd} = 0.968$; for the case that the contribution of inelastic collisions with $N_{int} = 20$ is taken into account, $\tilde{\rho}_{Rd} = 0.926$. This result indicates that the contribution of internal degrees of freedom leads to a decrease in $\tilde{\rho}_{Rd}$.

Similar values of $\tilde{T}_{Rd}$ and $\tilde{\rho}_{Rd}$ were obtained for other values of T_g and R_d relevant for Diesel engine conditions ($T_g = 750$ K and $R_d = 20$ μm) and the values of T_s in the range from 300 K to the critical temperature. The corresponding values of $\tilde{T}_{Rd}$

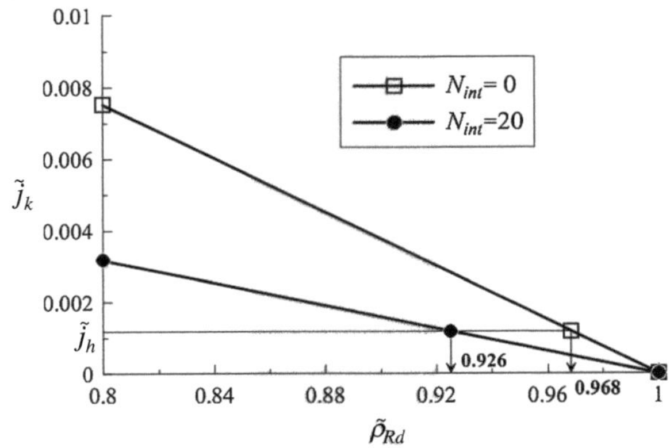

Fig. 6.14 Plots of $\tilde{j}_k \equiv j_k/(\rho_s\sqrt{R_v T_s})$ versus $\tilde{\rho}_{Rd}$ for $N_{int} = 0$ and 20, and the plot of $\tilde{j}_h \equiv j_h/(\rho_s\sqrt{R_v T_s})$ versus $\tilde{\rho}_{Rd}$ for $T_s = 600$ K, $T_g = 1{,}000$ K, $R_{d0} = 5$ μm, and $\tilde{T}_{Rd} = 1.026$. The intersections between the plots of $\tilde{j}_k$ and $\tilde{j}_h$ give the required values of $\tilde{\rho}_{Rd}$. Reprinted from [128]. Copyright Elsevier (2013)

and $\tilde{\rho}_{Rd}$ were used for the analysis of heating and evaporation of n-dodecane droplets in realistic Diesel engine-like conditions [128]. The results are presented in Sect. 6.6.

For realistic $\beta_m < 1$, the values of $\tilde{q}_k$ were practically indistinguishable from those predicted by the model for $\beta_m = 1$. The plots of $\tilde{j}_k$ versus $\tilde{\rho}_{Rd}$ for $\tilde{T}_{Rd} = 1.026$, $N_{int} = 20$, and $\beta_m = 1$ and 0.36 are shown in Fig. 6.15. The plot for $\beta_m = 1$ is identical to the one shown in Fig. 6.14, but presented in a wider range of $\tilde{\rho}_{Rd}$. The value of $\beta_m = 0.36$ is close to that predicted for $T_s = 600$ K (see (6.49). As follows from Fig. 6.15, the linear dependence of $\tilde{j}_k$ on $\tilde{\rho}_{Rd}$ is maintained for $\beta_m = 0.36$, but the values of $\tilde{j}_k$ predicted by the model with $\beta_m = 0.36$ are lower than those predicted by the model with $\beta_m = 1$. We follow the same procedure of calculating $\tilde{\rho}_{Rd}$ as described earlier for $\beta_m = 1$ and shown in Fig. 6.14. As one can see from Fig. 6.15, this leads to the prediction of lower values of $\tilde{\rho}_{Rd}$ for $\beta_m = 0.36$ than for $\beta_m = 1$. Thus the reduction of β_m is expected to lead to the enhancement of the kinetic effects.

6.5 Kinetic Boundary Condition

As mentioned at the very beginning of Sect. 6.4, the model described in Sect. 6.2 is based on two serious assumptions. Firstly, the contribution of inelastic collisions was ignored, and secondly, the evaporation coefficient was assumed equal to 1. The model described in Sect. 6.4 is expected to be able to relax the first assumption. To relax the second assumption, the values of the evaporation coefficient need to be calculated, using the molecular dynamics technique. These values of the evaporation coefficient would allow us to formulate the kinetic boundary conditions at the surface

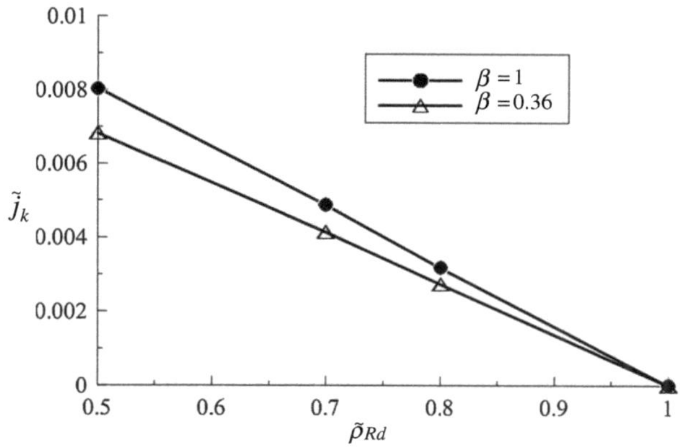

Fig. 6.15 Plots of $\tilde{j}_k$ versus $\tilde{\rho}_{\mathrm{Rd}}$ for $N_{\mathrm{int}} = 20$ and $\beta_m = 1$ and 0.36 versus $\tilde{\rho}_{\mathrm{Rd}}$ for $T_s = 600$ K, $T_g = 1{,}000$ K, $R_{d0} = 5\ \mu$m, and $\tilde{T}_{\mathrm{Rd}} = 1.026$. Reprinted from [128]. Copyright Elsevier (2013)

of the droplet in the form

$$f^{\mathrm{out}} = \beta_m f^e + (1 - \beta_m) f^r \quad (v_x > 0), \tag{6.40}$$

where f^{out} is the overall distribution function of molecules leaving the interface from the liquid phase, β_m is the evaporation coefficient, f^e is the distribution function of evaporated molecules, f^r is the distribution function of reflected molecules, and v_x is the velocity component normal to the interface. In the equilibrium state, the evaporation and condensation coefficients are equal.

An overview of the background of molecular dynamics simulation research is presented in Sect. 6.5.1. A new model, specifically designed to perform the molecular dynamics simulation of n-dodecane is presented and discussed in Sect. 6.5.2. In Sect. 6.5.3, the results of application of the latter model to calculation of evaporation coefficient are shown.

6.5.1 Molecular Dynamics Simulations (Background)

The first step in a molecular dynamics (MD) simulation is the choice of the model that best represents the interaction between molecules. Such a model has to take into account the attraction between molecules when the distance between them is large enough and the repelling forces when molecules approach close to one another. The model almost universally used in MD simulations is based on the so-called Lennard-Jones 12-6 potential, whose analytical form can be represented as [20]

$$V(R_{ij}) = 4\varepsilon_{ij}\left[\left(\frac{\sigma_{ij}}{R_{ij}}\right)^{12} - \left(\frac{\sigma_{ij}}{R_{ij}}\right)^{6}\right], \tag{6.41}$$

where R_{ij} is the distance between molecules i and j, ε_{ij} and σ_{ij} are the minimal energy and the zero energy separation distance relative to the pair of molecules (a different form of this potential was used in [169, 171]).

For cross interactions between different species, mixing rules, such as the Lorenz–Berthelot rules, are commonly used [4, 30]:

$$\varepsilon_{ij} = \sqrt{\varepsilon_i \varepsilon_j}, \tag{6.42}$$

$$\sigma_{ij} = (\sigma_i + \sigma_j)/2, \tag{6.43}$$

where the parameters ε_i and σ_i refer to specific molecules. Data about them is widely available [20, 60, 94]. For example, for nitrogen molecules [30], $\varepsilon_{N_2} = 95.9 k_B$ J and $\sigma_{N_2} = 0.371$ nm, where $k_B = 1.38066 \times 10^{-23}$ J/K is the Boltzmann constant. When these constants are unknown, they can be estimated from the properties of a fluid at the critical point, a liquid at the normal boiling point, and a solid at the melting point [20]. At $R_{ij} > R_m$, where R_m corresponds to the value of R_{ij} when $V = -\varepsilon_{ij}$, the value of V decreases rapidly with increasing R_{ij}. At $R_{ij} = 3\sigma_{ij}$, the absolute value of V is less than $0.01\varepsilon_{ij}$ [20]. This allows us to ignore the interaction between the molecules separated by more than about $R_{ij} = 3\sigma_{ij}$. The actual threshold values were taken to be equal to $R_{ij} = 3\sigma_{ij}$ [30], $R_{ij} = 3.5\sigma_{ij}$ [8], $R_{ij} = 2.5\sigma_{ij}$, $R_{ij} = 5\sigma_{ij}$, and $R_{ij} = 10\sigma_{ij}$ [166]. In [8], this threshold value of R_{ij} was accounted for by representing the expression for the potential in the form

$$V(R_{ij}) = \begin{cases} 4\varepsilon_{ij}\left[\left(\frac{\sigma_{ij}}{R_{ij}}\right)^{12} - \left(\frac{\sigma_{ij}}{R_{ij}}\right)^{6}\right] + V_{\text{shift}} & \text{if} \quad 0 < R_{ij} \le R_{\text{cut}} \\ 0 & \text{if} \quad R_{ij} > R_{\text{cut}}, \end{cases} \tag{6.44}$$

where

$$V_{\text{shift}} = -4\varepsilon_{ij}\left[\left(\frac{\sigma_{ij}}{R_{\text{cut}}}\right)^{12} - \left(\frac{\sigma_{ij}}{R_{\text{cut}}}\right)^{6}\right].$$

Once the values of $V(R_{ij})$ have been calculated, the actual trajectory of the ith molecule is calculated from Newton's law [166]:

$$m_i \frac{\mathrm{d}^2 R_i}{\mathrm{d}t^2} = \sum_{j=1}^{j=N} \nabla V(R_{ij}), \tag{6.45}$$

where R_i and m_i are the position and mass of the ith molecule, the summation is performed for all molecules in a sphere of radius R_{cut}.

Potentially, this approach could allow us to solve the problem of droplet evaporation in a self-consistent way without any additional assumptions. Even the formalism of the kinetic theory could be made redundant if this approach were fully implemented. This approach, however, has a number of serious limitations which severely limit the range of its applicability. Firstly, the number of molecules which can be analysed is limited by available computing power. It seems that the largest number of molecules ever simulated, 24,000, was reported in [37]. Secondly, the molecules analysed by this approach should be nearly spherical. Ideal candidates for this would be monoatomic molecules, such as Ar or Xe [30, 66, 164, 166], although the model has been applied to N_2 as well [30]. The applicability of this approach to simulating evaporation of Diesel fuel becomes highly questionable. Even if we assume that the latter is n-dodecane $C_{12}H_{26}$, this large molecule does not allow approximation as a sphere.

More advanced molecular dynamics investigations were reported by Tsuruta and Nagayama [163]. In that paper, two models for intermolecular potential were used: the Carravetta–Clementi model [24] and the extended simple point charge model [15]. In both models, the intermolecular interactions were treated as a combination of the short-range pairwise potential of atoms and the long-range Coulombic interaction. The predictions of the extended simple point charge model were shown to be in better agreement with experimental data. It was pointed out that the translational motion is of primary importance for the evaporation/condensation process, whereas the effects of the rotational motion are insignificant. The evaporation/condensation coefficients calculated using this approach were shown to be in good agreement with the values of this coefficient predicted by the transition state theory earlier developed in [102] (see Sect. 6.5.3).

Among previous molecular dynamics simulations on n-alkane liquid–vapour interfaces focused on interfacial properties such as phase equilibria, interface tension, and thermodynamic parameters, we can mention [5, 57, 62, 107, 108, 141, 177]. However, none of these studies focused on the molecular dynamics analysis of the vapour–liquid equilibria of n-dodecane (the closest approximation to Diesel fuel) and the estimation of the evaporation/condensation coefficient for this substance.

An alternative approach to molecular dynamics simulation of droplet evaporation processes could be based on quantum-mechanical methods (see, e.g., [54, 55]). The actual calculation of β_m for evaporating fuel droplets using this approach is not a trivial problem, and it has not yet been performed to the best of my knowledge (V. Gun'ko, private communication). The development of the evaporation models is closely linked with the development of solvation models. The latter are reviewed in [31]. The paper [31] also provides a method of estimating the differences in the free energies of molecules in the liquid and gaseous states.

The evaporation rates of various organics from Diesel fuel nanodrops were analysed using the solvation model (to estimate changes in the Gibbs free energy on transfer of a molecule from a liquid medium into the gas phase) and the kinetic gas theory (to estimate the collision rate of molecules with the clusters/nanodrops in the gas phase) [52]. The evaporation rate studied in [52] referred to the actual rate of molecules leaving the surfaces of the droplets; in the equilibrium state, this rate is equal to the rate of condensing molecules, and it cannot be used for the estimate of the reduction of droplet radii, as discussed in previous chapters. The application of the kinetic gas theory restricted the analysis to molecular clusters and nanodroplets. The solvation models SMD (HF/6-31g(d,p) and DFT/6-31g(d,p)) were used to calculate the Gibbs free energy of solvation, ΔG_s (in the n-dodecane, tetraline, benzene, and isopropyltoluene solution), and to analyse the effects of surroundings on the evaporation rate of normal, iso-, and cycloalkanes, 1–3 ring substituted aromatics, substitutes tetralines and indane (all cyclic compounds were with aliphatic C_1–C_6 side groups) considered as approximations of the components (in the C_8–C_{27} range) of Diesel fuel. An increase in the molecular size of alkanes from n-octane (C_8) to n-heptacosane (C_{27}) (considered as solutes) led to a decrease by several orders of magnitude in the evaporation rate of the molecules from n-dodecane (solvent) nanodrops at $T = 300$–650 K. This difference decreased with increasing temperature. An increase in the aromaticity of both solvents and solutes resulted in a strong decrease in the evaporation rate of all compounds. The evaporation rates for cycloalkanes were found to be between the values for normal (or isoalkanes) and 2–3 ring aromatics of similar molecular weight. The evaporation rates of the components of Diesel fuel were estimated using the results of quantum-chemical and classical kinetic gas theory calculations and temperature-dependence of the density and evaporation enthalpy of these components. This hybrid approach allowed the authors of [52] to study the effects on the evaporation rates of (i) various surroundings for molecules in Diesel fuel nanodrops, (ii) the size and type of evaporated molecules, and (iii) nanodrop size. This approach was used to estimate the influence of changes in contributions of various organics on the evaporation rate of nanodrops (e.g., normal, iso-, mono- and bicycloalkanes, monoaromatics and polyaromatics, cycloalkanoaromatics (indanes, tetralines)) [52].

The results of further analysis of some of the problems considered in [52] are presented in [53]. In that paper, the evaporation rate (γ) of n-alkane molecules in the C_8–C_{27} range from molecular clusters and nanodroplets was analysed using the quantum-chemical solvation model (SMD) and kinetic gas theory, assuming that the system is in a state of thermodynamic equilibrium (evaporation and condensation rates are equal). The droplet size, liquid density, evaporation enthalpy, and Gibbs free energy of evaporation were calculated at 300–640 K. The quantum-chemical calculations (SMD/HF and SMD/B3LYP methods with the 6-31G(d,p) basis set) were used to estimate changes in the Gibbs free energy during the transfer of a molecule from a liquid medium (clusters or nanodroplets) into the gas phase. The kinetic gas theory was used to estimate the collision rate of molecules/clusters/nanodroplets in the gas phase. This rate depends on partial pressures, temperature, sizes, and masses of molecules and clusters/nanodroplets. An increase in the molecular size of evaporated alkanes from octane to heptacosane resulted in a strong decrease in the values of γ. Preliminary estimates of the evaporation/condensation coefficient, based on the direct analysis of the collisions of individual molecules with molecular clusters, were presented.

6.5.2 United Atom Model

The models described in Sect. 6.5.1 cannot be directly applied to heavy hydrocarbon fuel evaporation and condensation analysis, since these fuels have chain-like molecules and cannot be simulated by these models. A number of approaches have been developed to describe the dynamics of complex hydrocarbon molecules such as n-dodecane, including the OPLS (Optimised Potential for Liquid Simulation), originally suggested by Jorgensen et al. [68], and the de Pablo and Toxvaerd models. These approaches are reviewed by Smit et al. [143], who also suggested their new approach, based on the OPLS, which was claimed to be more accurate than those suggested earlier. All these approaches are based on the observation that the C–H bond in complex hydrocarbon molecules is much shorter and much stronger than the C–C bond, and also stronger than the van der Waals forces between molecules. Thus the methyl (CH_3) or methylene (CH_2) groups can be regarded as separate atom-like structures in a relatively simple united atom model (UAM) when these groups can be treated as separate atoms (cf. Simon et al. [140]). The underlying physics of all these approaches is essentially the same, but they differ by the values of energy parameters ε for CH_3 and CH_2, diameters of these groups, and bond bending and torsion potentials. Smit et al. [143] applied all the above-mentioned approaches to complex hydrocarbons to determine their vapour–liquid coexistence curves, using a simulation based on the Gibbs-ensemble technique and the configuration-bias Monte Carlo method. All approaches based on the UAM model gave nearly identical results at standard conditions but predicted critical temperatures which could differ by up to 100 K (the critical temperature is mainly controlled by the ratio $\varepsilon_{CH_3}/\varepsilon_{CH_2}$). It was claimed that the new approach developed by Smit et al. [143] gave a good description of the phase behaviour of this curve over a large temperature range. Simon et al. [140] applied the Toxvaerd model to the molecular dynamic simulation of n-octane. Other applications of the OPLS approach are discussed in [69].

The focus of this section is on an overview of the united atom model (UAM) to the study of the n-dodecane ($C_{12}H_{26}$) liquid–vapour interfaces and a description of the simulation method, following [23, 167].

The essence of the united atom model applied to n-dodecane, in which methyl (CH_3) and methylene (CH_2) groups are regarded as united atoms, is schematically illustrated in Fig. 6.16. The interactions between the molecules of n-dodecane are modelled using the Optimised Potential for Liquid Simulation (OPLS) [23, 167], and the chain-like molecules are modelled by the united atom model, including the methyl (CH_3) and methylene (CH_2) groups. The essence of this model is schematically illustrated in Fig. 6.17.

The nonbonded interactions between atoms are described by the truncated Lennard-Jones (L-J) 12-6 potential given by (6.41). The energy parameters of CH_2 and CH_3 groups (atoms) are $\varepsilon_{CH_2}/k_B = 47$ K and $\varepsilon_{CH_3}/k_B = 114$ K respectively, and the energy parameter between CH_2 and CH_3 groups is estimated as $\varepsilon_{CH_3-CH_3}/k_B = \sqrt{\varepsilon_{CH_2}\varepsilon_{CH_3}}/k_B = 73.2$ K (k_B is the Boltzmann constant). The diameters of the methylene and methyl groups are assumed to be equal and are estimated as $\sigma = 3.93 \times 10^{-10}$ m. The L-J 12-6 interaction is truncated at 13.8×10^{-10} m. Note that the above-mentioned Lennard-Jones parameters for several alkanes were calculated from accurate full-dimensional intermolecular potentials (see [67] for the details).

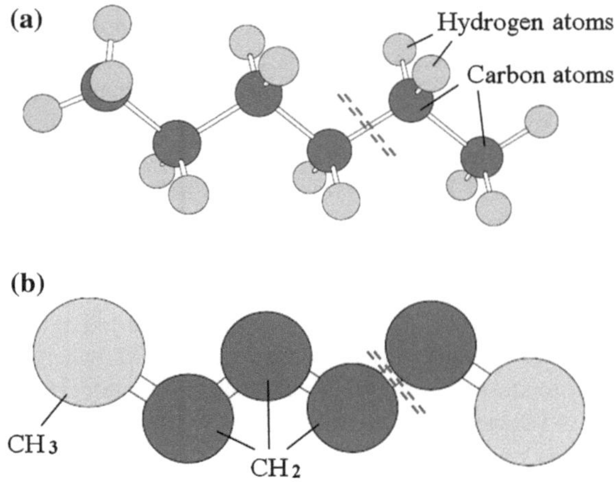

Fig. 6.16 Schematic presentation of an *n*-dodecane molecule (**a**) and its presentation using the united atom model (**b**). The bending angles between neighbouring bonds (~114°) (zigzag structure of the molecule) are taken into account. Reprinted from Journal of Chemical Physics [23]. Copyright AIP (2011)

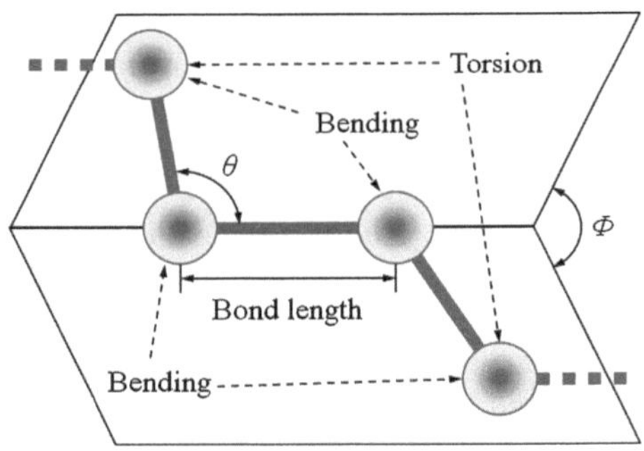

Fig. 6.17 Schematic representation of the bonds and the interactions between them for a portion of the *n*-dodecane chain consisting of four united atoms. Reprinted from Journal of Chemical Physics [23], Copyright AIP (2011)

The interactions within the chains include bond bending and torsion with the bond length constrained at 1.53×10^{-10} m. The bond-bending potential between the three atoms is estimated as

$$u^{\text{bend}}(\theta) = \frac{1}{2} k_\theta \left(\theta - \theta_0\right)^2, \tag{6.46}$$

where the bending coefficient is estimated as $k_\theta / k_B = 62{,}500\,\text{K/rad}^2$, and the equilibrium angle is $\theta_0 = 114°$ (see Fig. 6.17). The values of the bending coefficient for some other complex molecules are given in [170].

The torsion potential between the two atoms with two atoms between them is estimated as

$$u^{\text{tors}}(\varphi) = c_0 + 0.5\, c_1\,(1 + \cos\varphi) + 0.5\, c_2\,(1 - \cos 2\varphi) + 0.5\, c_3\,(1 + \cos 3\varphi)\,, \tag{6.47}$$

where $c_0/k_B = 0\,\text{K}$, $c_1/k_B = 355\,\text{K}$, $c_2/k_B = -68.19\,\text{K}$, $c_3/k_B = 791.3\,\text{K}$, and φ is the dihedral angle with 180° for the equilibrium state.

A system of 720 n-dodecane molecules (8640 CH_2 and CH_3 groups) was considered in a three-dimensional rectangular simulation box of $L_x \times L_y \times L_z = 64.24\sigma \times 16.48\sigma \times 16.48\sigma$, where σ is the reduced length for liquid temperatures 400, 450, 500, and 550 K [167]. This size of box corresponded to $25.25 \times 6.48 \times 6.48\,\text{nm}^3$. These molecules were initially oriented along the x-axis and placed in the middle of the simulation box. They had zigzag configurations, and the number of molecules chosen was 5 in the x direction and 12 in both y and z directions.

The equations of motion of the atoms were integrated using the Verlet leapfrog method [4]. The bond lengths were constrained by the SHAKE scheme [4]. The time steps in all simulations were taken equal to 5 fs. Periodic boundary conditions were applied in all directions [167].

The system was relaxed with a constraint of fixed homogeneous and isotropic temperature. The molecules began to relocate within the liquid phase and to evaporate gradually. Typically, the system required 15,000 ps to reach an equilibrium state. Finally, the liquid film was sandwiched between the layers of the vapour phases. Then data were sampled for another 5,000 ps. The positions of the two liquid–vapour interfaces were identified by their density profiles. The interface parameters, such as density and evaporation/condensation coefficient, were obtained by averaging them over these 5,000 ps.

A molecular dynamics study on transport properties and structure at the liquid–vapour interfaces of alkanes, based on essentially the same model as described above, was presented in [28]. The study of these authors was focused on decane, tetracosane, and hexatriacontane. In [96], molecular dynamics simulations were used for the analysis of droplet breakup.

6.5.3 Evaporation Coefficient

The simplest approach to the estimate of the evaporation/condensation coefficient can be based on transition state theory [102]. In this theory, the condensation/evaporation process at the liquid–vapour interface is considered a kind of chemical reaction, and the general theory of rate processes [49] is applied. As a result, the following expression for the evaporation/condensation coefficient was derived [102]:

$$\beta_m = \left[1 - \left(\frac{V^l}{V^g}\right)^{1/3}\right] \exp\left[-\frac{1}{2}\frac{1}{\left(\frac{V^l}{V^g}\right)^{1/3} - 1}\right], \tag{6.48}$$

where V^l and V^g are the specific volumes of liquid and gas respectively.

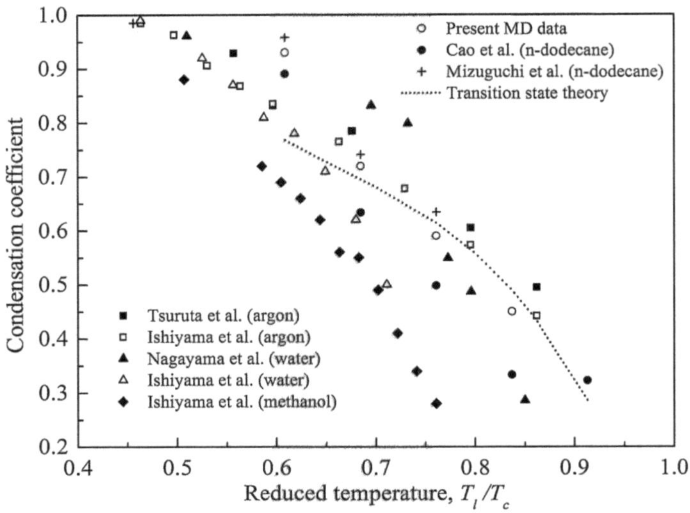

Fig. 6.18 The predicted values of the condensation coefficient versus reduced temperature as predicted by the present molecular dynamics (MD) simulation for *n*-dodecane (present MD data), inferred from published data for *n*-dodecane (Cao et al. [23]; Mizuguchi et al. [99]), argon (Tsuruta et al. [164]; Ishiyama et al. [64]), water (Ishiyama et al. [65]; Nagayama et al. [163]), and methanol (Ishiyama et al. [65]), and predicted by the transition state theory. Reproduced with modifications from Fig. 10 of [167]

Keeping in mind $V^l \ll V^g$ for *n*-dodecane, except when the temperatures are close to the critical temperature, we can expect, based on expression (6.48), that β_m should be close to 1, in agreement with the assumption made in Sect. 6.2. This, however, needs to be justified more rigorously, based on molecular dynamics simulations. This is done later in this section.

Using the united atom and the numerical models described in the previous section, the evaporation/condensation coefficient was estimated from (6.5). The equality of evaporation and condensation coefficients in the equilibrium allows us to focus our analysis on one of these coefficients. The condensation coefficient was chosen. The results were compared with the predictions of the transition state theory, which predicts the values of β_m given by (6.48).

The condensation coefficient of *n*-dodecane was calculated for liquid temperatures $T_l = 400, 450, 500$, and 550 K [167]. The plots of β_m versus the reduced temperature for *n*-dodecane are shown in Fig. 6.18 (filled circles). As follows from this figure, the condensation coefficient decreases from about 0.93 at 400 K to about 0.45 at 550 K. The condensation coefficients of *n*-dodecane reported in [23] using MD simulation in the system with 400 molecules are also shown in Fig. 6.18 (open circles). The values of the condensation coefficient predicted by the transition state theory (6.48) were obtained using the densities of the liquid and vapour phases obtained in our MD simulations (dotted line). As follows from this figure, there is a general agreement between the present results, the predictions of the transition state theory, and the prediction of the results of MD simulations for n-dodecane reported in [99], although the reasons for noticeable deviation between these results are still to be investigated. One of the reasons could be linked to the limited number of molecules used in calculations performed in [167]. Also, the united atom model used in these calculations might be less accurate at reduced temperatures close to 1. The condensation coefficients of other substances using MD simulations reported by other authors are also shown in Fig. 6.18 (filled [164] and open [64] squares for argon, filled [65] and open [163] triangles for water, and filled diamonds for methanol [65]).

As follows from Fig 6.18, in all cases the condensation coefficient decreases when the liquid temperature increases. This is consistent with the results for the interface thickness presented in [167]. The thicker interfaces at higher temperatures are expected to reduce the values of the evaporation/condensation coefficients.

The condensation coefficient for complex polyatomic molecules, like n-dodecane, would be expected to be lower than for simpler molecules due to the constraint imposed by the rotational motion of molecules in the liquid phase [47]. However, in MD simulations, the results of which are shown in Fig. 6.18, it was found that the rotational energy has no noticeable effect on this coefficient, and its value is mainly controlled by the translational energy, in agreement with previously reported results in [23, 163]. On the other hand, based on the transition state theory, we can expect that this coefficient is close to unity at low temperatures when $(V^g/V^l)^{(1/3)}$ is large and that it decreases with increasing temperature due to the decrease of the ratio $(V^g/V^l)^{(1/3)}$. This trend is consistent with the prediction of MD simulations shown in Fig. 6.18.

The values of MD calculations of β_m for n-dodecane, inferred from Fig. 6.18, can be approximated as

$$\beta(m)(T_s) = 7 \times 10^{-6}\, T_s^2 - 9.8 \times 10^{-3}\, T_s + 3.7215. \tag{6.49}$$

This equation was used in the calculations whose results are presented in Sect. 6.6, following [128]. These values of $\beta(m)$ were substituted into (6.40) to obtain the value of f^{out} [128].

Note that another boundary condition used in the kinetic modelling is that the distribution of molecules leaving the droplet surface is close to Maxwellian. The validity of this assumption was investigated in [168]. It was shown that the distribution functions of evaporated and reflected molecules for the velocity component normal to the surface deviate considerably from the Maxwellian, while the distribution function for all molecules leaving this surface (evaporated and reflected) is close to Maxwellian. The role of these effects in kinetic modelling has not yet been investigated.

6.6 Results of the Kinetic Calculations

The results presented in this section are mainly reproduced from [128].

6.6.1 Results for $\beta_m = 1$

The algorithm described in Sect. 6.4 was applied to computation of the heating and evaporation of Diesel fuel droplets in a hot gas at two temperatures (750 and 1,000 K). The initial droplet temperature and gas pressure in all cases were assumed equal to 300 K and 30 bar respectively. The initial droplet radii were assumed to be equal to 5 μm. Droplets were assumed to be stationary, but the effect of swelling was taken into account. The calculations were performed using the Infinite Thermal Conductivity (ITC) and Effective Thermal Conductivity (ETC) models for the liquid phase, the Abramzon and Sirignano model for the gas phase (see Sect. 4.2.2), and kinetic models taking and not taking into account the effects of inelastic collisions. The value of the evaporation coefficient was assumed equal to 1 in all cases.

The results of calculation of the radii and surface temperatures of droplets immersed into gas with temperature 750 K are shown in Figs. 6.19 and 6.20 respectively. As follows from Fig. 6.20, at the initial stage of droplet heating and evaporation, the kinetic effects on temperature are negligible, but the difference in temperatures predicted by the hydrodynamic models taking and not taking into account the effects of liquid finite thermal conductivity is quite noticeable. This is consistent with results reported earlier in [123]. Initially, the droplet surface temperature predicted by the ETC model is larger than that predicted by the ITC model, as expected. This larger droplet surface temperature predicted by the ETC model leads to a reduction in the heat flux supplied to the droplet, which eventually leads to a situation in which the droplet surface temperature predicted by the ETC model becomes smaller than that predicted by the ITC model. This happens at times close to 0.4 ms. Smaller temperatures predicted by the ETC model at this stage lead to slightly slower evaporation rate and longer evaporation time of droplets, as shown in Fig. 6.19.

As follows from Fig. 6.19, the kinetic models taking into account inelastic collisions predict longer evaporation times compared with the hydrodynamic ETC and ITC models, in agreement with the predictions of the model not taking into account the effects of inelastic collisions, described earlier (see [119]). The kinetic model taking into account inelastic collisions predicts longer evaporation times compared with the model ignoring this effect, in agreement with the prediction of Fig. 6.14. Similar enhancement of the kinetic effects due to the contribution of inelastic collisions can be observed from the temperature curves shown in Fig. 6.20.

As in the case of gas temperature equal to 750 K, for gas temperature equal to 1,000 K at the initial stage of droplet heating and evaporation, the kinetic effects on temperature are negligible, but the difference in temperatures predicted by the hydrodynamic models taking and not taking into account the effects of liquid finite thermal

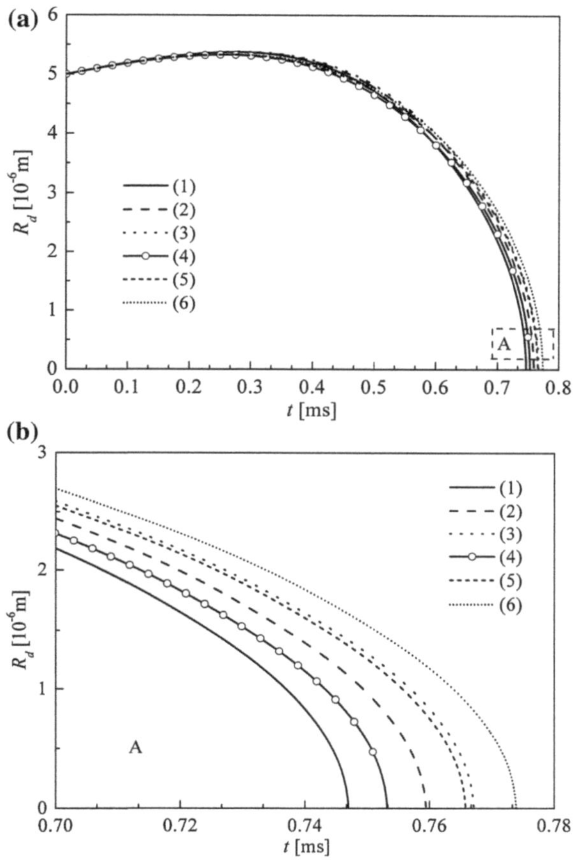

Fig. 6.19 **a** Plots of R_d versus time t for an n-dodecane droplet, predicted by the hydrodynamic ITC model (*curves* 1), the kinetic ITC model, ignoring the effects of inelastic collisions (*curves* 2), the kinetic ITC model, taking into account the effects of inelastic collisions (*curves* 3), the hydrodynamic ETC model (*curves* 4), the kinetic ETC model, ignoring the effects of inelastic collisions (*curves* 5), and the kinetic ETC model, taking into account the effects of inelastic collisions (*curves* 6). **b** Zoomed part of **a** showing the plots at the final stage of the evaporation process. All plots are shown for $T_g = 750$ K, $T_{d0} = 300$ K, $R_{d0} = 5$ μm. Reprinted from [128]. Copyright Elsevier (2013)

conductivity is noticeable (see [128] for the details). In the latter case, initially the droplet surface temperature predicted by the ETC model is larger than that predicted by the ITC model, as expected, but at times larger than about 0.25 ms, the surface temperature predicted by the ETC model becomes smaller than that predicted by the ITC model. In contrast to the case shown in Fig. 6.19, in the case of gas temperature equal to 1,000 K, the net effects of the increase of the temperature predicted by the ETC model before 0.25 ms and its decrease at times after 0.25 ms lead to shorter evaporation time predicted by the ETC model compared with the ITC model.

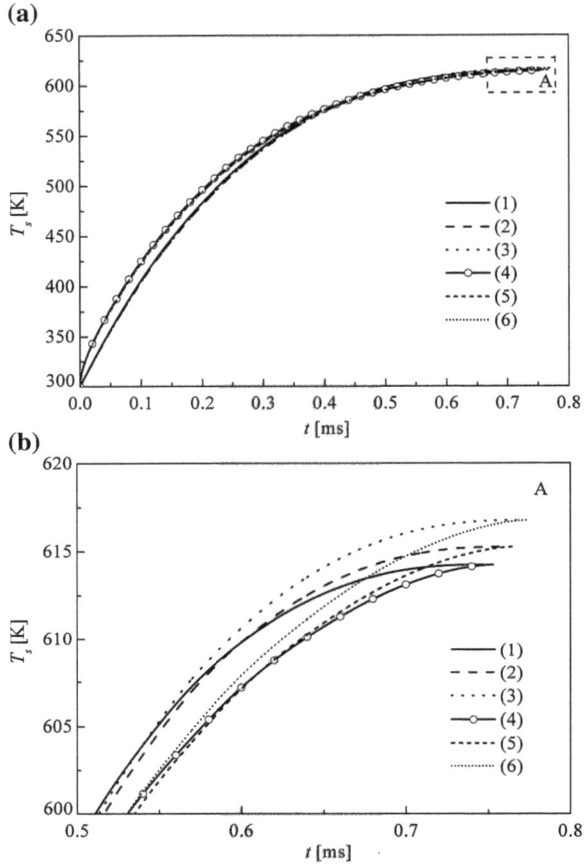

Fig. 6.20 The same as Fig. 6.19 but for T_s. Reprinted from [128]. Copyright Elsevier (2013)

As in the case shown in Figs. 6.19 and 6.20, for gas temperature equal to 1,000 K, the kinetic models taking and not taking into account inelastic collisions predict longer evaporation times and higher temperatures at the final stage of droplet evaporation than the hydrodynamic ETC and ITC models. Also, as in the case shown in Figs. 6.19 and 6.20, the kinetic model taking into account inelastic collisions predicts longer evaporation times and higher temperatures at the final stage of droplet evaporation than the model ignoring this effect. These kinetic effects are stronger when gas temperature is equal to 1,000 K than in the case shown in Figs. 6.19 and 6.20.

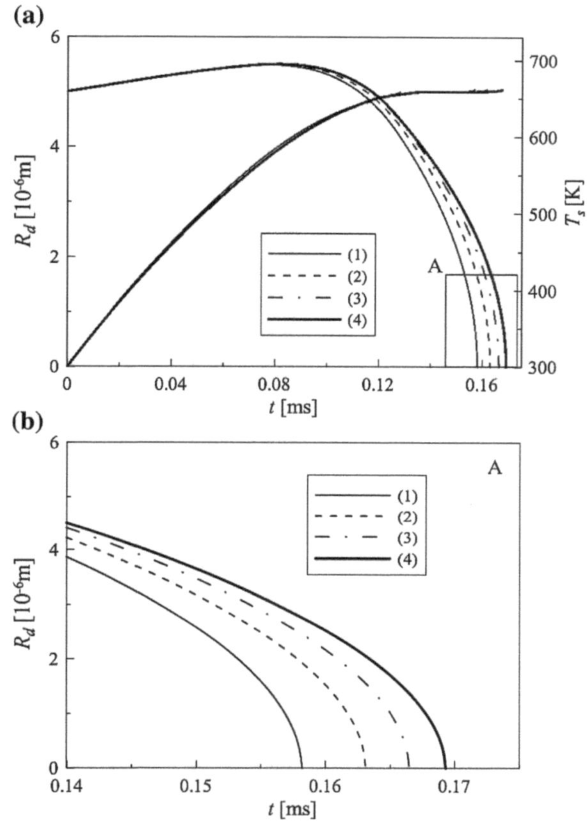

Fig. 6.21 **a** Plots of R_d and T_s versus time t for an n-dodecane droplet predicted by the hydrodynamic ITC model (*curves* 1), the kinetic ITC model, ignoring the effects of inelastic collisions and nonunity of β (*curves* 2), the kinetic ITC model, taking into account the effects of inelastic collisions but ignoring the nonunity of β (*curves* 3), and the kinetic ITC model, taking into account the effects of inelastic collisions and the nonunity of β, which is calculated based on (6.49) (*curves* 4). **b** Zoomed part of **a** showing the values of R_d at the final stage of the evaporation process. All plots are shown for $T_g = 1,500$ K, $T_{d0} = 300$ K, $R_{d0} = 5$ μm. Reprinted from [128]. Copyright Elsevier (2013)

6.6.2 Results for $\beta_m < 1$

As already mentioned, in our earlier kinetic models we assumed that $\beta_m = 1$. This was justified by the fact that no reliable experimental data or theoretical predictions of β_m were available at that time. The situation changed after the publication of papers [23, 167], where a new model for molecular dynamic simulations of n-dodecane was suggested. Using this model, the values of the evaporation coefficient β_m were calculated, and the temperature-dependence of β_m was approximated by (6.49). For $T_s = 600$ K, Eq. (6.49) predicts $\beta_m = 0.36$. This value of β_m was used for the plot presented in Fig. 6.15.

In what follows, the effects of $\beta_m < 1$ on n-dodecane droplet heating and evaporation are presented for the ITC model only. These effects for the ETC model are similar to those for the ITC model.

The results of calculation of the radii and surface temperatures of droplets with initial radii equal to 5 μm immersed in gas with temperature 1,500 K using the ITC model with $\beta_m = 1$ and β_m predicted by (6.49) are shown in Fig. 6.20. As one can see from this figure, the effect of $\beta_m < 1$ is quite visible. This effect leads to increased droplet evaporation time, thus enhancing the kinetic corrections to the prediction of the hydrodynamic model.

In the case of gas temperature equal to 1,000 K, the effect of $\beta_m < 1$ on droplet radius and temperature is small and can be safely ignored in most practical applications. For gas temperature equal to 750 K, this effect is even smaller than in the case of gas temperature equal to 1,000 K. This means that in these cases, the results of the analysis for $\beta_m = 1$ presented in the previous section are applicable to the realistic case $\beta_m < 1$.

6.7 Kinetic Modelling in the Presence of Three Components

One of the important limitations of the approaches described above is that they are based on the assumption that Diesel fuel can be approximated by n-dodecane. The more detailed analysis of the composition of Diesel fuel described in the previous chapter showed that it includes hundreds of various hydrocarbon components. These components can be subdivided into two main groups: alkanes and aromatics. The assumption that n-dodecane can approximate alkanes is a reasonable one (see [52]), while aromatics can be approximated by p-dipropylbenzene [52]. In this case, a more accurate approximation of Diesel fuel than that based on its approximation by n-dodecane would be a mixture of n-dodecane and p-dipropylbenzene. Mass fractions of n-dodecane in this mixture vary from 0.8 to 0.7 [52].

To improve the accuracy of kinetic modelling of Diesel fuel droplet heating and evaporation, one would naturally expect that this modelling should take into account the contributions of both n-dodecane and p-dipropylbenzene. We cannot, however, perform this modelling before a kinetic algorithm of the solution of the system of three Boltzmann equations for n-dodecane, p-dipropylbenzene and air taking into account the effects of inelastic collisions between respective molecules has been developed. Preliminary results of the development of such an algorithm and its testing are presented below, following [136].

The evolution of the molecular velocity distribution functions of air $f_a \equiv f_a(\mathbf{r}, t, \mathbf{v})$ (or any other neutral gas, including nitrogen), n-dodecane $f_n \equiv f_n(\mathbf{r}, t, \mathbf{v})$, and p-dipropylbenzene $f_p \equiv f_p(\mathbf{r}, t, \mathbf{v})$ is controlled by the corresponding Boltzmann equations:

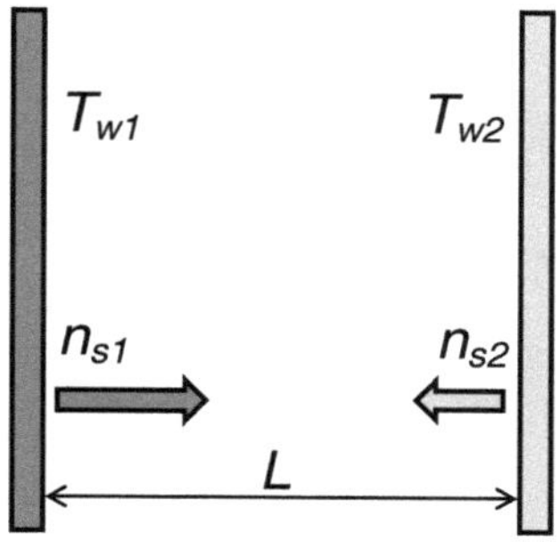

Fig. 6.22 A schematic representation of the setup used for the analysis of mass/heat transfer processes in up to three component mixtures. The distance between the walls L is assumed to be equal to 10 mean free molecular paths of n-dodecane molecules ℓ at the temperature of the second wall (T_{w2}). Reprinted from [136]. Copyright Elsevier (2014)

$$\left.\begin{aligned} \frac{\partial f_a}{\partial t} + \mathbf{v}_a \frac{\partial f_a}{\partial \mathbf{r}} &= J_{aa} + J_{an} + J_{ap} \\ \frac{\partial f_n}{\partial t} + \mathbf{v}_n \frac{\partial f_n}{\partial \mathbf{r}} &= J_{na} + J_{nn} + J_{np} \\ \frac{\partial f_p}{\partial t} + \mathbf{v}_p \frac{\partial f_p}{\partial \mathbf{r}} &= J_{pa} + J_{pn} + J_{pp} \end{aligned}\right\}, \tag{6.50}$$

where $J_{\alpha\beta}$ ($\alpha = a, n, p$; $\beta = a, n, p$) are collision integrals taking into account the contribution of the collisions between molecules. Triple collisions are ignored. All collision integrals $J_{\alpha\beta}$ are calculated taking into account the contribution of internal degrees of freedom (inelastic collisions) as described earlier.

The preliminary testing of the numerical algorithm for the solution of system (6.50) was performed for the case of two parallel walls kept at constant temperatures $T_{w1} = 1$ and $T_{w2} = 1$, as shown in Fig. 6.22. The distance between the walls L was equal to 10 mean free paths for n-dodecane ℓ_n at temperature T_{w2}. Space between the walls was filled with the following:

(1) a mixture of n-dodecane, p-dipropylbenzene and nitrogen (three component mixture);
(2) a mixture of n-dodecane and nitrogen (two component mixture);
(3) pure n-dodecane or p-dipropylbenzene (single component).

Initially, the molar fractions of all components were assumed to be equal to 1. It is also assumed that the distribution functions of n-dodecane and p-dipropylbenzene at both walls are half-Maxwellian with densities $n_{s1} = 2$ and $n_{s2} = 1$ describing evaporated species. Molecules of n-dodecane and p-dipropylbenzene hitting the walls were removed from the domain (they were condensed), while for nitrogen molecules, it was assumed that the flux of incoming molecules was equal to the flux of molecules reflected from the walls. Note that although, strictly speaking, the above-mentioned combination of parameters might not be realistic for practical applications, it turned out to be useful for testing the code.

Table 6.9 Chemical formulas, molar masses, and molecular diameters of *n*-dodecane, *p*-dipropylbenzene, and nitrogen

Component	Chemical formula	Molar mass (kg/kmol)	Molecular diameter (Å)
n-dodecane	$C_{12}H_{26}$	170.3	7.12
p-dipropylbenzene	$C_{12}H_{18}$	162.27	6.73
nitrogen	N_2	28.97	3.617

The chemical formulas, molar masses, and molecular diameters of *n*-dodecane, *p*-dipropylbenzene, and nitrogen are given in Table 6.9.

The space grid in the numerical analysis was taken equal to $0.5\,\ell_n$, and $12 \times 12 \times 12$ velocity grid points were used.

The predicted distributions of the number density of *n*-dodecane for three and two component mixtures and one component substance versus the distance from the first plate are shown in Fig. 6.23 (the distance x is normalised by ℓ_n). Two cases were considered: (a) the contribution of internal degrees of freedom (corresponding to rotational and vibrational movements of the constituent molecules) is ignored; (b) the numbers of these degrees for *n*-dodecane and *p*-dipropylbenzene are equal to 20, while nitrogen has 2 internal degrees of freedom. As follows from our previous analysis, only the first 20 internal degrees of freedom of such complex molecules as *n*-dodecane contribute to the results predicted by the kinetic calculations (the actual number of internal degrees of freedom of *n*-dodecane, taking into account translational, rotational, and vibrational movements of the constituent molecules, is expected to be much more than 20). It is anticipated that the same conclusion should remain valid for *p*-dipropylbenzene.

As follows from Fig. 6.23, the density distributions for three-component and two-component mixtures are very close. They coincide within the accuracy of plotting when the contribution of internal degrees of freedom is taken into account (see Fig. 6.23b). Both these distributions differ considerably from the distribution for the one-component substance. In all cases, the expected concentration jumps near the walls are clearly seen. These jumps are noticeably larger for the one-component substance than for three- or two-component mixtures. The physical nature of these density jumps has been widely discussed in the literature, beginning with the pioneering monograph [46]. In [118], it was shown that the introduction of these jumps into a conventional CFD code (FLUENT) enables this code to be used for modelling rarefied gases.

The contribution of internal degrees of freedom has little effect on the distribution of number densities of *n*-dodecane. It leads to an increase in number density jumps for three- and two-component mixtures and a slight decrease in these jumps for the case of a single component.

The distributions of the number densities of *p*-dipropylbenzene for a three-component mixture and a one-component substance are shown in Fig. 6.24. Comparing Figs. 6.23 and 6.24, one can see that the distributions of *n*-dodecane and

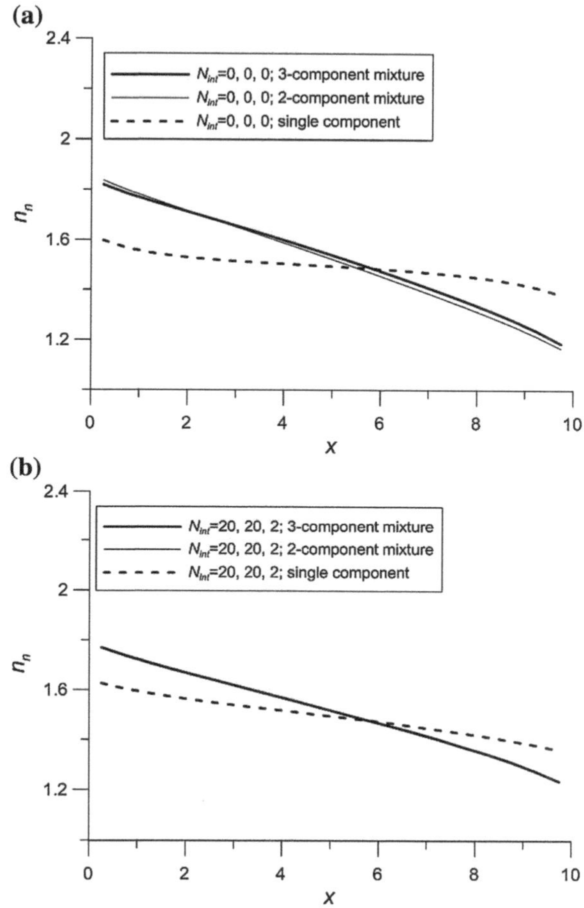

Fig. 6.23 Plots of n-dodecane number density n_n versus the distance from the first plate x predicted by the model taking into account the contributions of three components, n-dodecane, p-dipropylbenzene, and nitrogen (*thick solid curves*); two components, n-dodecane and nitrogen (*thin solid curves*); and a single component (n-dodecane) (*dashed curves*). The contribution of internal degrees of freedom is either **a** ignored or **b** taken into account assuming that $N_{\text{int}} = 20$ for n-dodecane and p-dipropylbenzene and $N_{\text{int}} = 2$ for nitrogen. Reprinted from [136]. Copyright Elsevier (2014)

p-dipropylbenzene are very close, which is related to the closeness of masses and sizes of the corresponding molecules (see Table 6.9).

The distributions of the number density of nitrogen for three- and two-component mixtures are shown in Fig. 6.25. Comparing Fig. 6.25 with Figs. 6.23 and 6.24, one can see that in contrast to n-dodecane and p-dipropylbenzene, the number density of nitrogen increases from the first plate to the second one. This means that nitrogen molecules are expected to move from the region of high concentration of n-dodecane and p-dipropylbenzene to the region of low concentration of these molecules, which

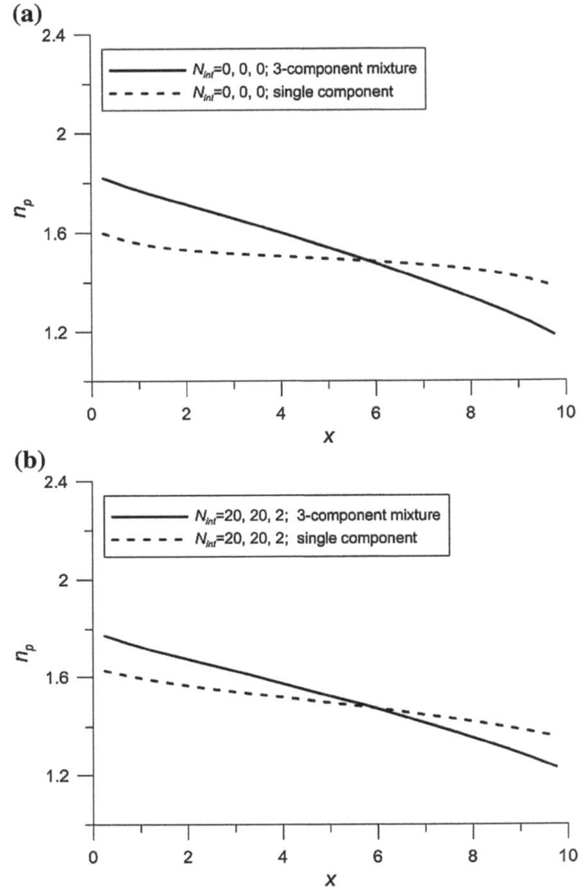

Fig. 6.24 The same as Fig. 6.23 but for *p*-dipropylbenzene, taking into account the contributions of three components, *n*-dodecane, *p*-dipropylbenzene, and nitrogen (*thick solid curves*), and a single component (*p*-dipropylbenzene) (*dashed curves*). Reprinted from [136]. Copyright Elsevier (2014)

is consistent with the corresponding thermodynamics equations for the mixtures. In contrast to the case shown in Fig. 6.23, the distribution of nitrogen is rather different for three- and two-component mixtures. Since the number density of nitrogen at the walls is not specified, we cannot infer any information about the number density jump in the vicinity of the walls for nitrogen. As in the cases shown in Figs. 6.23 and 6.24, the contribution of internal degrees of freedom produces only minor effects on the number density of nitrogen.

The number fluxes of *n*-dodecane and *p*-dipropylbenzene for the three-component mixture and single components versus the number of internal degrees of freedom N_{int} are shown in Fig. 6.26. It is assumed that N_{int} for *n*-dodecane and *p*-dipropylbenzene are equal and that N_{int} for nitrogen is equal to 2, except when N_{int} for *n*-dodecane and *p*-dipropylbenzene are equal to 0. In the latter case, N_{int} for nitrogen was taken equal to 0 as well. As one can see from this figure, although the number densities of *n*-dodecane and *p*-dipropylbenzene are weak functions of the number of internal

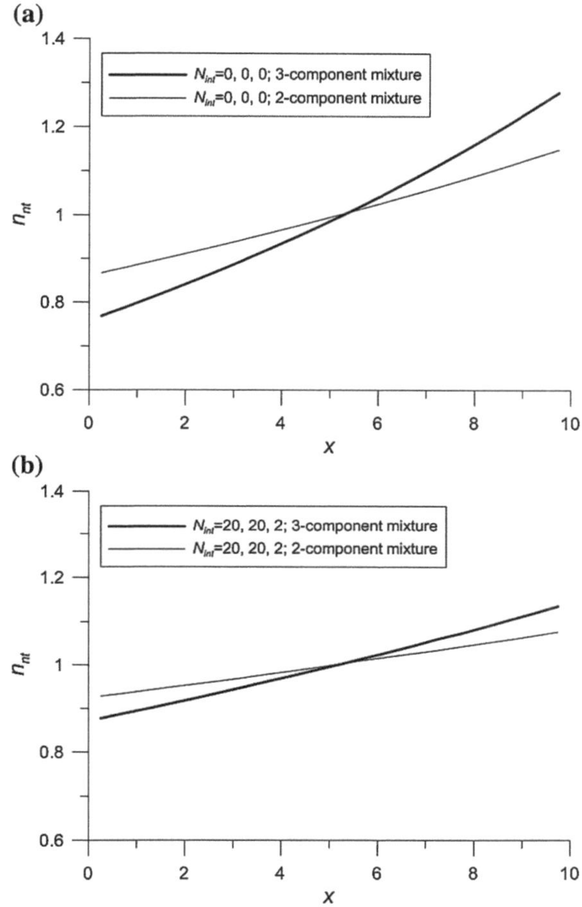

Fig. 6.25 The same as Figs. 6.23 and 6.24 but for nitrogen, taking into account the contributions of three components, *n*-dodecane, *p*-dipropylbenzene, and nitrogen (*thick solid curves*); and two components, *n*-dodecane and nitrogen (*thin solid curves*). Reprinted from [136]. Copyright Elsevier (2014)

degrees of freedom, the dependence of number fluxes of these components on this number is rather strong, at least at $N_{\text{int}} \leq 20$. At larger N_{int}, these fluxes do not depend on N_{int}, in agreement with our earlier results. Mass fluxes of *n*-dodecane and *p*-dipropylbenzene are close for the three-component mixture, but are noticeably different for the case of single components. Mass fluxes of *n*-dodecane and *p*-dipropylbenzene are both much larger for the single components than for the three-component mixture, as expected.

Functionality testing similar what has been presented above was performed for pure heat transfer and combined heat and mass transfer between the walls [136]. The results obtained in this case were similar to those described above for the case of pure

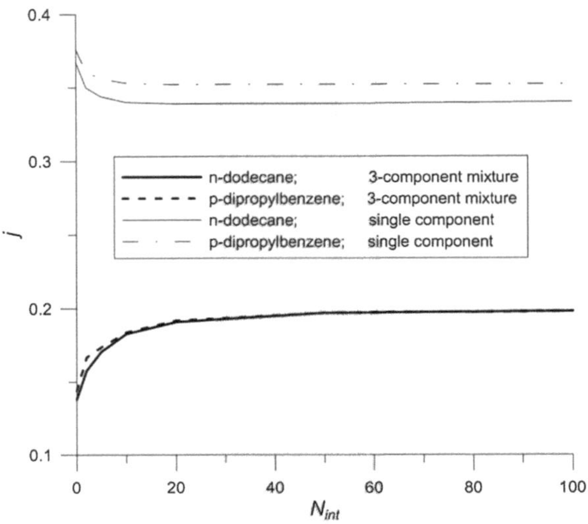

Fig. 6.26 Plots of number fluxes of *n*-dodecane and *p*-dipropylbenzene versus the number of internal degrees of freedom N_{int} predicted by the model taking into account the contributions of three components, *n*-dodecane, *p*-dipropylbenzene, and nitrogen, and the models of single components (*n*-dodecane or *p*-dipropylbenzene). Reprinted from [136]. Copyright Elsevier (2014)

mass transfer. In the case of pure heat transfer between the walls, the distributions of temperatures of various components between the walls were shown to be qualitatively similar. It was pointed out that the velocity distribution functions of *n*-dodecane remain almost Maxwellian at various positions between the walls when both heat and mass transfer processes are taken into account. Noticeable deviations from the Maxwellian distributions were observed for the corresponding distribution functions of nitrogen.

It was concluded that the predictions of the new kinetic algorithm are qualitatively consistent with the predictions of more basic models of the phenomena for which it was tested. It is ready for the analysis of a wider range of heat mass transfer problems in which the contribution of three components in the mixture is essential, including the problem of Diesel fuel droplet heating and evaporation.

References

1. Abramov, A. A., Kogan, M. N., & Makashev, N. K. (1981). Numerical analysis of the processes in strongly non-equilibrium Knudsen layers. *Reports of the Academy of Sciences of USSR. Mechanics of Liquids and Gases* No. 3, pp. 72–81 (in Russian).
2. Akhiezer, A. I., Akhiezer, I. A., Polovin, R. V., Sitenko, A. G., & Stepanov, K. N. (1975). *Plasma electrodynamics*. Oxford: Pergamon.
3. Alexandrov, A. F., Bogdankevich, L. S., & Rukhadze, A. A. (1978). *Fundamentals of plasma electrodynamics*. Moscow: Vysshaya Shkola (in Russian).
4. Allen, M. P., & Tidesley, D. J. (1984). *Computer simulation of liquids*. Oxford: Oxford University Press.

5. Amat, M. A., & Rutledge, G. C. (2010). Liquid-vapor equilibria and interfacial properties of n-alkanes and perfluoroalkanes by molecular simulation. *Journal of Chemical Physics, 132.11*, 114704. doi:10.1063/1.3356219.
6. Anisimov, S. I. (1968). Evaporation of metals under the influence of laser radiation. *Journal of Experimental and Theoretical Physics, 54*, 339–342 (in Russian).
7. Anisimov, S. I., Imas, Ya. A., Romanov, G. S., & Khodyko, Yu. V. (1970). Effect of high intensity radiation on metals. Moscow: Nauka Publishing House (in Russian).
8. Anisimov, S. I., Dunikov, D. O., Zhakhovskii, V. V., & Malyshenko, S. P. (1999). Properties of a liquid-gas interface at high-rate evaporation. *Journal of Chemical Physics, 110*, 8722–8729.
9. Aoki, K., Takata, S., & Kosuge, S. (1998). Vapor flows caused by evaporation and condensation on two parallel plane surfaces: effect of the presence of a noncondensable gas. *Physics of Fluids, 10*, 1519–1533.
10. Aoki, K., Bardos, C., & Takata, S. (2003). Knudsen layer for gas mixtures. *Journal of Statistical Physics, 112*, 629–655.
11. Aristov, V. V., & Tcheremissine, F. G. (1992). *Direct numerical solution of the Boltzmann equation*. Moscow: Computer Centre of Russian Academy of Sciences.
12. Bakhvalov, N. S., Zhidkov, N. P., & Kobelkov, G. M. (1989). *Numerical methods*. Moscow: Nauka Publishing House (in Russian).
13. Banasiak, J., & Groppi, M. (2003). Solvability of linear kinetic equations with multi-energetic inelastic scattering. *Reports on Mathematical Physics, 52*, 235–253.
14. Benedetto, D., & Caglioti, E. (1999). The collapse phenomenon in one-dimensional inelastic point particle systems. *Physica D: Nonlinear Phenomena, 132*, 457–475.
15. Berendsen, H. J. C., Grigera, J. R., & Straatsma, T. P. (1987). The missing term in effective pair potential. *Journal of Physics and Chemistry, 91*, 6269–6271.
16. Bhatnagar, P. L., Gross, E. P., & Krook, M. (1954). A model for collision processes in gases. 1. Small amplitude processes in charged and neutral one component systems. *Physical Review, 94*, 511–525.
17. Biben, T., Martin, Ph. A., & Piasecki, J. (2002). Stationary state of thermostated inelastic hard spheres. *Physica A: Statistical Mechanics and its Applications, 310*, 308–324.
18. Bird, G. A. (1976). *Molecular gas dynamics*. Oxford: Oxford University Press.
19. Bird, G. A. (1994). *Molecular gas dynamics and the direct simulation of gas flows*. Oxford: Oxford University Press.
20. Bird, R. B., Stewart, W. E., & Lightfoot, E. N. (2002). *Transport phenomena*. Chichester: John Wiley & Sons.
21. Borgnakke, C., & Larsen, P. S. (1975). Statistical collision model for Monte Carlo simulation of polyatomic gas mixture. *Journal of Computational Physics, 18*, 405–420.
22. Buet, C., Cordier, S., & Degond, P. (1998). Regularized Boltzmann operators. *Computers and Mathematics with Applications, 35*, 55–74.
23. Cao, B.-Y., Xie, J.-F., & Sazhin, S. S. (2011). Molecular dynamics study on evaporation and condensation of n-dodecane at liquid-vapour phase equilibria. *Journal of Chemical Physics, 134*, 164309–164309-9. doi:10.1063/1.3579457.
24. Carravetta, V., & Clementi, E. (2011) (1984). Water-water interaction potential: an approximation of the electron correlation contribution by a functional of the SCF density matrix. *Journal of Chemical Physics, 81*, 2646–2651. doi:10.1063/1.447973.
25. Chapman, S., & Cowling, T. G. (1990). *The mathematical theory on non-uniform gases*. Cambridge: Cambridge University Press.
26. Cercignani, C. (1981). Strong evaporation of a polyatomic gas. In S. S. Fisher (Ed.), *Rarefied gas dynamics*, part 1 (pp. 305–310). New York: AIAA.
27. Chernyak, V. (1995). The kinetic theory of droplet evaporation. *Journal of Aerosol Science, 26*, 873–885.
28. Chilukoti, H. K., Kikugawa, G., & Ohara, T. (2013). A molecular dynamics study on transport properties and structure at the liquid-vapor interfaces of alkanes. *International Journal of Heat and Mass Transfer, 59*, 144–154.

29. Cipola, J. W., Lang, J. H., & Loyalka, S. K. (1974). Kinetic theory of evaporation and condensation. *Proceedings of the 8th International Symposium on Rarefied Gas Dynamics* (pp. 1773–1776). New York-London: Academic Press.
30. Consoline, L., Aggarwal, S. K., & Murad, S. (2003). A molecular dynamics simulation of droplet evaporation. *International Journal of Heat and Mass Transfer*, *43*, 3179–3188.
31. Cramer, C. J., & Truhlar, D. G. (1999). Implicit solvation models: equilibria, structure, spectra, and dynamics. *Chemical Physics*, *99*, 2161–2200.
32. Doi, T. (2011). Numerical analysis of the time-dependent energy and momentum transfers in a rarefied gas between two parallel planes based on the linearized Boltzmann equation. *ASME Journal of Heat Transfer*, *133*, 022404.
33. Dombrovsky, L. A., & Sazhin, S. S. (2003). A parabolic temperature profile model for heating of droplets. *ASME Journal of Heat Transfer*, *125*, 535–537.
34. Dombrovsky, L. A., Sazhin, S. S., Sazhina, E. M., Feng, G., Heikal, M. R., Bardsley, M. E. A., et al. (2001). Heating and evaporation of semi-transparent diesel fuel droplets in the presence of thermal radiation. *Fuel*, *80*, 1535–1544.
35. Dongari, N., & Agrawal, A. (2012). Modeling of Navier-Stokes equations for high Knudsen number gas flows. *International Journal of Heat and Mass Transfer*, *55*, 4352–4358.
36. Lu, G., Duan, Y.-Y., Wang, X.-D., & Lee, D.-J. (2012). Internal flow in evaporating droplet on heated solid surface. *International Journal of Heat and Mass Transfer*, *54*, 4437–4447.
37. Dunikov, D. O., Malyshenko, S. P., & Zhakhovskii, V. V. (2000). Liquid-gas interface during transient condensation: molecular dynamics investigation. In A. Leontiev & A. Klimenko, (Eds.), *Proceedings of 3rd Russian National Heat and Mass Transfer Conference* (Vol. 4, pp. 257–260). Moscow: Moscow Power Engineering Institute Publishing House (in Russian).
38. Dunikov, D. O., Malyshenko, S. P., & Zhakhovskii, V. V. (2001). Corresponding states law and molecular-dynamics simulations of the Lennard-Jones fluid. *Journal of Chemical Physics*, *115*, 6623–6631.
39. Elliott, R. J. (1995) Plasma kinetic theory. In R. Rendy (Ed.), *Plasma physics: an introductory course* (pp. 29–53). Cambridge: Cambridge University Press.
40. Elperin, T., & Krasovitov, B. (1995). Radiation, thermal diffusion and kinetic effects in evaporation and combustion of large and moderate size droplets. *International Journal of Heat and Mass Transfer*, *38*, 409–418.
41. Feng, Z.-G., & Michaelides, E. E. (2012). Heat transfer from a nano-sphere with temperature and velocity discontinuities at the interface. *International Journal of Heat and Mass Transfer*, *55*, 6491–6498.
42. Ferrari, L., & Carbognani, A. (1998). Differential kinetic equations for a Rayleigh gas with inelastic collisions. *Physica A: Statistical Mechanics and its Applications*, *251*, 452–468.
43. Flynn, P. F., Durrett, R. P., Hunter, G. L., zur Loye, A. O., Akinyemi, O. C., Dec, J. E. et al. (1999). Diesel combustion: an integrated view combining laser diagnostics, chemical kinetics, and empirical validation. *SAE report 1999–01-0509*.
44. Fournier, N., & Mischler, S. (2005). A spatially homogeneous Boltzmann equation for elastic, inelastic and coalescing collisions. *Journal de mathématiques pures et appliquées*, *84*, 1173–1234.
45. Frezzotti, A. (1986). Kinetic theory study of the strong evaporation of a binary mixture. In V. Boffi & C. Cercignani (Eds.), *Rarefied gas dynamics* (Vol. 2, pp. 313–322). Stutgart: Teubner.
46. Fuchs, N. A. (1959). *Evaporation and droplet growth in gaseous media*. London: Pergamon Press.
47. Fujikawa, S., & Maerefat, M. (1990). A study of the molecular mechanism of vapor condensation. *JSME International Journal Series II*, *33*, 634.
48. Furioli, G., Pulvirenti, A., Terraneo, E., & Toscani, G. (2010). Convergence to self-similarity for the Boltzmann equation for strongly inelastic Maxwell molecules. *Annales de l'Institut Henri Poincaré AN*, *27*, 719–737.
49. Gladstone, G., Laidler, K. J., & Eyring, H. (1941). *The theory of rate processes*. New York: McGraw-Hill.
50. Gombosi, T. I. (1994). *Gaskinetic theory*. Cambridge: Cambridge University Press.

51. Gumerov, N. A. (2000). Dynamics of vapour bubbles with non-equilibrium phase transitions in isotropic acoustic fields. *Physics of Fluids, 12*, 71–88.
52. Gun'ko, V. M., Nasiri, R., Sazhin, S. S., Lemoine, F., & Grisch, F. (2013). A quantum chemical study of the processes during the evaporation of real-life Diesel fuel droplets. *Fluid Phase Equilibria, 356*, 146–156.
53. Gun'ko, V. M., Nasiri, R., & Sazhin, S. S. (2014). A study of the evaporation and condensation of n-alkane clusters and nanodroplets using quantum chemical methods. *Fluid Phase Equilibria, 366*, 99–107.
54. Gun'ko, V. M., & Turov, V. V. (1999). Structure of hydrogen bonds and ^{1}H NMR spectra of water at the interface of Oxides. *Langmuir, 15*, 6405–6415.
55. Gun'ko, V. M., Zarko, V. I., Leboda, R., Marciniak, M., Janusz, W., & Chibowski, S. (2000). Highly dispersed X/SiO_2 and C/X Si O_2 (X = Aluminia, Titania, Aluminia/Titania) in the gas liquid media. *Journal of Colloid and Interface Science, 230*, 396–409.
56. Hanchak, M. S., Briones, A. M., Ervin, J. S., & Byrd, L. W. (2013). One-dimensional models of nanoliter droplet evaporation from a hot surface in the transition regime. *International Journal of Heat and Mass Transfer, 57*, 473–483.
57. Harris, J. G. (1992). Liquid-vapor interfaces of alkane oligomers: structure and thermodynamics from molecular dynamics simulations of chemically realistic models. *Journal of Physics and Chemistry, 96*, 5077–5086.
58. Harvie, D. J. E., & Fletcher, D. F. (2001). A simple kinetic theory treatment of volatile liquid-gas interfaces. *ASME Journal of Heat Transfer, 123*, 486–491.
59. Hertz, H. (1882). Über die Verdunstung der Flüssigkeiten, inbesondere des Quecksilbers, im lufteeren Raume. *Annalen der Physik und Chemie, 17*, 177–200.
60. Hirschfelder, J. O., Curtiss, C. F., & Bird, R. B. (1967). *Molecular theory of gases and liquids* (4th ed.). New York: John Wiley & Sons.
61. Hirth, J. P., & Pound, G. M. (1963). *Condensation and evaporation: nucleation and growth kinetics*. Oxford-London-Paris-Frankfurt: Pergamon Press.
62. Ibergay, C., Ghoufi, A., Goujon, F., Ungerer, P., Boutin A., Rousseau, B. et al. (2007). Molecular simulations of the $n-$alkane liquid-vapor interface: interfacial properties and their long range corrections. *Physical Review E, 75*, 051602 (18 pages).
63. Imamuro, T., & Sturtevant, B. (1990). Numerical study of discrete-velocity gases. *Physics of Fluids, 2*, 2196–2203.
64. Ishiyama, T., Yano, T., & Fujikawa, S. (2004). Molecular dynamics study of kinetic boundary conditions at an interface between argon vapor and its condensed phase. *Physics of Fluids, 16*(8), 2899–2907. doi:10.1063/1.1763936.
65. Ishiyama, T., Yano, T., & Fujikawa, S. (2004). Molecular dynamics study of kinetic boundary conditions at an interface between a polyatomic vapor and its condensed phase. *Physics of Fluids, 16*(12) 4713–4726. doi:10.1063/1.1811674.
66. Ishiyama, T., Fujikawa, S., Kurz, W., & Lauterborn, W. (2013). Nonequilibrium kinetic boundary condition at the vapor-liquid interface of argon. *Physical Review E, 88*, 042406.
67. Jasper, A. W., & Miller, J. A. (2014). Lennard-Jones parameters for combustion and chemical kinetics modeling from full-dimensional intermolecular potentials. *Combustion and Flame, 161*, 101–110.
68. Jorgensen, W. L., Madura, J. D., & Swenson, C. J. (1984). Optimized intermolecular potential functions for liquid hydrocarbons. *Journal of the American Chemical Society, 106*, 6638–6646.
69. Jorgensen, W. L., & Tirado-Rives, J. (1988). The OPLS potential functions for proteins. Energy minimizations for crystals of cyclic peptides and crambin. *Journal of the American Chemical Society, 110*, 1657–1666.
70. Kennaird, E. H. (1938). *Kinetic theory of gases*. New York: McGraw-Hill.
71. Kharchenko, V., Balakrishnan, N., & Dalgarno, A. (1998). Thermalization of fast nitrogen atoms in elastic and inelastic collisions with molecules of atmospheric gases. *Journal of Atmospheric and Solar-Terrestrial Physics, 60*, 95–106.

72. Knight, S. J. (1979). Theoretical modelling of rapid surface vaporization with back pressure. *AIAA Journal*, *17*, 519–523.
73. Knudsen, M. (1915). Die Maximale Verdampfungsgeschwindigkeit des Quecksilbers. *Annals of Physics*, *47*, 697–708.
74. Koffman, L. D., Plesset, M. S., & Lees, L. (1984). Theory of evaporation and condensation. *Physics of Fluids*, *27*, 876–880.
75. Kogan, M. N. (1969). *Rarefied gas dynamics*. New York: Plenum.
76. Kogan, M. N., & Makashev, N. K. (1971). On the role of the Knudsen layer in the theory of heterogeneous reactions and flows with the surface reactions. *Reports of the Academy of Sciences of USSR. Mechanics of Liquids and Gases*, No. 6, pp. 3–11 (in Russian).
77. Korabelnikov, A. V., Nakoryakov, V. E., & Shreiber, I. R. (1981). Taking nonequilibrium vaporization into account in problems of vapour-bubble dynamics. *High Temperatures*, *19*, 586–590.
78. Korobov, N. M. (1989). *Trigonometric sums and their applications*. Moscow: Nauka Publishing House (in Russian).
79. Koura, K. (1998). Improved null-collision technique in the direct simulation Monte Carlo method: application to vibrational relaxation of nitrogen. *Computers and Mathematics with Applications*, *35*, 139–154.
80. Kremer, G. M., Silva, A. W., & Alves, G. M. (2010). On inelastic reactive collisions in kinetic theory of chemically reacting gas mixtures. *Physica A: Statistical Mechanics and its Applications*, *389*, 2708–2718.
81. Kryukov, A. P., & Levashov, V. Yu. (2011). About evaporation-condensation coefficients on the vapor-liquid interface of high thermal conductivity matters. *International Journal of Heat and Mass Transfer*, *54*, 3042–3048.
82. Kryukov, A. P., Levashov, V. Yu., & Sazhin, S. S. (2004). Evaporation of diesel fuel droplets: kinetic versus hydrodynamic models. *International Journal of Heat and Mass Transfer*, *47*, 2541–2549.
83. Kryukov, A. P., Levashov, V. Yu., & Shishkova, I. N. (2004). Non-equilibrium recondensation in a dusty medium. In S. A. Zhdanok et al. (Eds.), *Proceedings of the 5th Minsk International Heat and Mass Transfer Forum*, May 24–28 (Vol. 2, pp. 138–139). Minsk: Research Institute of Heat and Mass Transfer Publishing House (in Russian).
84. Kryukov, A. P., Levashov, V. Yu., & Shishkova, I. N. (2005). Condensation in the presence of a non-condensable component. *Journal of Engineering Physics and Thermophysics*, *78*, 15–21 (in Russian).
85. Kryukov, A. P., Levashov, V. Yu., Shishkova, I. N., & Yastrebov, A. K. (2005). *The numerical solution of the Boltzmann equation for engineering applications*. Moscow: Moscow Power Engineering Institute Publishing House (in Russian).
86. Kryukov, A. P., Levashov, V. Yu., & Shishkova, I. N. (2006). Flows of vapour-gas mixtures in micro- and nano-systems in the presence of evaporation and condensation. *Proceedings of the 5th National Russian Heat Transfer Conference* (Vol. 78, Issue No. 3, pp. 164–167), Moscow (in Russian).
87. Kucherov, R. Ya., & Rickenglaz, L. E. (1959). On hydrodynamics boundary conditions for modelling vaporization and condensation processes. *Journal of Experimental and Theoretical Physics*, *37*, 125–126 (in Russian).
88. Kucherov, R. Ya., & Rickenglaz, R. E. (1960). On the measurement of the coefficient of condensation. *Reports of the Academy of Science of USSR (Doklady)*, *133*, 1130–1131 (in Russian).
89. Kucherov, R. Ya., Rickenglaz, R. E., & Tzulaya, T. S. (1963). Kinetic theory of additional condensation in the presence of the small temperature jump. *Soviet Physics—Technical Physics*, *7*, 1027–1030.
90. Labuntsov, D. A. (1967). Analysis of the evaporation and condensation processes. *High Temperatures*, *5*, 579–585.
91. Labuntsov, D. A., & Kryukov, A. P. (1977). Processes of intensive evaporation. *Thermal Power Energy (Teploenergetika)*, No. 4, 8–11 (in Russian).

92. Labuntsov, D. A., & Kryukov, A. P. (1979). Analysis of intensive evaporation and condensation. *International Journal of Heat and Mass Transfer*, *22*, 989–1002.
93. Lambiotte, R., Ausloos, M., Brenig, L., & Salazar, J. M. (2007). Energy and number of collision fluctuations in inelastic gases. *Physica A: Statistical Mechanics and its Applications*, *375*, 227–232.
94. Lee, L. L. (1988). *Molecular nonideal fluids*. Boston: Butterworths.
95. Lifshitz, E. M., & Pitaevski, L. P. (1979). *Physical kinetics*. Moscow: Nauka Publishing House (in Russian).
96. Ludwig, K. F., & Micci, M. (2011). Molecular dynamics simulations of Rayleigh and first wind-induced breakup. *Atomization and Sprays*, *21*, 275–281.
97. Malyshenko, S. P., & Dunikov, D. O. (2002). On the surface tension corrections in non-uniform and nonequilibrium liquid-gas systems. *International Journal of Heat and Mass Transfer*, *45*, 5201–5208.
98. Marek, R., & Straub, J. (2001). Analysis of the evaporation coefficient and the condensation coefficient of water. *International Journal of Heat and Mass Transfer*, *44*, 39–53.
99. Mizuguchi, H., Nagayama, G., & Tsuruta, T. (2010). Molecular dynamics study on evaporation coefficient of biodiesel fuel. In *Seventh International Conference on Flow Dynamics*, Sendai, Japan, pp. 386–387.
100. Murakami, M., & Oshima, K. (1974). Kinetic approach to the transient evaporation and condensation problem. In M. Becker & M. Fiebig (Eds.), *Rarefied gas dynamics*. Pors-Wahn: DFVLR Press, paper F6.
101. Muratova, T. M., & Labuntsov, D. A. (1969). Kinetic analysis of the evaporation and condensation processes. *Thermal Physics of High Temperatures*, *7*, 959–967 (in Russian).
102. Nagayama, G., & Tsuruta, T. (2003). A general expression for the condensation coefficient based on transition state theory and molecular dynamic simulation. *Journal of Chemical Physics*, *118*, 1392–1399.
103. Neizvestny, A. I., & Onishenko, L. I. (1979). Experimental determination of the condensation coefficient for distilled water. *Physics of Atmosphere and Ocean*, *15*, 1052–1078 (in Russian).
104. Nigmatilin, R. I. (1991). *Dynamics of multiphase media* (Vol. 1). New York: Hemisphere Publishing Corporation.
105. Odukoya, A., & Naterer, C. F. (2013). Droplet evaporation and de-pinning in rectangular microchannels. *International Journal of Heat and Mass Transfer*, *56*, 127–137.
106. Peeters, P., Luijten, C. C. M., & van Dongen, M. E. H. (2001). Transitional droplet growth and diffusion coefficients. *International Journal of Heat and Mass Transfer*, *44*, 181–193.
107. Petrilla, B. A. (2008). Droplet vaporization of N-heptane using molecular dynamics. *Master's Thesis in Aerospace Engineering*. USA: The Pennsylvania State University.
108. Petrilla, B. A., Trujillo, M. F., & Micci, M. (2010). N-heptane droplet vaporization using molecular dynamics. *Atomization and Sprays*, *20*, 581–593.
109. Popov, S. P., & Tcheremissine, F. G. (1999). Conservative method for the solution of the Boltzmann equation for centrally symmetrical interaction potentials. *Computational Mathematics and Mathematical Physics*, *39*, 163–176.
110. Rogier, F., & Schneider, J. A. (1994). A direct method for solving the Boltzmann equation. *Transport Theory and Statistical Physics*, *23*, 1–3.
111. Rose, J. (1998). Interphase matter transfer, the condensation coefficient and dropwise condensation. In J. S. Lee (Ed.), *Heat Transfer 1998. Proceedings of 11th International Heat Transfer Conference*, August 23–28, 1998, Kyongji, Korea (Vol. 1, pp. 89–104). Seoul: The Korean Society of Mechanical Enginers.
112. Rose, J. W. (2000). Accurate approximate equations for intensive sub-sonic evaporation. *International Journal of Heat and Mass Transfer*, *43*, 3869–3875.
113. Santos, A. (2003). Transport coefficients of d-dimensional inelastic Maxwell models. *Physica A: Statistical Mechanics and its Applications*, *321*, 442–466.
114. Santos, A. (2006). A simple model kinetic equation for inelastic Maxwell particles. In M. S. Ivanov & A. K. Rebrov (Eds.), *Proceedings of the 25th International Symposium on Rarefied Gas Dynamics*, July 21–28th 2006, St Petersburg, Russia (pp. 191–196). Novosibirsk: Publishing House of the Siberian Branch of the Russian Academy of Sciences, ISBN 978.

115. Sazhin, S. S. (1978). Cyclotron whistler-mode instability in a collisional plasma. *Geomagnetic research* (Vol. 23, pp. 105–107). Moscow: Soviet Radio (in Russian).
116. Sazhin, S. S. (1993). *Whistler-mode waves in a hot plasma*. Cambridge: Cambridge University Press.
117. Sazhin, S. S. (2006). Advanced models of fuel droplet heating and evaporation. *Progress in Energy and Combustion Science*, *32*(2), 162–214.
118. Sazhin, S. S., & Serikov, V. V. (1997). Rarefied gas flows: hydrodynamic versus Monte Carlo modelling. *Planetary and Space Science*, *45*, 361–368.
119. Sazhin, S. S., & Shishkova, I. N. (2009). A kinetic algorithm for modelling the droplet evaporation process in the presence of heat flux and background gas. *Atomization and Sprays*, *19*, 473–489.
120. Sazhin, S. S., Wild, P., Leys, C., Toebaert, D., & Sazhina, E. M. (1993). The three temperature model for the fast-axial-flow CO_2 laser. *Journal of Physics D: Applied Physics*, *26*, 1872–1883.
121. Sazhin, S. S., Wild, P., Sazhina, E. M., Makhlouf, M., Leys, C., & Toebaert, D. (1994). The three dimensional modelling of the processes in the fast-axial-flow CO_2 laser. *Journal of Physics D: Applied Physics*, *27*, 464–469.
122. Sazhin, S. S., Feng, G., Heikal, M. R., Goldfarb, I., Goldshtein, V., & Kuzmenko, G. (2001). Thermal ignition analysis of a monodisperse spray with radiation. *Combustion and Flame*, *124*, 684–701.
123. Sazhin, S. S., Kristyadi, T., Abdelghaffar, W. A., & Heikal, M. R. (2006). Models for fuel droplet heating and evaporation: comparative analysis. *Fuel*, *85*, 1613–1630.
124. Sazhin, S. S., Shishkova, I. N., Kryukov, A. P., & Heikal, M. R. (2007). Evaporation of droplets into a background gas: kinetic modelling. *International Journal of Heat and Mass Transfer*, *50*, 2675–2691.
125. Sazhin, S. S., Shishkova, I. N., Kryukov, A. P., Levashov, V. Yu., & Heikal, M. R. (2008). Evaporation of droplets into a background gas in the presence of heat flux: kinetic and hydrodynamic modelling. *Proceedings of the 19th International Symposium on Transport Phenomena*, 17–20 August, 2008, Reykjavik, Iceland, paper 48.
126. Sazhin, S. S., Shishkova, I. N., Kryukov, A. P., Levashov, V. Yu., & Heikal, M. R. (2009). A Simple algorithm for kinetic modelling of Diesel fuel droplet evaporation. *Progress in Computational Heat and Mass Transfers; Proceedings of 6th ICCHMT*, May 18–21, 2009, Guangzhou, China, paper 151 (pp. 386–391).
127. Sazhin, S. S., Shishkova, I. N., & Heikal, M. (2010). Kinetic modelling of fuel droplet heating and evaporation: Calculations and approximations. *International Journal of Engineering Systems Modelling and Simulation*, *2*, 169–176.
128. Sazhin, S. S., Xie, J.-F., Shishkova, I. N., Elwardany, A. E., & Heikal, M. R. (2013). A kinetic model of droplet heating and evaporation: effects of inelastic collisions and a non-unity evaporation coefficient. *International Journal of Heat and Mass Transfer*, *56*, 525–537.
129. Sazhina, E. M., Sazhin, S. S., Heikal, M. R., Babushok, V. I., & Johns, R. (2000). A detailed modelling of the spray ignition process in Diesel engines. *Composites Science and Technology*, *160*, 317–344.
130. Schrage, R. W. (1953). *A theoretical study of interphase mass transfer*. New York: Columbia University Press.
131. Shankar, P. N., & Marble, F. M. (1971). Kinetic theory of transient condensation and evaporation at a plane surface. *Physics of Fluids*, *14*, 510–516.
132. Sharipov, F., & Kalempa, D. (2005). Velocity slip and temperature jump coefficients for gaseous mixtures. IV. Temperature jump coefficient. *International Journal of Heat and Mass Transfer*, *48*, 1076–1083.
133. Shen, C. (1983). The concentration-jump coefficient in a rarefied binary gas mixture. *Journal of Fluid Mechanics*, *137*, 221–231.
134. Shidlovskiy, V. P. (1967). *Introduction to the dynamics of rarefied gases*. New York: American Elsevier Publishing Company.

135. Shishkova, I. N., & Sazhin, S. S. (2006). A numerical algorithm for kinetic modelling of evaporation processes. *Journal of Computational Physics*, *218*, 635–653.
136. Shishkova, I. N., & Sazhin, S. S. (2014). A solution of the Boltzmann equations in the presence of three components and inelastic collisions. *International Journal of Heat and Mass Transfer*, *71*, 26–34.
137. Shishkova, I. N., Sazhin, S. S., & Xie, J.-F. (2013). A solution of the Boltzmann equation in the presence of inelastic collisions. *Journal of Computational Physics*, *232*, 87–99.
138. Shusser, M., Ytrehus, T., & Weihs, D. (2000). Kinetic theory analysis of explosive boiling of a liquid droplet. *Fluid Dynamics Research*, *27*, 353–367.
139. Sibold, D., & Urbassek, H. M. (1991). Kinetic study of evaporating flows from cylindrical jets. *Physics of Fluids*, *3*, 870–878.
140. Simon, J.-M., Kjelstrup, S., Bedeaux, D., & Hafskjold, B. (2004). Thermal flux through a surface of n-octane. A non-equilibrium molecular dynamics study. *Journal of Physical Chemistry*, *108*, 7186–7195.
141. Simon, J.-M., Bedeaux, D., Kjelstrup, S., Xu, J., & Johannessen, E. (2006). Integral relations verified by non-equilibrium molecular dynamics. *Journal of Physical Chemistry*, *110*, 18528–18536.
142. Sizhuk, A., & Yezhov, S. (2006). The dynamic theory for the inelastically colliding particles. *Journal of Molecular Liquids*, *127*, 84–86.
143. Smit, B., Karaborni, S., & Siepmann, J. I. (1995). Computer simulations of vapour-liquid phase equilibria of n-alkanes. *Journal of Chemical Physics*, *102*, 2126–2140.
144. Sone, Y. (2000). Kinetic theoretical studies of the half-space problem of evaporation and condensation. *Transport Theory and Statistical Physics*, *29*, 227–260.
145. Sone, Y., Aoki, K., Sugimoto, H., & Yamada, T. (1988). Steady evaporation and condensation on a plane condensed phase. *Theoretical and Applied Mechanics (Bulgaria)*, *19*, 89–93.
146. Sone, Y., & Onishi, Y. (1973). Kinetic theory of evaporation and condensation. *Journal of Physical Society of Japan*, *35*, 1773–1776.
147. Sone, Y., & Onishi, Y. (1978). Kinetic theory of evaporation and condensation—hydrodynamic equation and slip boundary condition. *Journal of Physical Society of Japan*, *44*, 1981–1994.
148. Sone, Y., & Sugimoto, H. (1990). Strong evaporation from a plane condensed phase. In G. E. A. Meier & P. A. Thompson (Eds.), *Adiabatic waves in liquid-vapour systems, Proceedings of IUTAM Symposium*, Gottingen, Germany 1989 (pp. 293–304). Berlin: Springer-Verlag.
149. Sone, Y., & Sugimoto, H. (1993). Kinetic theory analysis of steady evaporating flows from a spherical condensed phase into a vacuum. *Physics of Fluids*, *5*, 1491–1511.
150. Sone, Y., & Sugimoto, H. (1995). Evaporation of rarefied gas from a cylindrical condensed phase into a vacuum. *Physics of Fluids*, *7*, 2072–2085.
151. Sone, Y., Sugimoto, H., & Aoki, K. (1999). Cylindrical Couette flows of a rarefied gas with evaporation and condensation: reversals and bifurcation of flows. *Physics of Fluids*, *11*, 476–490.
152. Stoecklin, T., & Voronin, A. (2007). H—N_2 inelastic collision dynamics on new potential energy surface. *Chemical Physics*, *331*, 385–395.
153. Sugimoto, H., & Sone, Y. (1992). Numerical analysis of steady flows of a gas evaporating from its cylindrical condensed phase on the basis of kinetic theory. *Physics of Fluids*, *4*, 419–440.
154. Taguchi, S., Aoki, K., & Takata, S. (2003). Vapor flows condensing at incidence onto a plane condensed phase in the presence of noncondensable gas. I. Subsonic condensation. *Physics of Fluids*, *15*, 689–705.
155. Taguchi, S., Aoki, K., & Takata, S. (2004). Vapor flows condensing at incidence onto a plane condensed phase in the presence of noncondensable gas. I. Supersonic condensation. *Physics of Fluids*, *16*, 79–92.
156. Takata, S., & Aoki, K. (1999). Two-surface problems of a multi-component mixture of vapors and noncondensable gases in the continuum limit in the light of kinetic theory. *Physics of Fluids*, *11*, 2743–2756.

157. Tcheremissine, F. G. (1997). Conservative method for the solution of the Boltzmann equation for centrally symmetrical interaction potentials. *Reports of Russian Academy of Science, 357*, 53–56 (in Russian).
158. Tcheremissine, F. G. (1998). Conservative evaluation of Boltzmann collision integral in discrete ordinates approximation. *Computers and Mathematics with Applications, 35*, 215–221.
159. Tcheremissine, F.G. (2000). Discrete approximation and examples of the solution of the Boltzmann equation. In *Computational dynamics of rarefied gases* (pp. 37–74). Moscow: Computer Centre of Russia Academy of Sciences.
160. Thomas, J. P., Chang, T. S., & Siewert, S. E. (1974). Reverse temperature gradient in the kinetic theory of evaporation. *Physical Review Letters, 33*, 680–682.
161. Tilinin, I. S. (1996). Impact-parameter dependence of inelastic energy losses in slow atom-atom collisions. *Nuclear Instruments and Methods in Physics Research B, 115*, 102–105.
162. Titarev, V. A., & Shahov, E. M. (2002). Heat loss and evaporation from a flat surface after a rapid increase of its temperature. *Reports of the Russian Academy of Sciences: Mechanics of Liquids and Gases* No. 1, pp. 141–153 (in Russian).
163. Tsuruta, T., & Nagayama, G. (2004). Molecular dynamics studies on the condensation coefficient of water. *Journal of Physical Chemistry B, 108*, 1736–1743.
164. Tsuruda, T., Tanaka, H., & Masuika, T. (1999). Condensation/evaporation coefficient and velocity distribution at liquid-vapour interface. *International Journal of Heat and Mass Transfer, 42*, 4107–4116.
165. Versteeg, H. K., & Malalasekera, W. (2007). *An introduction to computational fluid dynamics* (2nd ed.). Harlow (UK): Longman.
166. Walther, J. H., & Koumoutsakos, P. (2001). Molecular dynamics simulation of nanodroplet evaporation. *ASME Journal of Heat Transfer, 123*, 741–748.
167. Xie, J.-F., Sazhin, S. S., & Cao, B.-Y. (2011). Molecular dynamics study of the processes in the vicinity of the n-dodecane vapour/liquid interface. *Physics of Fluids, 23*, 112104 (11 pages). doi:10.1063/1.3662004.
168. Xie, J.-F., Sazhin, S. S., & Cao, B.-Y. (2012). Molecular dynamics study of condensation/evaporation and velocity distribution of n-dodecane at liquid-vapour phase equilibria. *Journal of Thermal Science and Technology, 7*, 288–300.
169. Xu, X., Cheng, C., & Chowdhury, I. H. (2004). Molecular dynamics study of phase change mechanisms during femtosecond laser ablation. *ASME Journal of Heat Transfer, 126*, 727–734.
170. Yanagihara, H., Stanković, I., Blomgren, F., Rosén, A., & Sakata, I. (2014). A molecular dynamics simulation investigation of fuel droplet in evolving ambient conditions. *Combustion and Flame, 161*, 541–550.
171. Yang, T. H., & Pan, C. (2005). Molecular dynamics simulation of a thin water layer evaporation and evaporation coefficient. *International Journal of Heat and Mass Transfer, 48*, 3516–3526.
172. Young, J. B. (2011). Calculation of Knudsen layers and jump conditions using the linearised G13 and R13 moment methods. *International Journal of Heat and Mass Transfer, 54*, 2902–2912.
173. Young-ping, P. (1971). Application of kinetic theory to the problem of evaporation and condensation. *Physics of Fluids, 14*, 306–311.
174. Ytrehus, T. (1977). Theory and experiments on gas kinetics in evaporation. In J. L. Potter (Ed.), *Rarefied gas dynamics*, part 2 (pp. 1197–1212). New York: AIAA.
175. Ytrehus, T., & Aukrust, T. (1986). Mott-Smith solution for weak condensation. In V. Boffi & C. Cercignani (Eds.), *Rarefied gas dynamics* (Vol. 2, pp. 271–280). Stuttgart: Teubner.
176. Ytrehus, T., & Ostmo, S. (1996). Kinetic approach to interphase processes. *International Journal of Multiphase Flow, 22*, 133–135.
177. Zahn, D. (2008). Length-dependent nucleation mechanisms rule the vaporization of n-alkanes. *Chemical Physics Letters, 467*, 80–83.
178. Zenevich, V. A., Billing, G. D., & Jolicard, G. (1999). Vibrational-rotational energy transfer in H_2–H_2 collisions II. The relative roles of the initial rotational excitation of both diatoms. *Chemical Physics Letters, 312*, 530–535.

Chapter 7
Heating, Evaporation and Autoignition of Sprays

Abstract Possible approaches to coupled solutions of the equations describing heating and evaporation of fuel droplets leading to autoignition of a fuel vapour/air mixture are discussed. Firstly, the focus is on a brief overview of various approaches to the modelling of autoignition (in the presence or in the absence of droplets). Then a simplified model for coupling between droplet heating, evaporation, and the autoignition of the fuel vapour/air mixture, based on the integral manifold method, is presented and discussed. This is followed by a description of the approach to the numerical solution of the coupled system of ordinary differential equations (ODEs) describing droplet heating and evaporation and the autoignition process based on the dynamic decomposition technique. As in the original integral manifold method, the formal approach to decomposition of the system of ODEs in the latter method is based on the division of system variables into 'fast' and 'slow'. In contrast to the original version of the integral manifold method, however, linearised variations of slow variables during the time evolution of the fast variables are taken into account as a first-order approximation to the fast manifold.

In many practical engineering applications, heating and evaporation of fuel droplets leads to autoignition of a fuel vapour/air mixture. Temperature rise of the fuel vapour/air mixture during the autoignition process leads to the enhancement of droplet heating and evaporation processes. Hence, the analysis of fuel droplet heating and evaporation in realistic engine conditions should be coupled with the analysis of autoignition. Possible approaches to coupled solutions of the equations describing all these processes are discussed later in this chapter (Sects. 7.2 and 7.3). In the next section, the focus is on a brief overview of various approaches to the modelling of autoignition (in the presence or in the absence of droplets).

S. Sazhin, *Droplets and Sprays*, DOI: 10.1007/978-1-4471-6386-2_7,

7.1 Autoignition Modelling

Autoignition can be defined as an onset of combustion in a reactive medium raised above a certain temperature and pressure [63, 106]. Sometimes, autoignition is called spontaneous ignition, self-ignition, or homogeneous ignition [63]. It can be triggered by rapid compression of the fuel-oxidiser mixture, as in conventional Diesel engines, or by pressure waves due to very fast heat release, as in spark ignition engines.

In early approaches, autoignition was modelled as a one-step chemical reaction [29, 115]. In the simplest case, the chemical power released in the gas phase per unit volume was calculated based on the Arrhenius approximation [63]:

$$P_{\rm ch} = M_f Q_f C_f A \exp\left(-\frac{E}{R_u T_g}\right), \tag{7.1}$$

where M_f is the molar mass of fuel, Q_f is the specific combustion energy, C_f is the fuel vapour molar concentration, A is the preexponential factor (in 1/s), E is the activation energy (in J/kmol). Equation (7.1) is based on the assumption that there is no deficiency of oxygen in the system. The simplicity of this equation and its straightforward generalisations (see [63]) makes it particularly attractive for qualitative analysis of the autoignition process and its analytical study (see [17, 38–40, 71, 93]).

In many realistic situations, however, the autoignition chemistry is much more complicated, and its modelling, based on (7.1), can lead to misleading results. An in-depth analysis of autoignition models is a vast topic, the coverage of which would require a separate monograph or review paper, similar to [2]. In what follows, only a brief summary of autoignition models is presented, mainly with regard to their application to modelling the autoignition process in Diesel engines.

There are two main parameters which characterise the autoignition process: the initial temperature T_{g0} above which autoignition can develop and the time delay $\tau_{\rm del}$ before the start of autoignition [13]. Both these parameters strongly depend on pressure, fuel composition, condition of internal surface of the engine, and other factors [7, 8, 30, 72, 74, 109]. The detailed kinetic mechanism (DKM) of the autoignition process includes up to about 1000 chemical reactions and hundreds of species [15, 26, 31, 61, 73, 83, 86, 125]. Additional problems in the construction of DKMs are due to the scarcity or lack of kinetic data on many of the reactions involved. Even quantum chemistry methods cannot ensure accurate calculation of rate constants due to the fact that the decisive factors in this calculation are the small differences between the high energy levels of reacting molecules [5]. The incorporation of this model into CFD codes for the modelling of the autoignition process would be an almost impossible task. This problem has stimulated the development of a number of reduced chemical models (e.g., [6, 7, 10, 47–50, 53, 84, 89, 111]). The one suggested in [47] was successfully used in [110] for modelling autoignition of isolated n-heptane droplets. The general approach to simplification of DKMs was discussed by Maas and Pope in [68]. Poppe et al. [85] considered a model based on 30 reactions and 21 species. Basevich and Frolov [6] went even further and reduced the number of reactions to 21 and the number of species to 13. However, even this simplification turned out to be insufficient to model the chemical processes in combination with the flow and heat/mass transfer calculations in realistic three-dimensional geometries. Further simplification of this model led to a four-step model with adjusted rate coefficients as suggested by Müller et al. [75]. The practical applicability of this model was questioned by Griffiths in the accompanying discussion of that paper. He drew attention to the fact that according to this model, ignition was brought about entirely by thermal feedback, while the chain branching process was ignored. This is difficult to justify in a model intending to capture the key features of the autoignition process.

A number of authors have considered turbulent autoignition models in an attempt to take into account the contribution of flow characteristics [11, 12, 14, 35, 56, 65, 116, 123, 124]. Although these models are important for an in-depth understanding of the process, the complexity of turbulence modelling requires considerable simplification of the chemical kinetics. For example, the model for turbulent combustion used in [116, 124] is based on the four-step model suggested by Müller et al. [75]. In more recent papers, however, the results of turbulent ignition modelling alongside complex chemistry are described [82].

A group of researchers from Shell Research Ltd. attempted to capture the essential features of the process rather than to construct a chemically and physically rigorous model [54]. They achieved this by introducing generic species with chemical reaction constants deduced from experimental data and developed a model based on the eight-step chain branching reaction scheme incorporated into four processes:

Initiation:

$$RH + O_2 \rightarrow 2R^*$$

Propagation:

$$\begin{aligned} R^* &\rightarrow R^* + P \\ R^* &\rightarrow R^* + B \\ R^* &\rightarrow R^* + Q \\ R^* + Q &\rightarrow R^* + B \end{aligned}$$

Branching:

$$B \rightarrow 2R^*$$

Termination:

$$\begin{aligned} R^* &\rightarrow \text{out} \\ 2R^* &\rightarrow \text{out} \end{aligned}$$

Here RH represents hydrocarbon fuel (C_nH_{2m}), R^* is the radical, B is the branching agent, Q is the intermediate agent, and P is the product, consisting of CO, CO_2, and H_2O. Following Benson [8], the branching agent is related to hydroperoxide (RO_2H) at low temperatures and to hydrogen peroxide (H_2O_2) at high temperatures. The intermediate species can be generally related to aldehydes (RCHO).

This model was validated and further developed by a number of researchers [28, 55, 62, 76, 92, 98, 99, 101, 112] and is now the most widely used in automotive applications. It is known as the Shell model.

In [92], the equations of the Shell model were reexamined with a view to their more effective implementation into a computational fluid dynamics code. A simplification of the solution procedure without compromising accuracy was achieved by replacing time as an independent variable with the fuel depletion, which is the difference between the initial fuel concentration and the current one. All other variables used in this model, including temperature, concentration of oxygen, radicals, intermediate and branching agents, were expressed as functions of fuel depletion. The equations for the temperature and concentration of the intermediate agent were of first order and allowed analytical solutions. The concentrations of oxygen and fuel were related via an algebraic equation, which was solved in a straightforward way. In this case, the numerical solution of five coupled first-order ordinary differential equations was reduced to the solution of only two coupled first-order differential equations for the concentration of radicals and branching agent. It was possible to rearrange these equations even further so that the equation for the concentration of the radicals was uncoupled from the equation for the branching agent. In this case, the equation for the concentration of radicals became a second-order ordinary differential equation. This equation was solved analytically in two limiting cases and numerically in the general case. The solution to the first-order ordinary differential equation for the concentration of the branching agent and the solution to the first-order differential equation for time were represented in the form of integrals containing the concentration of the radicals obtained earlier. This approach allowed the central processing unit (CPU) time to be more than halved and made the calculation of the autoignition process using the Shell model considerably more effective.

The applications of the Shell model to modelling autoignition in gasoline and Diesel engines were described in [98]. The complexities of modelling autoignition in Diesel sprays was highlighted. In contrast to autoignition in gasoline engines, autoignition of Diesel fuel sprays takes place at a wide range of equivalence ratios and temperatures. This makes it necessary to impose flammability limits to restrict the range of equivalence ratios in which the autoignition model is active. The autoignition chemical delay for n-dodecane was shown to be much less than the physical delay due to the droplet transit time, atomization, heating, evaporation, and mixing (see the analysis later in this section). This enabled the authors of [98] to use the less accurate than DCM but more computer efficient Shell model for the analysis of Diesel fuel chemical autoignition. Since experimental data for the chemical autoignition delay for n-dodecane were not available, the investigation of the applicability of the Shell model to modelling the Diesel fuel autoignition process was based on data for n-heptane. The ignition time delays for premixed n-heptane predicted by calculations using the kinetic rate parameters corresponding to the primary reference fuel, RON70, showed good agreement with experimental results when A_{f4} (preexponential factor in the rate of production of the intermediate agent) was chosen in the range between 3×10^6 and 6×10^6. It was pointed out that the difference between the end-of-compression temperature as predicted by the adiabatic law and the actual end-of-compression temperature, taking into account the exothermic reactions at the end of compression, needs to be accounted for. The relation between the two temperatures was approximated by a linear function. It was assumed that this approach can be extended to n-dodecane.

Based on the analysis of the autoignition in a monodisperse spray in conditions relevant to Diesel engines, it was shown that in the case of droplets with an initial radius (R_{d0}) about or greater than 6 μm, the physical ignition delay (due to droplet heating and evaporation) dominates over the chemical ignition delay (time required for chemical reactions), while for the smaller droplets with $R_{d0} \leq 2.5\,\mu$m, the opposite is true [99]. The start of the autoignition process was predicted near the periphery of both monodisperse and polydisperse sprays, in agreement with current understanding of this phenomenon (a more rigorous approach to the analysis of evaporating polydisperse sprays using the direct quadrature-based sectional method of moments is described in [51]). The autoignition stage of the polydisperse Diesel combustion predicted by the Shell autoignition model in combination with conventional models for droplet heating and evaporation (ignoring the temperature gradient inside droplets) was shown to agree with available experimental data for a medium-duty truck Diesel engine [99]. Since most of the droplets in Diesel engines have radii about or greater than 6 μm, accurate modelling of droplet heating and evaporation seems to be more important than accurate modelling of the chemical autoignition process. This provides additional support for the application of the Shell model for modelling the latter process when compared with more accurate, but more complicated models, such as DCM. Also, this provides a stimulus for the development of more accurate models for droplet heating and evaporation. Note that the chemical ignition delay is strongly influenced by the droplet surface temperature, which is largely controlled by the selected model of droplet heating and evaporation (see Chaps. 4 and 5).

Analysis of the problem of combustion of fuel sprays and droplets, following the autoignition process, is beyond the scope of this monograph (see [25, 32, 33, 36, 64, 88, 117, 120, 121] and the references therein). Also, the analysis of numerous papers focused on experimental studies of spray autoignition and combustion in internal combustion engines is beyond the scope of this monograph (e.g., [3]).

As mentioned earlier, taking into account finite thermal conductivity in droplets is expected to increase noticeably the accuracy of modelling of the autoignition process in Diesel engines [9]. The reasons for this effect were investigated more closely in [96]. In that paper, the effect of the temperature gradient inside fuel droplets on droplet evaporation, breakup, and the ignition of fuel vapour/air mixture was investigated based on a zero-dimensional code. This code took into account the coupling between the liquid and gas phases and described the autoignition process based on the Shell model. The effect of temperature gradient inside droplets was investigated by comparing the predictions of the Effective Thermal Conductivity (ETC) model and the Infinite Thermal Conductivity (ITC) model, both of which were implemented in this code. It was pointed out that in the absence of breakup, the influence of the temperature gradient in droplets on droplet evaporation in a realistic Diesel engine environment is generally small (a few percent). In the presence of the breakup process, however, the temperature gradient inside the droplets can lead to a significant decrease in evaporation time. This was attributed to the fact that the effect of temperature gradient inside droplets leads to a substantial increase in droplet surface temperature at the initial stages of its heating. This is translated into a decrease in surface tension and the threshold radii of the unstable droplets. Even in the absence of break-up, the effect of the temperature gradient inside the droplets was shown to lead to a noticeable decrease in the total autoignition delay. In the presence of breakup, this effect is enhanced substantially, leading to more than halving of the total ignition delay. It was recommended that the effect of the temperature gradient inside droplets be taken into account in CFD codes describing droplet breakup and evaporation processes and the ignition of the evaporated fuel/air mixture.

The problem of modelling the autoignition in sprays is different from that focused on the interaction between sprays and developed flames. The latter problem has been considered in a number of papers, including [57, 118].

The results reported in [9, 96] are based on the coupled solutions of ordinary differential equations for droplets and chemical reactions. A more detailed review of the mathematical aspects of these coupled solutions is presented in the next section.

7.2 Coupled Solution: A Simplified Model

In this section, the interaction between droplet heating, evaporation, and the autoignition of the fuel vapour/air mixture is illustrated using a rather simplistic model, following [93, 97]. In what follows, a brief summary of its physical and mathematical formulations are given.

7.2.1 Physical Model

Spray is assumed to be monodisperse, and the effects of droplet motion are ignored. The first assumption can be justified by the fact that a well-defined maximum in droplet distribution by radii is usually observed, although in this case, some important effects linked with the polidispersity of realistic sprays are bound to be overlooked [45, 46]. The second assumption is not justified near the nozzle, but it is widely accepted in the region away from the nozzle where most droplets are entrained by the surrounding air [94]. The evaporation process is described by the hydrodynamic model based on the assumption that the fuel vapour near the droplet surfaces is always saturated [90]. The autoignition process is described in terms of a one-step Arrhenius reaction (see Eq. (7.1)). The radiative heat exchange between fuel droplets and ambient air is taken into account, assuming that droplets are grey opaque spheres and the radiative temperature is equal to the gas temperature (gas is optically thick). This approximation is applicable for relatively large droplets. In the general case, the effects of semitransparency of droplets would need to be taken into account [90]. Spalding mass and heat transfer numbers are assumed to be well below 1.

The temperature-dependence of thermal conductivity, diffusivity, and the mass transfer coefficient are taken into account as in [39].

The droplet temperatures are assumed to be equal and constant, and the temperature gradients inside droplets are ignored. This model is known as the first type of droplet-vaporization model, according to Sirignano's classification [103]. It yields the well-known d^2 law, whereby the droplet diameter squared decreases linearly with time. More advanced models of droplet heating and evaporation were discussed in Chap. 4.

We appreciate that although the above-mentioned restrictions of the model still allow us to provide a qualitatively correct description of the process of spray autoignition, many important details of this process become overlooked. This means that this approach cannot replace the conventional numerical analysis of this process (e.g., [90]), but it can effectively complement it.

The mathematical formulation of the physical model described above is presented in the next section.

7.2.2 Mathematical Formulation

Using the assumptions described in Sect. 7.2.1, the energy balance equation for the gas (air) is written as [93, 97]:

$$\rho_g c_{pg} \alpha_g \frac{\mathrm{d}T_g}{\mathrm{d}t} = \eta c'_{ff} M_f \alpha_g Q_f A \exp\left(-\frac{E}{R_u T_g}\right) - 4\pi R_d^2 n h \left(T_g - T_{s0}\right) - 4\pi R_d^2 n \sigma \varepsilon_d \left(T_g^4 - T_{s0}^4\right), \qquad (7.2)$$

where ρ_g is the gas density, c_{pg} the specific heat capacity of gas, α_g the nondimensional volumetric gas content, $\eta = c'_f/c'_{ff}$ the mole fraction of the combustible gas, c'_{ff} the overall fuel molar concentration (vapour molar concentration when all droplets are evaporated and no combustion processes are involved), c'_f the fuel molar concentration far from droplets, M_f the molar mass of fuel ($\rho_f = M_f c_f$), Q_f the specific heat of the reaction, A a constant preexponential rate factor (in s^{-1}), E the activation energy, R_u the universal gas constant, R_d the droplet radius, h the convection heat transfer coefficient, n the droplet number density, ε_d the emissivity of the droplet surfaces, σ the Stefan–Boltzmann constant, T_{s0} the temperature of droplet surfaces (assumed to be constant).

The choice of the normalizing factor c'_{ff} ensures that $\eta \leq 1$. The value of A is adjusted to give agreement between the reaction rate predicted by the one-step reaction and by a more realistic Shell autoignition model [93]. The equation for the radius of a droplet takes the form

$$\frac{\mathrm{d}R_{\mathrm{d}}}{\mathrm{d}t} = -\frac{h_m M_f (c'_{fs} - c'_f)}{\rho_l}, \tag{7.3}$$

where ρ_l is the density of the liquid fuel, c'_{fs} the saturated vapour molar concentration, and h_m the mass transfer coefficient.

The equation for the concentration of combustible gas is described as

$$M_f c'_{ff} \alpha_g \frac{\mathrm{d}\eta}{\mathrm{d}t} = -M_f c'_{ff} \eta \alpha_g A \exp\left(-\frac{E}{R_u T_g}\right) + 4\pi R_d^2 n h_m M_f \left(c'_{fs} - c'_f\right). \tag{7.4}$$

For the typical Diesel engine parameters, $t_d \ll t_g$, where t_d and t_g are the characteristic times of droplet and gas heating, and our assumption that droplet heating can be ignored [93] is justified. If we restrict the analysis to stationary droplets, the expressions for h and h_m become [93]:

$$h = \frac{k_{g0}}{\sqrt{T_{g0}}} \frac{\sqrt{T_g}}{R_d}, \tag{7.5}$$

$$h_m = \frac{D_{i0p}}{R_d} \left(\frac{T_g}{T_{g0}}\right)^{3/2}, \tag{7.6}$$

where $D_{i0p} = D_{i0} p_{\mathrm{atm}}/p$, k_g is the gas thermal conductivity, and D_i the binary diffusion coefficient; the temperature-dependence of the specific heat is ignored, while D_i is proportional to $T_g^{3/2}$ and inversely proportional to p [58]; the suffix $_0$ indicates values at T_{g0}, and p_{atm} is the atmospheric pressure.

Using (7.5) and (7.6), system (7.2)–(7.4) can be rewritten in the dimensionless form [93]

$$\frac{\gamma \frac{\mathrm{d}\theta}{\mathrm{d}\tau}}{1+\beta\theta} = \eta \exp\left(\frac{\theta}{1+\beta\theta}\right) - \varepsilon_1 r_d \left[\theta (1+\beta\theta)^{1/2} + \varepsilon_3 r_d \left((1+\beta\theta)^4 - 1\right)\right], \tag{7.7}$$

$$\frac{\mathrm{d}r_d^3}{\mathrm{d}\tau} = -\varepsilon_1 \varepsilon_2 r_d \left[\theta (1+\beta\theta)^{1/2} + \varepsilon_3 r_d \left((1+\beta\theta)^4 - 1\right)\right], \tag{7.8}$$

$$\frac{\mathrm{d}\eta}{\mathrm{d}\tau} = -\eta \exp\left(\frac{\theta}{1+\beta\theta}\right) + \varepsilon_1 \psi r_d \left[\theta (1+\beta\theta)^{1/2} + \varepsilon_3 r_d \left((1+\beta\theta)^4 - 1\right)\right], \tag{7.9}$$

where

$$\theta = (T_g - T_{s0}) T_{s0}^{-2} E R_u^{-1}, r_d = R_d/R_{d0}, \tau = t/t_r, t_r = A^{-1} \exp(1/\beta),$$
$$\gamma = \rho_{g0} c_{pg} T_{go} \beta / \rho_{ff} Q_f, \beta = R_u T_{s0}/E, \psi = Q_f/L, \rho_{ff} = c'_{ff} M_f,$$
$$\varepsilon_1 = 4\pi R_{d0} n k_{g0} T_{s0}^{3/2} \beta (\rho_{ff} \alpha_g Q_f A \sqrt{T_{g0}})^{-1} \exp(1/\beta), \varepsilon_2 = Q_f \rho_{ff} \alpha_g (\rho_l L \alpha_l)^{-1},$$
$$\varepsilon_3 = T_{s0}^{5/2} T_{g0}^{1/2} R_{d0} \sigma_1 / (k_{g0} \beta), \sigma_1 = \sigma \varepsilon_1, \alpha_l = (4/3)\pi R_d^3 n.$$

The parameters β and γ are conventional parameters in the Semenov theory of thermal explosion [102], and their physical meanings are well known; β is the reduced initial temperature, and γ represents the final dimensionless adiabatic temperature of the thermally insulated system after explosion. Characteristic values of both these parameters are small compared with unity for most gaseous mixtures, due to the high exothermicity of the chemical reaction and the high activation energy. The parameters ε_1, ε_2 describe the interaction between gaseous and liquid phases. The parameter ε_3 describes the impact of thermal radiation and represents the ratio of radiative and convective heat transfer coefficients [39]. The initial conditions for (7.7)–(7.9) are

$$\theta(0) = \theta_0 = \theta(T_{g0}), \quad \eta(0) = \eta_0 = c'_{f0}/c'_{ff}, \quad r_d(0) = 1. \tag{7.10}$$

7.2.3 Analysis

Introducing new variables

$$p = \ln(1 + \beta\theta), \quad q = r_d^3, \quad s = \frac{\psi r^3}{\varepsilon_2} + \eta,$$

following [93], system (7.7)–(7.9) is simplified to

$$\frac{\gamma}{\beta}\frac{dp}{d\tau} = \left(s - \frac{\psi}{\varepsilon_2}q\right)\exp\left(\frac{e^p - 1}{\beta e^p}\right) - \varepsilon_1 q^{1/3} F(p, q), \tag{7.11}$$

$$\frac{ds}{d\tau} = -\left(s - \frac{\psi}{\varepsilon_2}q\right)\exp\left(\frac{e^p - 1}{\beta e^p}\right), \tag{7.12}$$

$$\nu\frac{dq}{d\tau} = -\varepsilon_1 q^{1/3} F(p, q), \tag{7.13}$$

where $\nu = 1/\varepsilon_2$,

$$F(p, q) = \left(\frac{e^p - 1}{\beta}\right) e^{p/2} + \varepsilon_3 q^{1/3}\left(e^{4p} - 1\right).$$

For typical values of parameters for a Diesel engine, the inequalities $\nu \ll \gamma/\beta < 1$ hold ($\gamma/\beta = 0.268$, $\nu = 0.0058$ for $T_{s0} = 600$ K and $\nu = 0.0084$ for $T_{s0} = 300$ K [93, 97]). This allows the differences in the values of ν, γ/β and unity to be taken into account and different scales of characteristic times to be introduced in (7.11)–(7.13).

Using the condition $\nu \ll \gamma/\beta < 1$, the authors of [93] applied conventional integral manifold theory (see [37, 44, 60, 104, 105, 108, 122]) to investigate this system. This theory is essentially focused on the analysis of the following autonomous system of equations:

$$\left.\begin{aligned} \dot{x} &= f(x, y, \varepsilon) \\ \varepsilon\dot{y} &= g(x, y, \varepsilon) \end{aligned}\right\}, \tag{7.14}$$

where $0 < \varepsilon \ll 1$, $x \in \mathscr{R}^m$, $y \in \mathscr{R}^n$, in $\mathscr{R}^{m+n} = \mathscr{R}^m \times \mathscr{R}^n$.

A smooth surface $y = h(x, \varepsilon)$ is a slow invariant manifold of system (7.14) if every trajectory $x = x(t, \varepsilon)$, $y = y(t, \varepsilon)$ of system (7.14) that has at least one common point $x = x_0$, $y = y_0$ with the surface $y = h(x, \varepsilon)$, i.e., $y_0 = h(x_0, \varepsilon)$, lies entirely on this surface, i.e., $y(t, \varepsilon) = h(x(t, \varepsilon), \varepsilon)$. Finding this manifold makes it possible not only to reduce the dimension of the system under consideration, but also to clarify the physical background of the processes described by system (7.14). One of the main restrictions in finding this manifold lies in the requirement that the function $g(x, y, \varepsilon)$ satisfy the Lipschitz condition

$$\|g(x_1, y_1) - g(x_2, y_2)\| \leq L(\|x_1 - x_2\| + \|y_1 - y_2\|), \tag{7.15}$$

where (x_1, y_1), (x_2, y_2) are arbitrary arguments from the domain under consideration and L is a positive constant.

One can easily see that the right-hand side of (7.13) does not satisfy the Lipschitz condition (7.15). Hence, the approach used in [93] was, strictly speaking, not mathematically correct. In [97], this system was analysed using a different approach, based on the Tikhonov theorem (see Appendix L).

Equations (7.11)–(7.12) correspond to (L.1) (in the space $\mathscr{R}^2$), while (7.13) corresponds to (L.2). The boundary layer equation corresponding to (7.13) can be represented as

$$\frac{\mathrm{d}q}{\mathrm{d}\bar{\tau}} = -\varepsilon_1 q^{1/3} F(p, q), \tag{7.16}$$

where $\bar{\tau} = \tau/\nu$, p can be considered a parameter in this equation. The validity of assumption (H1) in Appendix L follows from the fact that the function $q^{1/3}F(p, q)$ is Lipschitzian at $q > 0$.

Since $\nu \ll \gamma/\beta < 1$, we can expect that the variations of p and s are small compared with variations of q, at least until a steady state is reached. Hence, we can assume that $\dot{p} = 0$ and $\dot{s} = 0$, which means that p and s conserve their initial values:

$$p \equiv \ln(1 + \beta\theta_0), \quad s \equiv \nu\psi + \eta_0. \tag{7.17}$$

The validity of these assumptions was justified in [93], based on the numerical integration of (7.11)–(7.13) (see Fig. 2 of [93]).

Equations (7.17) allow us to rewrite (7.16) as

$$\frac{\mathrm{d}q}{\mathrm{d}\bar{\tau}} = -\mathscr{A} q^{1/3} - \mathscr{B} q^{2/3}, \tag{7.18}$$

where

$$\mathscr{A} = \varepsilon_1\theta_0 (1+\beta\theta_0)^{1/2}, \quad \mathscr{B} = \varepsilon_1\varepsilon_3 \left((1+\beta\theta_0)^4 - 1\right), \quad q(0) = 1.$$

The closeness of the solutions to (7.16) (in combination with (7.11)–(7.12)) and (7.18) was proven by direct integration of these equations.

Assumption (H2) follows from the fact that $q = 0$ is the only root of

$$-\mathscr{A} q^{1/3} - \mathscr{B} q^{2/3} = 0 \tag{7.19}$$

for $q \geq 0$. To check assumption (H3), we take function $V(q) = q^2/2$ and show that this is a Lyapunov function of (7.18) for the solution $q = 0$. Indeed, $V(q) > 0$ when $q \neq 0$, $V(q) = 0$ when $q = 0$, and

$$\frac{\mathrm{d}V}{\mathrm{d}\bar{\tau}} = -\mathscr{A} q^{4/3} - \mathscr{B} q^{5/3} < 0, \quad q > 0,$$

which implies that the function $V(q) = q^2/2$ is the Lyapunov function and solution $q = 0$ is asymptotically stable. The domain of attractivity is uniform in X, since the above-mentioned Lyapunov function exists for all $q > 0$. Hence, assumption (H3) is satisfied for $q \geq 0$.

After we substitute the solution $q = \phi(p, s) = 0$ into system (7.11)–(7.13), the latter can be rewritten as

$$\frac{\gamma}{\beta}\frac{\mathrm{d}p}{\mathrm{d}\tau} = s \exp\left(\frac{e^p - 1}{\beta e^p}\right), \tag{7.20}$$

$$\frac{\mathrm{d}s}{\mathrm{d}\tau} = -s \exp\left(\frac{e^p - 1}{\beta e^p}\right). \tag{7.21}$$

System (7.20) and (7.21) corresponds to (L.5). This system is Lipschitzian, and it has a unique solution [27]. Hence, assumption (H4) is satisfied. Assumption (H5) is satisfied, since every point $q_0 > 0$ belongs to the domain of attractivity of the steady state $q = 0$. This means that (L.8) and (L.9) are satisfied for $q > 0$ and $t > 0$, and the solution to (7.18) converges uniformly to $q = 0$ for $0 < t \leq T$. For $t > T$, $q = 0$ is the solution to this equation. Hence, the line $q = 0$ is a positively invariant manifold.

This allows us to perform the order reduction of the original non-Lipschitzian system (7.11)–(7.13) and reduce it to system (7.20) and (7.21) with positively invariant manifold $q = 0$.

The analysis of system (7.20) and (7.21) describing the ignition of the mixture of fuel vapour and air and equation

$$\nu\frac{\mathrm{d}q}{\mathrm{d}\tau} = -\mathscr{A} q^{1/3} - \mathscr{B} q^{2/3} \tag{7.22}$$

for the initial reduction of droplet radius was performed earlier in [93] without correct mathematical justification of the validity of these equations.

As follows from the analysis in [93], the ignition process can be subdivided into two stages: droplet evaporation and ignition of the gaseous mixture. The basic system of equations was eventually reduced to a system of just two equations: one for the normalised gas temperature and another for the normalised droplet radii. These equations were further simplified assuming that the changes of gas temperature are small, which is justified in the initial stages of the combustion process. Results predicted by the approximate analytical solution were compared with the results predicted by the CFD package VECTIS. The analytical solution was shown to underpredict the evaporation time. A considerably better agreement between the evaporation times predicted by VECTIS and the qualitative theory was achieved when the gas temperature was assumed to be equal to the local temperature in the vicinity of the droplets.

The example considered in this section allows us to understand the underlying physics of the process, but its applicability to realistic fuel spray systems is expected to be limited. The main problem with modelling realistic sprays lies in the large number of ordinary differential equations which need to be solved. These equations describe droplet heating and evaporation and chemical reactions in individual computational cells, the number of which can exceed hundreds if not thousands or even millions. Direct numerical solution of these equations does not seem to be feasible. Two possible approaches to their analysis are discussed in the following section.

7.3 Coupled Solution: Dynamic Decomposition

7.3.1 Decomposition Techniques

Decomposition of complex systems describing heating and evaporation of individual droplets (or droplet parcels) in sprays and the autoignition of fuel vapour/air mixture into simpler subsystems is de facto almost universally used in engineering and physics applications. It allows the numerical simulation to focus on the subsystems, thus avoiding substantial difficulties and instabilities related to numerical simulation of the original systems. Special rules are usually introduced to incorporate the results of numerical simulation of the subsystems into the general scheme of the simulation of the whole system. The hierarchy of the decomposition process has been in most cases fixed for the duration of the process.

As an example of such decomposition one can mention the solutions of ordinary and partial differential equations (ODEs and PDEs) describing spray dynamics in computational fluid dynamics (CFD) codes. Numerical spray modelling is traditionally based on the Lagrangian approach coupled with the Eulerian representation of the gas phase. This permits the decomposition of complicated and highly nonlinear systems of PDEs describing interactions between computational cells and the systems of ODEs that govern processes in individual computational cells, including liquid/gas phase exchange and chemical kinetics. The systems of ODEs are usually integrated using much shorter time steps δt (typically 10^{-6} s) than the global time steps used for calculating the gas phase Δt (typically 10^{-4} s). Thus the decomposition of ODEs and PDEs is de facto used, although its basis has not been rigorously investigated to the best of the author's knowledge [99, 113].

Further decomposition of the system of ODEs describing droplet dynamics inside individual computational cells is widely used. The simplest decomposition of this system of ODEs is based on the sequential solution of individual subsystems that this system comprises (split operator approach). In this approach, the solution of each individual subsystem for a given subset of variables is based on the assumption that all the other variables are fixed. The sequence of the solutions of individual subsystems is often chosen rather arbitrarily, and the results sometimes vary substantially depending on the order in which these subsystems are solved. In the case of a multiscale system, the reliability of this approach becomes questionable altogether, as shown later in this section.

To overcome these problems, the multiscale nature of the ODEs needs to be investigated before any attempt to solve them is made. This idea could be prompted by the approaches used in [91] for the analysis of the processes in CO_2 lasers and the one used in [92] for the analysis of equations describing the autoignition of Diesel fuel (the Shell model; see Sect. 7.1). Before solving a system of five stiff ODEs describing five temperatures in these lasers, the characteristic time scales of these equations were analysed [91]. It was shown that two of these equations describe rather rapid relaxation of two temperatures to the third one. This allowed the authors of [91] to replace the solution of five stiff equations by a system of just three nonstiff equations without any significant loss of accuracy. The approach used in [92] was different from that used in [91], but the ultimate result of reduction of the number of ODEs to be solved and elimination of the stiffness of the system of equations remained the same. In mathematical terms, in both papers the dimension of the ODE system was reduced. In other words, the system was decomposed into lower-dimensional subsystems.

A similar system decomposition into lower-dimensional subsystems was used in constructing reduced chemical mechanisms based on Intrinsic Low-Dimensional Manifolds (ILDM) (e.g., [59, 69, 87]) and Computational Singular Perturbation (CSP) (e.g., [52, 66, 70, 81, 114]). There are many similarities between these methods. They are based on a rigorous scale separation such that 'fast' and 'slow' subspaces of the chemical source term are defined and mechanisms of much reduced stiffness are constructed. These approaches, however, were developed with a view of application to modelling chemical kinetics. Their generalisation to general CFD codes has not been considered to the best of my knowledge.

A useful analytical tool for the analysis of stiff systems of ODEs, used for modelling of spray heating, evaporation, and ignition, could be based on the geometrical asymptotic approach to singularly perturbed systems (integral manifold method) as developed by Gol'dshtein and Sobolev [42, 43] for combustion applications (see also [37]). This approach overcomes some of the problems mentioned earlier and is essentially focused on systems of ordinary differential equations of the form (7.14). In practical implementations of the integral manifold method, a number of simplifying assumptions were made. These include the assumption that the slow variable is constant during the fast processes. This assumption opened the way to analytical study of the processes [17, 39, 41, 93] (see also the discussion in Sect. 7.2.3).

These approaches to decomposing systems of ODEs were developed and investigated with a view to application to rather special problems and were based on a number of assumptions. These include fixing the decomposition over the whole period of the process and not allowing its hierarchy to change with time. The underlying philosophy of these approaches, however, seems to be attractive for application to the analysis of a wide range of physical and engineering problems including spray modelling in general computational fluid dynamics (CFD) codes.

In what follows, an alternative approach to decomposition of the system of ODEs, following [18, 19, 21, 100], is described. In contrast to the approaches described above, the new method allows a change of the nature of the decomposition with time (dynamic decomposition).

7.3.2 Description of the Method

As in the original integral manifold method, the formal approach to decomposition of the system of ODEs in the new method is based on the division of system variables into 'fast' and 'slow'. This leads to the division of this system into 'fast' and 'slow' subsystems. In contrast to the original version of the integral manifold method, however, linearised variations of slow variables during the time evolution of the fast variables are taken into account as a first-order approximation to the fast manifold. The usefulness of this division depends on whether the 'fast' subsystem has lower dimension than the 'slow' subsystem. The procedure can be iterative and result in a hierarchical division of the original system. For example, the 'slow' subsystem can, in turn, be subdivided into 'slow' and 'very slow' subsystems.

The procedure suggested in [21] was initially focused on the simplest possible subdivision of the original system into two subsystems and applied to spray combustion modelling. Note that 'fast'–'slow' decomposition in this case was different for different phase space regions [39, 93] and for different time intervals. A wider range of application of this method is anticipated. The details of this method are described in the rest of this section, following [21].

Let us consider a system whose state is characterised by n dimensionless parameters, denoted by Z_n $(n = 1, 2, \ldots, n)$. The value of each of these parameters for a given place in space depends on time t, i.e., $Z_n = Z_n(t)$. This dependence can be found from the solution to the system of n equations, which can be represented in vector form as

$$\frac{\mathrm{d}\mathbf{Z}}{\mathrm{d}t} = \boldsymbol{\Phi}(\mathbf{Z}), \tag{7.23}$$

where

$$\mathbf{Z} = (Z_1, Z_2, \ldots, Z_n), \qquad \boldsymbol{\Phi} = (\Phi_1, \Phi_2, \ldots, \Phi_n).$$

In the general case, a rigorous coupled numerical solution could be found. This may be impractical when too many equations are involved. We believe that a more efficient approach to this problem could be based on reducing the number of dimensions of this system as discussed in Sect. 7.3.1. This could be based on organising (7.23) in terms of a decreasing parameter Y_i defined as

$$Y_i = \left| \frac{\Phi_i(t_k)}{Z_i(t_k)} \right|, \tag{7.24}$$

where $\Phi_i(t_k) \equiv \Phi_{ik} \equiv \Phi_i(Z_1(t_k), Z_2(t_k), \ldots Z_i(t_k), \ldots Z_n(t_k))$ and $Z_i(t_k) \equiv Z_{ik}$ are the right-hand sides of (7.23), and the values of Z_i are taken at the time t_k for the time step $\Delta t : t_k \rightarrow t_{k+1}$, $i = 1, 2, \ldots n$. For Z_{ik} close to zero, we have a special case which requires additional investigation. In practice, we can assume in most cases that Z_{ik} is a large number.

Note that the Taylor expansion of the right-hand side of (7.23) gives

$$\boldsymbol{\Phi}(\mathbf{Z}(t_{k+1})) \equiv \boldsymbol{\Phi}(\mathbf{Z}_{k+1}) = \boldsymbol{\Phi}(\mathbf{Z}_k) + \mathbf{D}\boldsymbol{\Phi}|_{\mathbf{Z}=\mathbf{Z}_k} (\mathbf{Z}_{k+1} - \mathbf{Z}_k) + o(\mathbf{Z}_{k+1} - \mathbf{Z}_k), \tag{7.25}$$

where $\mathbf{D}\boldsymbol{\Phi}$ is the Jacobian matrix of the vector field $\boldsymbol{\Phi}$.

In the general case, the value of $\boldsymbol{\Phi}(\mathbf{Z}_{k+1})$ is controlled mainly by $\boldsymbol{\Phi}(\mathbf{Z}_k)$. However, in the special case $\boldsymbol{\Phi}(\mathbf{Z}_k) = 0$, the second term on the right-hand side of (7.25) becomes dominant. Only in this or similar cases, and under certain conditions, can (7.23) be simplified to

$$\frac{\mathrm{d}Z_i(t_k)}{\mathrm{d}t} = \lambda_i(t_k) Z_i(t_k), \tag{7.26}$$

and the values of Y_i coincide with λ_i. The analysis of [18, 19, 21, 100], however, was focused not on this special case, but on the general case in which Y_i are not directly related to λ_i.

If Y_i is greater than a certain positive number $\alpha < 1$ chosen a priori, then the corresponding equation can be considered fast, and this should be solved rigorously. If the number of 'fast' equations is $f \neq 0$, then the system is said to be multiscale, and this procedure effectively reduces the dimension of the system (7.23) from n to f. This dimensional reduction is particularly attractive when f is small (1 or 2).

In an alternative approach, which turned out to be more useful than the previous one, we reorganise Y_i in descending order as

$$Y_{i_1} \geq Y_{i_2} \geq \cdots \geq Y_{i_j} \geq \cdots \geq Y_{i_n}. \tag{7.27}$$

If we are able to find $j = f$ such that

$$\frac{Y_{i_{f+1}}}{Y_{i_f}} < \epsilon, \tag{7.28}$$

where ϵ is another small parameter chosen a priori, then we can conclude that the system can be decomposed locally ($\Delta t : t_k \to t_{k+1}$) into two subsystems: 'fast' and 'slow'. Note that the subscript of i_j just indicates the order in which the parameters are organised (no summation over f on the right-hand side of (7.28) takes place).

The equations for these subsystems were represented in vector form as

$$\frac{d\mathbf{U}}{dt} = \boldsymbol{\Phi}_f \begin{pmatrix} \mathbf{U} \\ \mathbf{V} \end{pmatrix}, \tag{7.29}$$

$$\frac{d\mathbf{V}}{dt} = \boldsymbol{\Phi}_s \begin{pmatrix} \mathbf{U} \\ \mathbf{V} \end{pmatrix}, \tag{7.30}$$

where

$$\boldsymbol{\Phi}_f = (\Phi_{i_1}, \ldots, \Phi_{i_f}), \quad \boldsymbol{\Phi}_s = (\Phi_{i_{f+1}}, \ldots, \Phi_{i_n}),$$
$$\mathbf{U} = (Z_{i_1}, \ldots, Z_{i_f}), \quad \mathbf{V} = (Z_{i_{f+1}}, \ldots, Z_{i_n}).$$

The transformation from the original order of variables

$$\mathbf{Z} = (Z_1, Z_2, \ldots, Z_n)$$

to the new order of the same variables

$$\mathbf{Z}' = (Z_{i_1}, \ldots, Z_{i_f}, Z_{i_{f+1}}, \ldots, Z_{i_n})$$

was performed with the help of the transformation matrix $\mathbf{Q} = Q_{i,i_j}$, where $Q_{i,i_j} = 1$ with i is the original position of the variable, i_j is the final position of the variable, and $Q_{i,i_j} = 0$ for all other i and i_j. In this case, $\mathbf{Z}$ was represented as

$$\mathbf{Z} = \mathbf{Q}(\mathbf{Z}_0)\mathbf{Z}' = \mathbf{Q}(\mathbf{Z}_0) \begin{pmatrix} \mathbf{U} \\ \mathbf{V} \end{pmatrix} = \begin{pmatrix} \mathbf{Q}_f(\mathbf{Z}_0) & \mathbf{Q}_s(\mathbf{Z}_0) \end{pmatrix} \begin{pmatrix} \mathbf{U} \\ \mathbf{V} \end{pmatrix}, \tag{7.31}$$

where $\mathbf{Q}$ was calculated for the values of $\mathbf{Z}$ at the initial time t_0 or the beginning of the time step ($\mathbf{Z}_0$). The first f columns of the matrix $\mathbf{Q}$ refer to the fast subsystem, while the remaining $n - f$ columns refer to the slow subsystem. This is indicated by introducing the additional matrices $\mathbf{Q}_f$ and $\mathbf{Q}_s$.

Note that so far, the simplest form of matrices $\mathbf{Q}$ and $\mathbf{Q}^{-1}$ performing the change of the order of variables was considered. More complex forms of these matrices could potentially perform the decomposition of the originally nonmultiscale system into a multiscale one. Analysis of the latter decomposition, however, was beyond the scope of the analysis of [18, 19, 21, 100].

With a new small positive parameter $\varepsilon \ll 1$ having been introduced, and with the definitions of $\mathbf{Q}_f$ and $\mathbf{Q}_s$ in mind, the system of Eqs. (7.29) and (7.30) was rewritten in a form similar to that used in the Integral Manifold Method:

$$\varepsilon\frac{d\mathbf{U}}{dt} = \varepsilon\mathbf{Q}_f^{-1}(\mathbf{Z}_0)\boldsymbol{\Phi}\left(\mathbf{Q}(\mathbf{Z}_0)\begin{pmatrix}\mathbf{U}\\ \mathbf{V}\end{pmatrix}\right) \equiv \boldsymbol{\Phi}_{f\varepsilon}\begin{pmatrix}\mathbf{U}\\ \mathbf{V}\end{pmatrix}, \tag{7.32}$$

$$\frac{d\mathbf{V}}{dt} = \mathbf{Q}_s^{-1}(\mathbf{Z}_0)\boldsymbol{\Phi}\left(\mathbf{Q}(\mathbf{Z}_0)\begin{pmatrix}\mathbf{U}\\ \mathbf{V}\end{pmatrix}\right) \equiv \boldsymbol{\Phi}_s\begin{pmatrix}\mathbf{U}\\ \mathbf{V}\end{pmatrix}, \tag{7.33}$$

where $\boldsymbol{\Phi}_{f\varepsilon} = \varepsilon\boldsymbol{\Phi}_f$. In this representation, the right-hand sides of (7.32) and (7.33) are expected to be of the same order of magnitude over the specified period (time step).

Equations (7.32) and (7.33) were integrated over the time period $\Delta t : t_k \to t_{k+1}$. The zeroth-order solution to (7.33) is just a constant value of the slow variable: $\mathbf{V}_{k+1}^{C} = \mathbf{V}_k = (Z_{i_{f+1}k}, \ldots, Z_{i_n k})$, where the superscript ^{0} indicates the zeroth approximation, while the subscripts $_k$ and $_{k+1}$ indicate the point in time. The zeroth order for the fast variable was found from (7.32) with $\mathbf{V} = \mathbf{V}_k$. This was interpreted as the equation for the slow variable on the fast manifold. Thus, (7.32) (or (7.29)) was approximated by the following system:

$$\frac{d\mathbf{U}}{dt} = \boldsymbol{\Phi}_f\begin{pmatrix}\mathbf{U}\\ \mathbf{V_k}\end{pmatrix}. \tag{7.34}$$

The solution to (7.34) at $t = t_{k+1}$ ($\mathbf{U}_{k+1}^0$) is the zeroth-order approximation of the fast motion on the fast manifold at $t = t_k$. Note that the system of equations (7.34) can be stiff in the general case, but with a reduced level of stiffness, compared with the original system (7.23). Hence, the suggested method is expected to reduce the level of stiffness of the system and not to eliminate the stiffness altogether.

Under the same zeroth-order approximation, the slow variables would remain constant over the same time step. This assumption was used in the original formulation of the Method of Integral Manifolds [43]. This, however, might lead to an unphysical result in which slow variables would remain constant for every time $t > t_0$, whence the need to calculate slow variables using at least a first-order approximation. In the case in which ε is not asymptotically small, further, or higher-order, approximations need to be considered. In this case, the new time scale $\tau = 1/\varepsilon$ was introduced, and the slow and fast variables were represented as

$$\left.\begin{array}{l} V(\tau) = V^{(0)} + \varepsilon V^{(1)}(\tau) + \varepsilon^2 V^{(2)}(\tau) + \cdots \\ U(\tau) = U^{(0)}(\tau) + \varepsilon U^{(1)}(\tau) + \varepsilon^2 U^{(2)}(\tau) + \cdots \end{array}\right\}. \tag{7.35}$$

Substituting expressions (7.35) into (7.33) yielded

$$\frac{\mathrm{d}(V^{(0)} + \varepsilon V^{(1)}(\tau) + \varepsilon^2 V^{(2)}(\tau) + \cdots)}{\mathrm{d}\tau} = \varepsilon \boldsymbol{\Phi}_s \begin{pmatrix} U^{(0)}(\tau) + \varepsilon U^{(1)}(\tau) + \varepsilon^2 U^{(2)}(\tau) + \cdots \\ V^{(0)} + \varepsilon V^{(1)}(\tau) + \varepsilon^2 V^{(2)}(\tau) + \cdots \end{pmatrix}. \tag{7.36}$$

Equation (7.36) allowed the authors of [18, 19, 21, 100] to obtain the first-order solution for the slow variable in the form

$$V_{k+1} = V_{k+1}^{(0)} + \varepsilon V_{k+1}^{(1)} = V_k^{(0)} + \varepsilon \boldsymbol{\Phi}_s \begin{pmatrix} U_{k+1}^{(0)} \\ V_k^{(0)} \end{pmatrix} \Delta\tau. \tag{7.37}$$

Returning to the original variables, the expression for the value of $\mathbf{V}(t_{k+1}) \equiv \mathbf{V}_{k+1}$ was rewritten in the form

$$\begin{aligned} \mathbf{V}_{k+1} &= \mathbf{V}_k^{(0)} + \boldsymbol{\Phi}_s \begin{pmatrix} \mathbf{U}_{k+1}^{(0)} \\ \mathbf{V}_k^{(0)} \end{pmatrix} \Delta t \\ &= Z_{i_{f+1}k} + \Phi_{i_{f+1}} \left(\mathbf{U}_{k+1}^{(0)}, \mathbf{V}_k^{(0)}\right) \Delta t, \ldots Z_{i_n k} + \Phi_{i_{f+1}} \left(\mathbf{U}_{k+1}^{(0)}, \mathbf{V}_k^{(0)}\right) \Delta t. \end{aligned} \tag{7.38}$$

To increase the accuracy of the calculations, one could continue the process to take into account the first-order solution for the fast motion. Then the second-order solution for the slow motion could be obtained, etc.

Since the above-mentioned decomposition of the original system of equations is allowed to vary with time, it was suggested to call this the dynamic decomposition approach [21].

7.3.3 Application of the Method

In this section, the method described above is applied to a specific problem of modelling heating, evaporation, and combustion of an array of droplets, following [21]. A number of processes, including droplet dynamics, breakup, and coalescence, and the effects of temperature gradient inside droplets are ignored (see [90] and Chaps. 3 and 4). This is justified by the fact that the main emphasis of this section is on the investigation of a new method for the solution of systems of ODEs relevant to spray combustion modelling rather than providing a detailed analysis of the processes involved (see [21]).

7.3.3.1 Equations and Approximations

Equation (4.18) was used to describe the evolution of individual droplets, assuming, following [67], that $\varphi = 1$ (which implies that $B_T = B_M$). Also following [67], the expression for the partial fuel pressure at the surface of the droplet p_{fs} in the form

$$p_{fs} = \exp\left[a_f - \frac{b_f}{T_s - 43}\right] \tag{7.39}$$

was used, where a_f and b_f are constants to be specified for specific fuels, T_s is the surface temperature of fuel droplets in K; p_{fs} predicted by (7.39) is in kPa [21]. In the calculations, the values $a_f = 15.5274$ and $b_f = 5383.59$ recommended for Diesel fuel [67] were used.

Following [1, 34, 95], the following equation for stationary droplet temperature was used:

$$m_d c_l \frac{\mathrm{d}T_d}{\mathrm{d}t} = 4\pi R_d^2 h(T_g - T_d) - \dot{m}_d L + 4\pi R_d^2 \sigma \overline{Q}_a \theta_R^4, \tag{7.40}$$

where h was obtained from the equation

$$\mathrm{Nu} = \frac{2hR_d}{k_g} = 2\frac{\ln(1 + B_T)}{B_T}, \tag{7.41}$$

θ_R is the radiative temperature, as calculated from the P-1 model (it was assumed that $\theta_R = T_{\mathrm{ext}}$ (external temperature)), $\overline{Q}_a$ is the average absorption efficiency factor, calculated from the equation

$$\overline{Q}_a = a_r\, R_d^{b_r}, \tag{7.42}$$

a_r and b_r are polynomials in θ_R, c_l is the liquid specific heat capacity.

Equation (7.40) is a generalisation of (3.34) to take into account the effects of droplet evaporation and radiative heating (see Sect. 3.2.3).

To simplify the problem further, the effect of thermal radiation was ignored, and the following simplified form of the temperature-dependence of L was used [67]:

$$L = L_{T_{\mathrm{bn}}} \left(\frac{T_{\mathrm{c}} - T_s}{T_{\mathrm{c}} - T_{\mathrm{bn}}}\right),$$

where $L_{T_{\mathrm{bn}}}$ is the value of L at the droplet boiling temperature T_{bn}, T_{c} is the critical temperature. Following [67], the following values of parameters were used: $T_{\mathrm{bn}} = 536.4$ K, $T_{\mathrm{c}} = 725.9$ K, and $L_{T_{\mathrm{bn}}} = 254{,}000$ J/kg. Also, it was assumed that liquid density is constant.

The equations written for individual droplets were applied to describe droplet parcels, following the conventional approach widely used in computational fluid dynamics codes.

The equation for fuel vapour density follows directly from the conservation of fuel vapour:

$$\alpha_g \frac{\mathrm{d}\rho_{fv}}{\mathrm{d}t} = -\alpha_g \mathrm{CT} + \left[\sum_i \dot{m}_{di} \Big/ \mathrm{V} \right], \tag{7.43}$$

where ρ_{fv} is the fuel vapour density, α_g is the volume fraction of gas assumed equal to 1, the summation is over all droplets in the volume V, CT is the chemical term describing fuel depletion (in $\mathrm{kg/(m^3 s)}$).

As in [119], the following expression of the rate of reaction was used:

$$k_{\mathrm{cr}} = A[\mathrm{fuel}]^a [\mathrm{O_2}]^b \exp\left[-E/(BT)\right], \tag{7.44}$$

where k_{cr} has units of mol/(cm^3 s), while the concentrations of fuel [fuel] and oxygen $[O_2]$ have units of mol/cm^3. The values of the coefficients for $C_{10}H_{22}$ were used (these are the closest to n-dodecane ($C_{12}H_{26}$)—the closest approximation for Diesel fuel) [119]:

$$A = 3.8 \times 10^{11} \ \frac{1}{\mathrm{s}} \left(\frac{\mathrm{mol}}{\mathrm{cm}^3}\right)^{1-a-b} = 2.137 \times 10^9 \ \frac{1}{\mathrm{s}} \left(\frac{\mathrm{kmol}}{\mathrm{m}^3}\right)^{1-a-b};$$

$$E = 30 \ \frac{\mathrm{kcal}}{\mathrm{mol}} = 1.255 \times 10^8 \ \frac{\mathrm{J}}{\mathrm{kmol}}; \qquad a = 0.25; \quad b = 1.5.$$

Using A in $\frac{1}{\mathrm{s}} \left(\frac{\mathrm{kmol}}{\mathrm{m}^3}\right)^{1-a-b}$ and E in $\frac{\mathrm{J}}{\mathrm{kmol}}$, it was written

$$\mathrm{CT} = A \ M_{O_2}^{-1.5} \ M_f^{0.75} \rho_{fv}^{0.25} \ \rho_{O_2}^{1.5} \exp\left[-E/(BT)\right], \tag{7.45}$$

where $M_{O_2} = 32\,\mathrm{kg/kmol}$, and $M_f = 170\,\mathrm{kg/kmol}$ are the molar masses of oxygen and fuel respectively in kg/kmol, ρ_{O_2} is the density of oxygen.

The single-step global reaction for n-dodecane combustion was written as

$$C_{12}H_{26} + 18.5O_2 \Longrightarrow 12\,CO_2 + 13H_20,$$

and the equation for density of oxygen was represented as

$$\frac{\mathrm{d}\rho_{O_2}}{\mathrm{d}t} = -18.5 \frac{M_{O_2}}{M_f} \mathrm{CT} = -3.48235 \ \ \mathrm{CT}. \tag{7.46}$$

The equivalence ratio, defined as

$$\varphi_R = \frac{\mathrm{Fuel/Air}}{(\mathrm{Fuel/Air})_{\mathrm{stoich}}} = \frac{\mathrm{Fuel/Oxygen}}{(\mathrm{Fuel/Oxygen})_{\mathrm{stoich}}} = \frac{18.5 \times 32}{170} \frac{\rho_{fv}}{\rho_{O_2}} = \frac{3.48\rho_{fv}}{\rho_{O_2}},$$

where $(\mathrm{Fuel}/\mathrm{Air})_{\mathrm{stoich}}$, is the stoichiometric ratio of the densities of fuel and air, was specified.

The condition of energy balance leads to the following equation for gas temperature:

$$c_{\mathrm{mix}}\rho_{\mathrm{mix}}\frac{\mathrm{d}T_g}{\mathrm{d}t} = \alpha_g Q_f \mathrm{CT} - \left[\sum_i m_{di} c_l \frac{\mathrm{d}T_{di}}{\mathrm{d}t} + \sum_i \dot{m}_{di} L + \sum_i \dot{m}_{di} c_{pF}(T_g - T_{di})\right] \Big/ \mathrm{V}, \tag{7.47}$$

where Q_f is the heat released per unit mass of burnt fuel vapour (in J/kg).

7.3.3.2 Application

Among various applications of the method described in Sect. 7.3.2 (e.g., [4]), the case described in [21] is considered in this section. In this case, the analysis was focused on polydisperse spray heating, evaporation, and ignition, using equations and approximations described above. Despite its simplicity, the model described above is capable of capturing the essential features of these processes. Three droplets with initial radii 5, 9, and 13 μm were considered. The initial temperatures of all droplets was taken to be equal to 400 K. The gas temperature was taken to be equal to 880 K [99]. The gas volume was chosen such that if the droplets are fully evaporated without combustion, then the equivalence ratio of fuel vapour/air mixture was assumed to be equal to 4. The initial density of oxygen was taken to be equal to 2.73 kg/m^3 (this corresponds to air pressure equal to 3 MPa). The initial mass fraction of fuel vapour was taken equal to zero. These values of the parameters can be considered an approximation of the actual conditions in Diesel engines [99].

The calculations were based on the equations described above and the equation for the density of the mixture ρ_{mix} of fuel vapour and air:

$$\frac{\mathrm{d}\rho_{\mathrm{mix}}}{\mathrm{d}t} = \sum_i \frac{\mathrm{d}\rho_i}{\mathrm{d}t}, \tag{7.48}$$

where ρ_i are the densities of the individual components.

The calculations were performed until the gas temperature reached 1100 K. At this temperature, the autoignition process was assumed to be complete [99]. Since the equations for droplet mass and temperature were solved for three droplets, the maximal number of equations to be solved was 10. Note that the density of the fuel vapour/air mixture could be derived algebraically from mass conservation. It was preferred, however, to solve the ODE so that the mass conservation in the system could be monitored as a validity check (see [21]).

Once the smaller droplets had evaporated, the number of equations was reduced. These coupled equations were solved using three approaches.

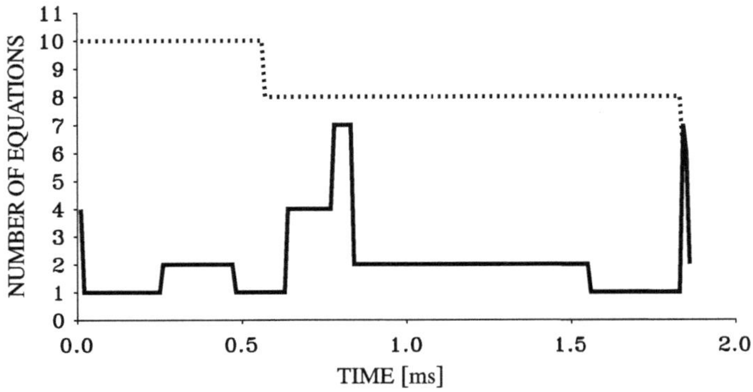

Fig. 7.1 Plots of the total number of equations solved (*dashed*) and the number of equations for fast variables (*solid*) for the values of parameters as specified in the text. The calculations continued until the autoignition process took place. Reprinted from Ref. [21]. Copyright Elsevier (2007)

Firstly, following widely used practice in CFD codes, the system of equations was divided into subsystems which were solved sequentially. This approach is widely referred to as the operator splitting technique (see [107]). These subsystems include equations for mass and temperature of each droplet (three subsystems) and equations for gas temperature, fuel, oxygen, and mixture density (additional fourth subsystem). When each of these subsystems was solved, the remaining variables were assumed to have constant values over the time step. This approach is referred to as the 'fixed decomposition approach' to distinguish it from the 'dynamic decomposition approach' discussed in Sect. 7.3.2. More specifically, at the first step, the equations for the mass and the temperature of each droplet were solved simultaneously (three systems of equations). Then the results were used to calculate the density of fuel and mixture and the temperature of gas without taking into account the chemical term (see (7.43), (7.48), and (7.47)). Next, the chemical term was calculated based on (7.45), and the result was used to calculate the density of oxygen (7.46). Finally, the values of the density of fuel and mixture and temperature of gas were updated using the chemical term. This approach is effectively equivalent to the simplest form of (A–B) splitting as described in [107].

Secondly, these equations were solved rigorously using the DLSODAR stiff solver from ODEPACK developed in the LLNL laboratory. This means that all equations were solved simultaneously in a coupled way.

The third approach was based on decomposing the original system following the procedure described in Sect. 7.3.2 with $\varepsilon = 0.25$. Note that this parameter is not related to the parameter ε used in (7.14). The total number of equations solved and the number of equations for fast variables could change with time as expected. The corresponding plots of the numbers of these equations are shown in Fig. 7.1. As follows from this figure, initially all 10 equations were solved, when the first or second approach was used. Then this number was reduced to 8 when the smallest droplet evaporated, and to 6 when the two smallest droplets evaporated. Initially, the

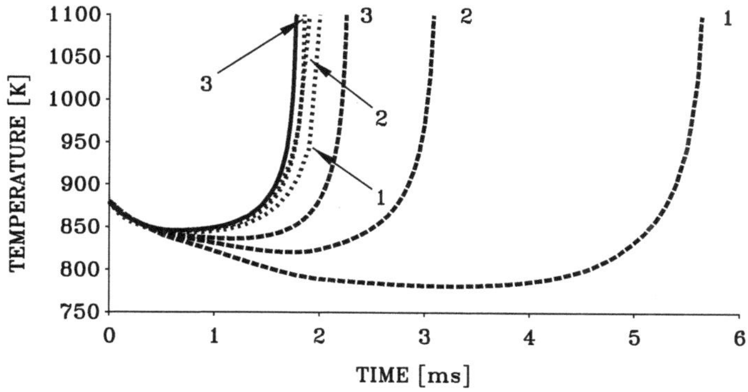

Fig. 7.2 Plots of gas temperature versus time, calculated using the first approach (fixed decomposition) (*dashed*), second approach (coupled solution of the full system of equations) (*solid*), and the third approach (dynamic decomposition) (*dotted*). Plots '1', '2', and '3' refer to calculations based on the time steps 10^{-4} s, 5×10^{-5} s, and 10^{-5} s respectively. The gas volume is chosen such that if the droplets are fully evaporated without combustion, then the equivalence ratio of fuel vapour/air mixture is equal to 4. Reprinted from Ref. [21]. Copyright Elsevier (2007)

number of equations for fast variables to be solved was equal to 4, then it dropped to just one equation describing fuel vapour density. Between about 0.25 and 0.5 ms, the number of equations for fast variables was equal to two (equations for fuel vapour density and the radius of the smallest droplet). Then again just the equation for fuel density was solved. Between about 0.6 and 0.8 ms, the number of fast equations to be solved is comparable with the total number of equations solved. During this period, decomposition of the system is not expected to be useful. After about 0.8 ms and until about 1.8 ms, only one equation (fuel density) or two equations (fuel density and the radius of the second droplet) were used. In this case, the decomposition technique described in Sect. 7.3.2 is expected to be particularly important.

The time evolution of gas temperature, calculated using the above-mentioned three approaches, is shown in Fig. 7.2 [21]. This figure indicates the total ignition delay time, i.e., the time required for the gas to reach 1,100 K. As follows from Fig. 7.2, the first approach appears to be very sensitive towards the time step chosen. If the time step 10^{-4} s is chosen, then the predicted total ignition delay is almost four times as great as that predicted based on the second approach (coupled solution of the whole system). If the time step is decreased to 5×10^{-5} and 10^{-5} s, then calculations using the first method appear to be more accurate than when this time step is equal to 10^{-4}, but still the accuracy of computations is hardly acceptable for practical applications. Even for the rather small time step 10^{-5} s, the predicted total ignition delay is more than 20 % greater than predicted by the rigorous coupled solution of this system of equations (second approach).

The application of the third approach to the solution of this system gives a rather different picture. Even in the case of the largest time step (10^{-4} s), the error of calculations of the total ignition time delay was just 13 %. In the case of smaller time steps, the time delay predicted by solving the decomposed system almost coincides with that obtained by rigorously solving the whole system with possible errors not exceeding 2 %. Essentially the same conclusion regarding the benefits of the third approach, based on the dynamic decomposition of the original system of equations, follows from similar figures showing the time evolution of the instantaneous equivalence ratio and the radius of the largest droplets. Also, a similar difference in the values of other variables predicted by the solutions of the system using the above-mentioned three approaches was observed.

Similar conclusions were obtained from the solution of equations for a different set of parameters, typical for the peripheral region of fuel sprays in Diesel engines (see [21]).

To compare the CPU efficiency of the new (dynamic decomposition) and conventional CFD (fixed decomposition) approaches, a series of runs for various time steps was performed for each case. A polydisperse spray including three droplet parcels, 10,000 droplets each, was injected at the start of the calculation. For fixed time steps, the CPU requirements of both approaches were about the same. As shown above, however, the accuracy of the new approach was always greater than that of the conventional approach. Thus a comparison of CPU requirements of both methods for fixed time steps would be misleading. An alternative approach needs to take into account the accuracy of calculations. As followed from our calculations, the autoignition delays predicted by the dynamic decomposition approach coincided with those predicted by the full coupled solution of the system of equations for the time step 10^{-6} s. This value of the autoignition delay was considered the true value for further comparisons. For example, for the second set of parameters, errors less than 1.5 % were achieved for time step 1.3×10^{-5} s for the conventional CFD approach, and for time step 2.4×10^{-5} s for the new approach. When this effect was taken into account, then in all cases under consideration, the CPU time for the new method was always smaller than that for the conventional approach. In some cases, the CPU reduction for the new approach was as high as a factor of 3. The CPU time was estimated based on the customised function DATE_AND_TIME. The standard function GETTIME did not give consistent results for small CPU times [21].

A number of additional tests were conducted to compare the performances of the standard fix decomposition approach (used in CFD codes) and the new dynamic decomposition method. Both programs were run sequentially on two workstations (Silicon Graphics, Intel 64-bit processor) and using two Fortran compilers (Intel—Fortran 95, GNU—Fortran 77) with the additional option of code optimisation. The above-mentioned customised function DATE_AND_TIME was used to estimate the CPU time required for system integration. It was found that for a given time step and for a low level of code optimisation, the CPU time is less for the standard fixed decomposition approach, whereas an increase in the level of optimisation leads to comparable times required for program execution. For example, for the first case for the time step $\Delta t = 5 \times 10^{-5}$, the average time of program execution with the code optimisation level set to O0 (i.e., no code optimization) was approximately equal to 0.12 s for the fixed decomposition approach and 0.21 s for the dynamic decomposition approach. For the optimisation level set to O2 (this option is the default one, it enables optimisations for speed, including global code scheduling, software pipelining, prediction, speculation, etc.), these times become 0.15 and 0.13 s respectively. Therefore, further improvements of realization of the dynamic decomposition method are possible and might be implemented, leading to optimising the realization structure of the code and numerics.

Note that the coupled solution of the system of equations using the stiff solver is always more accurate than the solutions based on dynamic and fixed decompositions, and is usually more CPU efficient. Hence, when the number of equations is relatively small (as in the cases considered in this section), there is no need to develop any decomposition technique in the first place. However, in realistic engineering calculations, when the number of droplet parcels could be in the tens of thousands [99], no stiff solver can cope with the full coupled system of ODEs describing them. The real competition in this case is between fixed and dynamic decomposition approaches as described above [21].

Note that for some problems, there is no marked difference in time scales between the original variables. This is a typical situation in modelling reacting flows due to strong coupling of thermochemical processes. In this case, the decomposition into fast and slow variables may fail. Also, the decomposition in the original coordinates can produce a relatively large subsystem (the number of governing equations of the reacting system can reach or exceed hundreds), and the modelling needs to be performed over very small time steps to capture the fine features of, for example, an autoignition process. For this type of problem, the application of the dynamic decomposition technique might be not practical.

A decomposition technique specifically designed for such systems was developed in [23]. This method represents the realisation of the general framework of Singularly Perturbed Vector Fields developed in [20, 22]. It allows finding the decomposition and approximating it efficiently for system reduction purposes. Then, depending on a particular asymptotic limit, either slow or fast subsystems are integrated according to the standard singular perturbation theory [37, 43]. In this technique, the same decomposition for the whole time interval under consideration, and for a given set of initial conditions, is applied. Hence it was called the Global Quasi-Linearisation (GQL) technique. Although this approach is less accurate than the dynamic decomposition technique, it can potentially be much more efficient for implementation into CFD codes due to the reduced number of dimensions involved. The efficiency of this approach was demonstrated for a relatively simple but typical combustion chemistry model for the carbon monoxide system [16].

The application of the Global Quasi-Linearisation technique to the analysis of cyclohexane/air mixture autoignition is discussed in [24]. That paper is focused on the application of this technique to a specific problem of cyclohexane/air autoignition in a rapid compression machine environment. A simplified autoignition mechanism including 50 species was used in the analysis based on the customised computational fluid dynamics (CFD) package FLUENT and the new GQL method. The results predicted by FLUENT were shown to be very close to the results predicted by the zero-dimensional code SPRINT, developed at the University of Leeds. The application of the GQL method showed that further substantial reduction of the mechanism is possible without significant loss of accuracy. The method was shown to be an efficient tool for identification and approximation of low-dimensional invariant system manifolds. The applicability of this approach to the problem of modelling spray heating, evaporation, and autoignition has not yet been investigated.

A probabilistic model of thermal explosion in polydisperse fuel sprays was suggested and developed in a series of papers by Nave et al. [77–80]. In this approach, the polydispersity was modelled using a probability density function (PDF) corresponding to the initial distribution of fuel droplet size. This approximation of polydisperse spray was believed to be more accurate than the traditional 'parcel' approximation, and it allowed an analytical treatment of the simplified model. Since the system of the governing equations represented a multiscale problem, the method of invariant (integral) manifolds was applied. An explicit expression for the critical condition for thermal explosion limit was derived analytically. Numerical simulations demonstrated a dependence of these thermal explosion conditions on the PDF type. Thus, this approach can be considered a generalisation of the thermal explosion conditions of the classical Semenov theory.

References

1. Abramzon, B., & Sirignano, W. A. (1989). Droplet vaporization model for spray combustion calculations. *International Journal of Heat and Mass Transfer*, *32*, 1605–1618.
2. Aggarwal, S. K. (1998). A review of spray ignition phenomena: present status and future research. *Progress in Energy and Combustion Science*, *24*, 565–600.
3. Aleiferis, P. G., Serras-Pereira, J., van Romunde, Z., Caine, J., & Wirth, M. (2010). Mechanisms of spray formation and combustion from a multi-hole injector with E85 and gasoline. *Combustion and Flame*, *157*, 735–756.
4. Amani, E., & Nobari, M. R. H. (2013). A calibrated evaporation model for the numerical study of evaporation delay in liquid fuel sprays. *International Journal of Heat and Mass Transfer*, *56*, 45–58.
5. Basevich, V. Y. (1990). Chemical kinetics in the combustion processes. In N. P. Cheremisinoff (Ed.), *Handbook of Heat and Mass Transfer, V. 4. Advances in Reactor Design and Combustion Science* (pp. 769–819). Houston: Gulf Publishing Company. NULL
6. Basevich, V. Y., & Frolov, S. M. (1994). A reduced kinetic scheme for autoignition modelling of iso-octane and *n*-heptane/air mixtures during the induction period for internal combustion engines. *Chemical Physics*, *13*, 146–156 (in Russian).
7. Basevich, V. Y., Beliaev, A. A., Branshtater, V., Neigauz, M. G., Tashl, R., & Frolov, S. M. (1994). Modelling of iso-octane and *n*-heptane autoignition with reference to IC engines. *Physics of Combustion and Explosion*, *30*, 15–24 (in Russian).
8. Benson, S. W. (1981). The kinetics and thermochemistry of chemical oxidation with application to combustion and flames. *Progress in Energy and Combustion Science*, *7*, 125–134.
9. Bertoli, C., & Migliaccio, M. (1999). A finite conductivity model for diesel spray evaporation computations. *International Journal of Heat and Fluid Flow*, *20*, 552–561.
10. Blin-Simiand, N., Rigny, R., Viossat, V., Circan, S., & Sahetchian, K. (1993). Autoignition of hydrocarbon/air mixtures in a CFR engine: experimental and modeling study. *Combustion Science and Technology*, *88*, 329–348.
11. Borghesi, G., Mastorakos, E., & Cant, R. S. (2013). Complex chemistry DNS of n-heptane spray autoignition at high pressure and intermediate temperature conditions. *Combustion and Flame*, *160*, 1254–1275.

12. Boudier, P., Henriot, S., Pinsot, T., & Baritaud, T. (1992). A model for turbulent flame ignition and propagation in spark ignition engines. In *Twenty-Fourth Symposium (International) on Combustion*/The Combustion Institute, pp. 503–510.
13. Brady, R. N. (1996). *Modern diesel technology*. London: Prentice-Hall.
14. Bruel, P., Rogg, B., & Bray, K. N. C. (1990). On auto-ignition in laminar and turbulent non-premixed systems. In *Twenty-Third Symposium (International) on Combustion*/The Combustion Institute, pp. 759–766.
15. Buda, F., Bounaceur, R., Warth, V., Glaude, P. A., Fournet, R., & Battin-Leclerc, F. (2005). Progress toward a unified detailed kinetic model for the autoignition of alkanes from C_4 to C_{10} between 600 K and 1200 K. *Combustion and Flame*, *142*, 170–186.
16. Bykov, V., & Maas, U. (2009). Investigation of the hierarchical structure of kinetic models in ignition problems. *Zeitschrift für Physikalische Chemie*, *223*, 461–479.
17. Bykov, V., Goldfarb, I., Gol'dshtein, V., & Greenberg, J. B. (2002). Thermal explosion in a hot gas mixture with fuel droplets: a two reactants model. *Combustion Theory and Modelling*, *6*, 1–21.
18. Bykov, V., Goldfarb, I., Goldshtein, V., Sazhin, S. S., & Sazhina, E. M. (2004). System decomposition technique: application to spray modelling in CFD codes. In *20th Annual Symposium of the Israeli Section of the Combustion Institute.* Book of Abstracts, (p. 16). Beer-Sheva, Israel: Ben-Gurion University.
19. Bykov, V., Goldfarb, I., Gol'dshtein, V., Sazhin, S., & Sazhina, E. (2005). An asymptotic approach to numerical modelling of spray autoignition. In *Proceedings of the Fourth Mediterranean Combustion Symposium*, MCS4, Paper V. 6 October 6–10. Lisbon, Portugal, (CD-ROM).
20. Bykov, V., Goldfarb, I., & Gol'dshtein, V. (2006). Singularly perturbed vector fields. *Journal of Physics: Conference Series*, *55*, 28–44.
21. Bykov, V., Goldfarb, I., Gol'dshtein, V., Sazhin, S., & Sazhina, E. (2007). System decomposition technique for spray modelling in CFD codes. *Computers and Fluids*, *36*, 601–610.
22. Bykov, V., & Gol'dshtein, V. (2008). On a decomposition of motions and model reduction. *Journal of Physics: Conference Series*, *138*, 012003.
23. Bykov, V., Gol'dshtein, V., & Maas, U. (2008). Simple global reduction technique based on decomposition approach. *Combustion Theory and Modelling*, *12*, 389–405.
24. Bykov, V., Griffiths, J. F., Piazzesi, R., Sazhin, S. S., & Sazhina, E. M. (2013). The application of the global quasi-linearisation technique to the analysis of the cyclohexane/air mixture autoignition. *Applied Mathematics and Computation*, *219*, 7338–7347.
25. Chen, C.-K., & Lin, T.-H. (2012). Streamwise interaction of burning drops. *Combustion and Flame*, *159*, 1971–1979.
26. Chevalier, C., Pitz, W. J., Warnatz, J., Westbrook, C. K., & Mellenk, H. (1992). Hydrocarbon ignition: automatic generation of reaction mechanisms and applications to modeling of engine knock. In *Twenty-Fourth Symposium (International) on Combustion*/The Combustion Institute, pp. 1405–1414.
27. Coddington, E. A., & Levinson, N. (1955). *Theory of ordinary differential equations*. New York: McGraw-Hill.
28. Cox, R. A., & Cole, J. A. (1985). Chemical aspects of the autoignition of hydrocarbon—air mixtures. *Combustion and Flame*, *60*, 109–123.
29. Crespo, A., & Liñan, A. (1975). Unsteady effects in droplet evaporation and combustion. *Combustion Science and Technology*, *11*, 9–18.
30. Crua, C., Kennaird, D. A., Sazhin, S. S., & Heikal, M. R. (2004). Diesel autoignition at elevated in-cylinder pressures. *International Journal of Engine Research*, *5*, 365–374.
31. Curran, H. J., Gaffuri, P., Pitz, W. J., & Westbrook, C. K. (1998). A comprehensive modeling study of n-heptane oxidation. *Combustion and Flame*, *114*, 149–177.
32. De, S., Lakshmisha, K. N., & Bilger, R. W. (2011). Modeling of nonreacting and reacting turbulent spray jets using a fully stochastic separated flow approach. *Combustion and Flame*, *158*, 1992–2008.

33. Demoulin, F. X., & Borghi, R. (2003). Modeling of turbulent spray combustion with application to diesel like experiment. *Combustion and Flame*, *129*, 281–293.
34. Dombrovsky, L. A., Sazhin, S. S., Sazhina, E. M., Feng, G., Heikal, M. R., Bardsley, M. E. A., et al. (2001). Heating and evaporation of semi-transparent Diesel fuel droplets in the presence of thermal radiation. *Fuel*, *80*, 1535–1544.
35. Dopazo, C., & O'Brien, E. E. (1974). An approach to the autoignition of a turbulent mixture. *Acta Astronautica*, *1*, 1239–1266.
36. Durand, P., Gorokhovski, M., & Borghi, R. (1999). An application of the probability density function model to diesel engine combustion. *Combustion Science and Technology*, *144*, 47–78.
37. Fenichel, N. (1979). Geometric singular perturbation theory for ordinary differential equations. *Journal of Differential Equations*, *31*, 53–98.
38. Goldfarb, I., Gol'dshtein, V., Kuzmenko, G., & Greenberg, J. B. (1998). On thermal explosion of a cool spray in a hot gas. In *Proceedings of the 27th International Symposium on Combustion* (Vol. 2, pp. 2367–2374). Colorado, USA.
39. Goldfarb, I., Gol'dshtein, V., Kuzmenko, G., & Sazhin, S. S. (1999). Thermal radiation effect on thermal explosion in gas containing fuel droplets. *Combustion Theory and Modelling*, *3*, 769–787.
40. Goldfarb, I., Sazhin, S., & Zinoviev, A. (2004). Delayed thermal explosion in flammable gas containing fuel droplets: asymptotic analysis. *International Journal of Engineering Mathematics*, *50*, 399–414.
41. Goldfarb, I., Gol'dshtein, V., & Maas, U. (2004). Comparative analysis of two asymptotic approaches based on integral manifolds. *IMA Journal of Applied Mathematics*, *69*, 353–374.
42. Gol'dshtein, V., & Sobolev, V. (1988). *Qualitative analysis of singularly perturbed systems*. Novosibirsk: Siberian Branch of USSR Academy of Science, Institute of Mathematics (in Russian).
43. Gol'dshtein, V., & Sobolev, V. (1992). Integral manifolds in chemical kinetics and combustion. In *Singularity Theory and Some Problems of Functional Analysis* (pp. 73–92). Providence: American Mathematical Society.
44. Gorban, A. N., Kazantzis, N., Kevrekidis, I. G., Öttinger, H. C., & Theodoropoulos, C. (Eds.). (2006). *Model reduction and coarse-graining approaches for multi-scale phenomena*. New York: Springer.
45. Gorokhovski, M., & Saveliev, V. L. (2003). Analyses of Kolmogorov's model of breakup and its application into Lagrangean computation of liquid sprays under air-blast atomization. *Physics of Fluids*, *15*, 184–192.
46. Greenberg, J. B. (2002). Stability boundaries of laminar premixed polydisperse spray flames. *Atomization and Sprays*, *12*, 123–144.
47. Griffiths, J. F. (1993). Kinetic fundamentals of alkane autoignition at low temperatures. *Combustion and Flame*, *93*, 202–206.
48. Griffiths, J. F. (1995). Reduced kinetic models and their application to practical combustion systems. *Progress in Energy and Combustion Science*, *21*, 25–107.
49. Griffiths, J. F., Jiao, Q., Schreiber, M., Meyer, J., & Knoche, K. F. (1992) Development of thermokinetic models for autoignition in a CFD code: Experimental validation and application of the results to rapid compression studies. In *Twenty-Fourth Symposium (International) on Combustion*/The Combustion Institute, pp. 1809–1815.
50. Griffiths, J., Piazzesi, R., Sazhina, E. M., Sazhin, S. S., Glaude, P. A., & Heikal, M. R. (2012). Cfd modelling of cyclohexane auto-ignition in an rcm. *Fuel*, *96*, 192–203.
51. Gumprich, W., & Sadiki, A. (2013) Direct quadrature based sectional method of moments for the simulation of evaporating polydisperse sprays. In *Proceedings of ILASS—Europe 2013, 25th European Conference on Liquid Atomization and Spray Systems* 1–4 September 2013 (paper 43). Chania, Greece.
52. Hadjinicolaou, M., & Goussis, D. M. (1999). Asymptotic solutions of stiff PDEs with the CSP method: the reaction diffusion equation. *SIAM Journal on Scientific Computing*, *20*, 781–910.
53. Halstead, M., Prothero, A., & Quinn, C. P. (1973). Modeling the ignition and cool-flame limits of acetaldehyde oxidation. *Combustion and Flame*, *20*, 211–221.

54. Halstead, M. P., Kirsch, L. J., & Quinn, C. P. (1977). The autoignition of hydrocarbon fuels at high temperatures and pressures—fitting of a mathematical model. *Combustion and Flame*, *30*, 45–60.
55. Hamosfakidis, V., & Reitz, R. D. (2003). Optimization of a hydrocarbon fuel ignition model for two single component surrogates of diesel fuel. *Combustion and Flame*, *132*, 433–450.
56. Hilbert, R., & Thevenin, D. (2002). Autoignition of turbulent non-premixed flames investigated using direct numerical simulations. *Combustion and Flame*, *128*, 22–37.
57. Hou, S.-S., Lin, J.-C., & Hsuan, C.-Y. (2014). The interaction between internal heat gain and heat loss on compound-drop spray flames. *Int. J Heat Mass Transfer*, *71*, 503–514.
58. Incroperra, F. P., & DeWitt, D. P. (1996). *Fundamentals of heat and mass transfer*. Chichester: Wiley.
59. Kaper, H. G., & Kaper, T. J. (2001). *Asymptotic analysis of two reduction methods for systems of chemical reactions*. Argonne National Lab, preprint ANL/MCS-P912-1001.
60. Knobloch, H. W., & Aulback, B. (1984). Singular perturbations and integral manifolds. *Journal of Mathematical and Physical Sciences*, *18*, 415–424.
61. Kojima, S. (1994). Detailed modeling of *n*-butane autoignition chemistry. *Combustion and Flame*, *99*, 87–136.
62. Kong, S.-C., Han, Z., & Reitz, R. D. (1996). *The development and application of a diesel ignition and combustion model for multidimensional engine simulation*. SAE Technical Paper 950278.
63. Kuo, K.-K. (1986). *Principles of combustion*. Chichester: Wiley.
64. Kuznetzov, V. R., & Sabel'nikov, V. A. (1990). *Turbulence and combustion*. New York: Hemisphere.
65. Lakshmisha, K. N., Zhang, Y., Rogg, B., & Bray, K. N. C. (1992). Modelling auto-ignition in a turbulent medium. In *Twenty-Fourth Symposium (International) on Combustion*/The Combustion Institute, pp. 421–428.
66. Lam, S. H., & Goussis, D. M. (1994). The GSP method for simplifying kinetics. *International Journal of Chemical Kinetics*, *26*, 461–486.
67. Lefebvre, A. H. (1989). *Atomization and sprays*. New York: Taylor & Francis.
68. Maas, V., & Pope, S. B. (1992). Implementation of simplified chemical kinetics based on intrinsic low-dimensional manifolds, In *Twenty-Fourth Symposium (International) on Combustion*/The Combustion Institute, pp. 103–112.
69. Maas, U., & Pope, S. B. (1992). Simplifying chemical kinetics: intrinsic low-dimensional manifolds in composition space. *Combustion and Flame*, *117*, 99–116.
70. Masias, A., Diamantis, D., Mastorakos, E., & Goussis, D. E. (1999). An algorithm for the construction of global reduced mechanisms with CSP data. *Combustion and Flame*, *117*, 685–708.
71. McIntosh, A. C., Gol'dshtein, V., Goldfarb, I., & Zinoviev, A. (1998). Thermal explosion in a combustible gas containing fuel droplets. *Combustion Theory and Modelling*, *2*, 153–165.
72. Minkoff, G. J., & Tipper, C. F. (1962). *Chemistry of combustion reactions*. London: Butterworth.
73. Minetti, R., Ribaucour, M., Carlier, M., Fittschen, C., & Sochet, L. R. (1994). Experimental and modelling study of oxidation and autoignition of butane at high pressure. *Combustion and Flame*, *96*, 201–211.
74. Minetti, R., Ribaucour, M., Carlier, M., & Sochet, L. R. (1996). Autoignition delays of a series of linear and branched chain alkanes in the intermediate range of temperatures. *Combustion Science and Technology*, *113–114*, 179–192.
75. Müller, U. C., Peters, N., & Liñan, A. (1992). Global kinetics for *n*-heptane ignition at high pressures. In *Twenty-Fourth Symposium (International) on Combustion*/The Combustion Institute, pp. 777–784.
76. Natarajan, B., & Bracco, F. V. (1984). On multidimensional modelling of auto-ignition in spark-ignition engines. *Combustion and Flame*, *57*, 179–197.
77. Nave, O., Bykov, V., & Gol'dshtein, V. (2010). A probabilistic model of thermal explosion in polydisperse fuel spray. *Applied Mathematics and Computation*, *217*, 2698–2709.

78. Nave, O., Bykov, V., Gol'dshtein, V., & Lehavi, Y. (2011). Numerical simulations applying to the analysis of thermal explosion of organic gel fuel in a hot gas. *Fuel*, *90*, 3410–3416.
79. Nave, O., & Gol'dshtein, V. (2011). The flammable spray effect on thermal explosion of a combustible gas-fuel mixture. *Combustion Science and Technology*, *183*, 519–539.
80. Nave, O., Gol'dshtein, V., & Dan, E. (2011). The delay phenomena in thermal explosion of polydisperse fuel spray—using the method of integral manifolds. *Atomization and Sprays*, *21*, 69–85.
81. Neophytou, M. K., Goussis, D. A., van Loon, M., & Mastorakos, E. (2004). Reduced chemical mechanism for atmospheric pollution using computational singular perturbation analysis. *Atmospheric Environment*, *38*, 3661–3673.
82. Neophytou, A., Mastorakos, E., & Cant, R. S. (2012). The internal structure of igniting turbulent sprays as revealed by complex chemistry DNS. *Combustion and Flame*, *159*, 641–664.
83. Pilling, M. J., Robertson, S. H., & Seakins, P. W. (1995). Elementary radical reactions and autoignition. *Journal of the Chemical Society, Faraday Transactions*, *91*, 4179–4188.
84. Pitsch, H., & Peters, N. (1998) *Investigation of the ignition process of sprays under diesel engine conditions using reduced n-heptane chemistry*. SAE paper 2464.
85. Poppe, C., Schreiber, M., & Griffiths, J. F. (1993). Modelling of n-heptane autoignition and validation of the results. In *Proceedings of Joint Meeting of British and German Sections of the Combustion Institute* (pp. 360–363), Cambridge.
86. Ranzi, E., Faravelli, T., Gaffuri, P., Pennati, G. C., & Sogaro, A. (1994). A wide range modeling of propane and *n*-butane oxidation. *Combustion and Flame*, *100*, 299–330.
87. Rhodes, C., Morari, M., & Wiggins, S. (1999). Identification of the low order manifolds: validating the algorithm of Maas and Pope. *Chaos*, *9*, 108–123.
88. Sabel'nikov, V., Gorokhovski, M., & Baricault, N. (2006). The extended IAM mixing model in the framework of the composition PDF approach; application to diesel spray combustion. *Combustion Theory and Modelling*, *10*, 155–169.
89. Sahetchian, K., Champoussin, J. C., Brun, M., Levy, N., Blin-Simiand, N., Aligrot, C., et al. (1995). Experimental study and modeling of dodecane ignition in a Diesel engine. *Combustion and Flame*, *103*, 207–220.
90. Sazhin, S. S. (2006). Advanced models of fuel droplet heating and evaporation. *Progress in Energy and Combustion Science*, *32*(2), 162–214.
91. Sazhin, S. S., Wild, P., Leys, C., Toebaert, D., & Sazhina, E. M. (1993). The three temperature model for the fast-axial-flow CO_2 laser. *Journal of Physics D: Applied Physics*, *26*, 1872–1883.
92. Sazhin, S. S., Sazhina, E. M., Heikal, M. R., Marooney, C., & Mikhalovsky, S. V. (1999). The Shell autoignition model: a new mathematical formulation. *Combustion and Flame*, *117*, 529–540.
93. Sazhin, S. S., Feng, G., Heikal, M. R., Goldfarb, I., Gol'dshtein, V., & Kuzmenko, G. (2001). Thermal ignition analysis of a monodisperse spray with radiation. *Combustion and Flame*, *124*, 684–701.
94. Sazhin, S. S., Feng, G., & Heikal, M. R. (2001). A model for fuel spray penetration. *Fuel*, *80*, 2171–2180.
95. Sazhin, S. S., Abdelghaffar, W. A., Sazhina, E. M., Mikhalovsky, S. V., Meikle, S. T., & Bai, C. (2004). Radiative heating of semi-transparent diesel fuel droplets. *ASME Journal of Heat Transfer*, *126*, 105–109. Erratum *126*, 490–491.
96. Sazhin, S. S., Abdelghaffar, W. A., Sazhina, E. M., & Heikal, M. R. (2005). Models for droplet transient heating: effects on droplet evaporation, ignition, and break-up. *International Journal of Thermal Sciences*, *44*, 610–622.
97. Sazhin, S. S., Shchepakina, E. A., & Sobolev, V. A. (2010). Order reduction of a non-Lipschitzian model of monodisperse spray ignition. *Mathematical and Computer Modelling*, *52*, 529–537.
98. Sazhina, E. M., Sazhin, S. S., Heikal, M. R., & Marooney, C. (1999). The Shell autoignition model: application to gasoline and Diesel fuels. *Fuel*, *78*, 389–401.

99. Sazhina, E. M., Sazhin, S. S., Heikal, M. R., Babushok, V. I., & Johns, R. (2000). A detailed modelling of the spray ignition process in Diesel engines. *Combustion Science and Technology*, *160*, 317–344.
100. Sazhina, E. M., Bykov, V., Goldfarb, I., Gol'dshtein, V., Sazhin, S. S., & Heikal, M. R. (2005). Modelling of spray autoignition by the ODE system decomposition technique. In *Proceedings of HEFAT2005 (4th International Conference on Heat Transfer, Fluid Mechanics and Thermodynamics)*. Cairo, Egypt. Paper number: SE2.
101. Schäpertöns, H., & Lee, W. (1985). *Multidimensional modelling of knocking combustion in SI engines*. SAE Technical Paper 850502.
102. Semenov, N. N. (1935). *Chemical kinetics and chain reactions*. Oxford: Oxford University Press.
103. Sirignano, W. A. (1999). *Fluid dynamics and transport of droplets and sprays*. Cambridge: Cambridge University Press.
104. Sobolev, V. A. (1984). Integral manifolds and decomposition of singularly perturbed systems. *System and Control Letters*, *5*, 169–179.
105. Sobolev, V. A., & Shchepakina, E. A. (2010). *Reduction of the models and critical phenomena in micro-kinetics*. Moscow: Fizmatlit. (in Russian).
106. Spalding, D. B. (1979). *Combustion and mass transfer*. New York: Pergamon Press.
107. Sportisse, B. (2000). An analysis of operator splitting techniques in the stiff case. *Journal of Computational Physics*, *161*, 140–168.
108. Strygin, V. V., & Sobolev, V. A. (1988). *Separation of motions by the integral manifolds method*. Moscow: Nauka (in Russian).
109. Suppes, G. J., Srinivasan, B., & Natarajan, V. P. (1995). *Autoignition of biodiesel, methanol, and 50:50 blend in a simulated diesel engine environment*. SAE paper 952758.
110. Tanabe, M., Kono, M., Sato, J., Koenig, J., Eigenbrod, C., Dinkelacker, F., et al. (1995). Two stage ignition of n-heptane isolated droplets. *Combustion Science and Technology*, *108*, 103–119.
111. Tanaka, S., Ayala, F., & Keck, J. C. (2003). A reduced chemical kinetic model for HCCI combustion of primary reference fuels in a rapid compression machine. *Combustion and Flame*, *133*, 467–481.
112. Theobald, M. A. (1986). *Numerical simulation of diesel autoignition*. PhD Thesis, MIT.
113. Utyuzhnikov, S. V. (2002). Numerical modeling of combustion of fuel-droplet-vapour releases in the atmosphere. *Flow, Turbulence and Combustion*, *68*, 137–152.
114. Valorani, M., & Goussis, D. M. (2001). Explicit time-scale splitting algorithm for stiff problems: auto-ignition of gaseous mixtures behind a steady shock. *Journal of Computational Physics*, *169*, 44–79.
115. Varshavski, G. A., Fedoseev, D. V., & Frank-Kamenetskii, A. D. (1968). Autoignition of a fuel droplet. In V. A. Fedoseev (Ed.). *Proceedings of the Sixths Conference on Evaporation, Combustion and Gas Dynamics in Disperse Systems*, October 1966 (pp. 91–95). Odessa: Odessa University Publishing House (in Russian).
116. Viggiano, A., & Magi, V. (2004). A 2-D investigation of n-heptane autoignition by means of direct numerical simulation. *Combustion and Flame*, *137*, 432–443.
117. Wang, C., Dean, A. M., Zhu, H., & Kee, R. J. (2013). The effects of multicomponent fuel droplet evaporation on the kinetics of strained opposed-flow diffusion flames. *Combustion and Flame*, *160*, 265–275.
118. Wang, H., Luo, K., & Fan, J. (2014). Effects of turbulent intensity and droplet diameter on spray combustion using direct numerical simulation. *Fuel*, *121*, 311–318.
119. Westbrook, C. K., & Dryer, F. L. (1981). Simplified reaction mechanism for the oxidation of hydrocarbon fuels in flames. *Combustion Science and Technology*, *27*, 31–43.
120. Wu, G., & Sirignano, W. A. (2010). Transient burning of a convective fuel droplet. *Combustion and Flame*, *157*, 970–981.
121. Wu, G., & Sirignano, W. A. (2011). Transient convective burning of interactive fuel droplets in double-layer arrays. *Combustion and Flame*, *158*, 2395–2407.

122. Zadiraka, K. V. (1957). On the integral manifold of a system of differential equations containing a small parameter. *Doklady Akademii Nauk SSSR*, *115*, 646–649 (in Russian).
123. Zhang, Y., Rogg, B., & Bray, K. N. C. (1993). Modeling of autoignition in nonpremixed turbulent systems: closure of the chemical-source terms. *Progress in Astronautics and Aeronautics*, *152*, 87–102.
124. Zhang, Y., Rogg, B., & Bray, K. N. C. (1995). 2-D simulation of turbulent autoignition with transient laminar flamelet source term closure. *Combustion Science and Technology*, *105*, 211–227.
125. Zhukov, V. P., Sechenov, V. A., & Starikovskii, A.Yu. (2005). Self-ignition of a lean mixture of n-pentane and air over a wide range of pressures. *Combustion and Flame*, *140*, 196–203.

Appendix A
Derivation of Formula (3.77)

In this appendix, the details of the derivation of Formula (3.77) are presented, following [2].

Introduction of the new variable $u = Tr = TR/R_d$ allows us to rewrite Eq. (3.76) as

$$\frac{\partial u}{\partial t} = \kappa_R \frac{\partial^2 u}{\partial r^2} + \tilde{P}(r) \tag{A.1}$$

with the boundary and initial conditions

$$\left.\begin{array}{ll} \frac{\partial u}{\partial r} + H(t)u = M(t) & \text{when } r = 1 \\ u = 0 & \text{when } r = 0 \\ u(t=0) = rT_0(rR_d) \equiv \tilde{T}_0(r) & \text{when } 0 \leq r \leq 1 \end{array}\right\} \tag{A.2}$$

where

$$H(t) = \frac{h(t)R_d}{k_l} - 1, \quad M(t) = \frac{h(t)T_{\text{eff}}(t)R_d}{k_l}, \quad \tilde{P}(r) = rP(r).$$

Assuming that $h(t) \equiv h = \text{const}$, we can write $H(t) \equiv h_0 = (hR_d/k_l) - 1 = \text{const}$. Introducing a new parameter $\mu_0(t) = hT_{\text{eff}}(t)R_d/k_l$, we can rewrite (A.2) as

$$\left.\begin{array}{ll} \frac{\partial u}{\partial r} + h_0 u = M(t) = \mu_0(t) & \text{when } r = 1 \\ u = 0 & \text{when } r = 0 \\ u(t=0) = rT_0(rR_d) \equiv \tilde{T}_0(r) & \text{when } 0 \leq r \leq 1 \end{array}\right\} \tag{A.3}$$

Recalling that $h_0 > -1$, let us look for the solution of Eq. (A.1) in the form

$$u(r,t) = \frac{1}{1+h_0} r\mu_0(t) + W(r,t). \tag{A.4}$$

On substituting the function (A.4) into (A.1), we obtain the equation for W,

S. Sazhin, *Droplets and Sprays*, DOI: 10.1007/978-1-4471-6386-2,

$$\frac{\partial W}{\partial t} = \kappa_R \frac{\partial^2 W}{\partial r^2} + \tilde{P}(r) - \frac{r}{1+h_0}\frac{\mathrm{d}\mu_0(t)}{\mathrm{d}t}, \tag{A.5}$$

with the boundary and initial conditions

$$\left.\begin{aligned} W|_{r=0} &= \left(\tfrac{\partial W}{\partial r} + h_0 W\right)\Big|_{r=1} = 0 \\ W|_{t=0} &= \tilde{T}_0(r) - \tfrac{r}{1+h_0}\mu_0(0). \end{aligned}\right\} \tag{A.6}$$

We look for the solution of (A.5) with the boundary and initial conditions (A.6) in the form

$$W(r,t) = \sum_{n=1}^{\infty} c_n(t) v_n(r), \tag{A.7}$$

where the functions $v_n(r)$ form the full set of nontrivial solutions to the equation

$$\frac{\mathrm{d}^2 v}{\mathrm{d}r^2} + \lambda^2 v = 0 \tag{A.8}$$

subject to the boundary conditions $v|_{r=0} = \left(\frac{\mathrm{d}v}{\mathrm{d}r} + h_0 v\right)\big|_{r=1} = 0$. The general solution $v(r) = A\cos\lambda r + B\sin\lambda r$ to (A.8) satisfies the boundary conditions when $A = 0$ and

$$\lambda\cos\lambda + h_0\sin\lambda = 0. \tag{A.9}$$

The solution to (A.9) gives a set of positive eigenvalues λ_n numbered in ascending order ($n = 1, 2, \ldots$). If $h_0 = 0$, then $\lambda_n = \pi(n - \frac{1}{2})$. Assuming that $B = 1$, the expressions for the eigenfunctions v_n can be written as

$$v_n(r) = \sin\lambda_n r \quad (n = 1, 2, \ldots).$$

The value of B is implicitly accounted for by the coefficients $c_n(t)$ in the series (A.7). The functions $v_n(r)$ form a full set of eigenfunctions which are orthogonal for $r \in [0, 1]$. The orthogonality of the functions v_n follows from the relation

$$\int_0^1 v_n(r) v_m(r)\mathrm{d}r = \delta_{nm}||v_n||^2, \tag{A.10}$$

where

$$\delta_{nm} = \begin{cases} 0 & n \neq m \\ 1 & n = m \end{cases}, \qquad ||v_n||^2 = \frac{1}{2}\left(1 - \frac{\sin 2\lambda_n}{2\lambda_n}\right) = \frac{1}{2}\left(1 + \frac{h_0}{h_0^2 + \lambda_n^2}\right).$$

The eigenvalue $\lambda_0 = 0$ describes the trivial eigenfunction $v_0(r) = 0$. The orthogonality of v_n allows us to expand the known functions in (A.5), (A.6) in the series

$$\tilde{P}(r) = \sum_{n=1}^{\infty} p_n v_n(r), \tag{A.11}$$

$$\tilde{T}_0(r) = \sum_{n=1}^{\infty} q_n v_n(r), \tag{A.12}$$

$$f(r) \equiv -r/(1+h_0) = \sum_{n=1}^{\infty} f_n v_n(r), \tag{A.13}$$

where:

$$\left.\begin{aligned} p_n = \tfrac{1}{\|v_n\|^2} \int_0^1 \tilde{P}(r) v_n(r) \mathrm{d}r, \qquad q_n = \tfrac{1}{\|v_n\|^2} \int_0^1 \tilde{T}_0(r) v_n(r) \mathrm{d}r, \\ f_n = \tfrac{1}{\|v_n\|^2} \int_0^1 f(r) v_n(r) \mathrm{d}r = -\tfrac{\sin\lambda_n}{\|v_n\|^2 \lambda_n^2}. \end{aligned}\right\} \tag{A.14}$$

On substituting series (A.7), (A.11), and (A.13) into (A.5), we obtain

$$\sum_{n=1}^{\infty} \left(\frac{\mathrm{d}c_n(t)}{\mathrm{d}t} + c_n(t)\kappa_R \lambda_n^2 \right) v_n(r) = \sum_{n=1}^{\infty} \left(p_n + f_n \frac{\mathrm{d}\mu_0(t)}{\mathrm{d}t} \right) v_n(r). \tag{A.15}$$

Both sides of (A.15) are Fourier series of functions $v_n(r)$. Two Fourier series are equal if and only if their coefficients are equal. This implies that

$$\frac{\mathrm{d}c_n(t)}{\mathrm{d}t} + c_n(t)\kappa_R \lambda_n^2 = p_n + f_n \frac{\mathrm{d}\mu_0(t)}{\mathrm{d}t}. \tag{A.16}$$

The initial condition for $c_n(t)$ follows from the initial condition for W:

$$c_n(0) = q_n + f_n \mu_0(0).$$

The solution to (A.16), subject to this initial condition, can be written as

$$c_n(t) = \frac{p_n}{\kappa_R \lambda_n^2} + \exp\left[-\kappa_R \lambda_n^2 t\right] \left(q_n + f_n \mu_0(0) - \frac{p_n}{\kappa_R \lambda_n^2} \right)$$
$$+ f_n \int_0^t \frac{\mathrm{d}\mu_0(\tau)}{\mathrm{d}\tau} \exp\left[-\kappa_R \lambda_n^2 (t-\tau)\right] \mathrm{d}\tau. \tag{A.17}$$

On substituting functions (A.17) and $v_n(r)$ into the series (A.7), we obtain the solution to (A.5) subject to (A.6) in the form

$$W(r,t)=\sum_{n=1}^{\infty}\left\{\frac{p_n}{\kappa_R\lambda_n^2}+\exp\left[-\kappa_R\lambda_n^2 t\right]\left(q_n+f_n\mu_0(0)-\frac{p_n}{\kappa_R\lambda_n^2}\right)\right.$$
$$\left.+f_n\int\limits_0^t\frac{\mathrm{d}\mu_0(\tau)}{\mathrm{d}\tau}\exp\left[-\kappa_R\lambda_n^2(t-\tau)\right]\mathrm{d}\tau\right\}\sin\lambda_n r, \tag{A.18}$$

where p_n, q_n, f_n, and λ_n are defined by (A.14) and (A.9), and

$$u(r,t)=\frac{1}{1+h_0}r\mu_0(t)+\sum_{n=1}^{\infty}\left\{\frac{p_n}{\kappa_R\lambda_n^2}+\exp\left[-\kappa_R\lambda_n^2 t\right]\left(q_n+f_n\mu_0(0)-\frac{p_n}{\kappa_R\lambda_n^2}\right)\right.$$
$$\left.+f_n\int\limits_0^t\frac{\mathrm{d}\mu_0(\tau)}{\mathrm{d}\tau}\exp\left[-\kappa_R\lambda_n^2(t-\tau)\right]\mathrm{d}\tau\right\}\sin\lambda_n r. \tag{A.19}$$

Recalling (A.14) and the definition of u, the final solution to (3.76) can be given as

$$T(r,t)=\frac{1}{r}\sum_{n=1}^{\infty}\left\{\frac{p_n}{\kappa_R\lambda_n^2}+\exp\left[-\kappa_R\lambda_n^2 t\right]\left(q_n-\frac{p_n}{\kappa_R\lambda_n^2}\right)\right.$$
$$-\frac{\sin\lambda_n}{||v_n||^2\lambda_n^2}\mu_0(0)\exp\left[-\kappa_R\lambda_n^2 t\right]$$
$$\left.-\frac{\sin\lambda_n}{||v_n||^2\lambda_n^2}\int\limits_0^t\frac{\mathrm{d}\mu_0(\tau)}{\mathrm{d}\tau}\exp\left[-\kappa_R\lambda_n^2(t-\tau)\right]\mathrm{d}\tau\right\}$$
$$\times\sin\lambda_n r+T_{\mathrm{eff}}(t). \tag{A.20}$$

This formula is identical to (3.77). In the limit of $P=0$, it reduces to (3.37). In deriving (A.20), we took into account that $T_{\mathrm{eff}}(t)=k_l\mu_0(t)/(hR_d)$.

If $T_0(r)$ is continuously twice differentiable, then the series in (A.18), (A.19), (A.20) converge absolutely and uniformly for all $t\geq 0$ and $r\in[0,1]$, since

$$|p_n|<\text{const},\qquad |\,q_n\,|<\frac{\text{const}}{\lambda_n^2},\qquad |\sin\lambda_n r\,|\leq 1,\qquad \exp\left(-\kappa_R\lambda_n^2 t\right)\leq 1,$$

and $\lambda_n^{-2}<n^{-2}$ for $n>1$. It can be shown that $\lambda_n>\pi(n-1)$. Hence for $n>1$

$$\lambda_n>n\pi\left(1-\frac{1}{n}\right)>n\pi/2>n. \tag{A.21}$$

Equation (A.20) [(or (3.77)] is expected to reduce to (3.35) in the limit $k_l \to \infty$. In what follows, it is shown that this indeed takes place, following [1]. Let us first restrict our analysis to the droplet's surface and represent (A.20) for $R = R_d$ in the form

$$T_s = T_\infty + (T_{s0} - T_\infty)(S_1 + S_2), \tag{A.22}$$

where

$$S_1 = \varepsilon \frac{\sin^2 \lambda_1}{\lambda_1^2 \, ||v_1||^2} \exp\left(-\kappa_R \lambda_1^2 t\right), \tag{A.23}$$

$$S_2 = \varepsilon \sum_{n=2}^{\infty} \frac{\sin^2 \lambda_n}{\lambda_n^2 ||v_n||^2} \exp\left(-\kappa_R \lambda_n^2 t\right), \tag{A.24}$$

$\varepsilon \equiv k_a/k_l \ll 1$. In this case, $h_0 = \varepsilon - 1$ in (A.3).

In what follows, the value of S_2 is estimated, and the expression for S_1 is rearranged in the limit $\varepsilon \to 0$. Since $\lambda_n \geq \lambda_2 > \pi/2$ for $n = 2, 3, ...$, we have

$$\left| \frac{\sin 2\lambda_n}{2\lambda_n} \right| \leq \frac{1}{2\lambda_n} \leq \frac{1}{2\lambda_2} < \frac{1}{\pi},$$

$$||v_n||^2 > \frac{1}{2}\left(1 - \frac{1}{\pi}\right) = \frac{\pi - 1}{2\pi}, \qquad \frac{1}{||v_n||^2} < \frac{2\pi}{\pi - 1}. \tag{A.25}$$

On the other hand, we can see from the solution to the equation for λ that

$$\lambda_n > \pi n/2, \qquad n = 2, 3, ...$$

This implies that

$$\left. \frac{1}{\lambda_n} \right|_{n \geq 2} < \frac{2}{\pi} \frac{1}{n}. \tag{A.26}$$

In view of inequalities (A.25) and (A.26), we can estimate the value of S_2 as

$$| S_2 | < \varepsilon \frac{2\pi}{\pi - 1} \left[\left(\frac{2}{\pi}\right)^2 \sum_{n=2}^{\infty} \frac{1}{n^2} \right] = \frac{8\varepsilon}{\pi(\pi - 1)} \left[\sum_{n=2}^{\infty} \frac{1}{n^2} \right] \to 0 \tag{A.27}$$

when $\varepsilon \to 0$ for all t.

We can see that for $\varepsilon \to 0$, $h_0 = \varepsilon - 1 \to -1 + 0$, and $\lambda_1 \to +0$. In this case, we can expand $\tan \lambda_1$ in a Taylor series and rewrite (A.9) as

$$1 + \frac{1}{3}\lambda_1^2 + \frac{2}{15}\lambda_1^4 + \cdots = \frac{1}{1 - \varepsilon} = \sum_{k=0}^{\infty} \varepsilon^k. \tag{A.28}$$

Let us look for the solution to (A.28) in the form of the series

$$\lambda_1^2 = c_1\varepsilon + c_2\varepsilon^2 + \cdots . \tag{A.29}$$

On substituting formula (A.29) into (A.28), we can write

$$1 + \frac{1}{3}\left(c_1\varepsilon + c_2\varepsilon^2 + \cdots\right) + \frac{2}{15}\left(c_1\varepsilon + c_2\varepsilon^2 + \cdots\right)^2 + \cdots = 1 + \varepsilon + \varepsilon^2 + \cdots \tag{A.30}$$

Equation (A.30) is satisfied when the coefficients of equal powers of ε are equal. This leads us to the following values of the first two coefficients: $c_1 = 3$ and $c_2 = -3/5$. In the limit $\varepsilon \to 0$, we can ignore all terms in (A.29) apart from the first one and write

$$\lambda_1^2 \approx 3\varepsilon. \tag{A.31}$$

On substituting expression (A.31) into expression (A.23) and keeping in mind that in the limit $\lambda_1 \to +0$, $(\sin(2\lambda_1))/(2\lambda_1) \approx 1 - (2\lambda_1)^2/6$ and $||\ v_1\ ||^2 \approx \lambda_1^2/3$, we obtain

$$S_1 = \exp\left(-\frac{3k_a t}{c_l \rho_l R_d^2}\right) = \exp\left(-\frac{3ht}{c_l \rho_l R_d}\right). \tag{A.32}$$

Recalling that $S_2 \to 0$ and having substituted expression (A.32) into expression (A.22), we obtain

$$T_s = T_\infty + (T_{s0} - T_\infty)\exp\left(-\frac{3ht}{c_l \rho_l R_d}\right). \tag{A.33}$$

The same equation could be obtained directly from the energy balance equation at the surface of the droplet, assuming that there is no temperature gradient inside the droplet [see (3.35)]:

$$\frac{4}{3}\pi R_d^3 \rho_l c_l \frac{\mathrm{d}T_s}{\mathrm{d}t} = 4\pi R_d^2 h(T_\infty - T_{s0}). \tag{A.34}$$

The analysis presented above can be generalised to the case $R \neq R_d$, and it can be shown that when $k_l \to \infty$ and $R \neq 0$, the temperature inside the droplets is equal to the surface temperature given by (A.33). The temperature at the point $R = 0$ is the same as at the points $R \neq 0$, since $\partial T/\partial R\ |_{R=0} = 0$. For practical applications, the series in expression (A.20) needs to be truncated. It was found that the best way to choose the number of terms in this series is to compare the surface temperature that it predicts at $t = 0$ and $R = R_d$ with T_{s0}. If the allowed error in the calculations is less than ε, then the number of terms N required must satisfy the following condition:

$$\Phi(N) = \left|\left(1 - \frac{T_\infty}{T_{s0}}\right)\left[1 - \frac{k_a}{k_l}\sum_{n=1}^{N}\frac{\sin\lambda_n^2}{\lambda_n^2 ||v_n||^2}\right]\right| < \varepsilon. \tag{A.35}$$

References

1. Sazhin, S. S., & Krutitskii, P. A. (2003). A conduction model for transient heating of fuel droplets. In H. G. W. Begehre, R. P. Gilbert, M. W. Wong (Eds.), *Progress in Analysis*. Proceedings of the 3d International ISAAC (International Society for Analysis, Applications and Computations) Congress (August 20–25, 2001, Berlin) (Vol. II, pp. 1231–1239). Singapore: World Scientific.
2. Sazhin, S. S., Krutitskii, P. A., Abdelghaffar, W. A., Mikhalovsky, S. V., Meikle, S. T., & Heikal, M. R. (2004). Transient heating of diesel fuel droplets. *International Journal Heat and Mass Transfer*, *47*, 3327–3340.

Appendix B
Derivation of Formula (3.95)

In this appendix, the details of the derivation of formula (3.95) are presented, following [2].

Introducing a new variable

$$u = \left(T - T_{g0}(R_g)\right) R,$$

we can simplify (3.76) and initial and boundary conditions [see (3.3) and (3.4)] to

$$\frac{\partial u}{\partial t} = \kappa \frac{\partial^2 u}{\partial R^2} + RP(t, R), \tag{B.1}$$

$$u|_{t=0} = -T_0 R \tag{B.2}$$

$$\begin{aligned} u|_{R=R_d^-} = u|_{R=R_d^+}, \quad & k_l \left[R_d u_R' - u \right]\Big|_{R=R_d^-} \\ = k_g \left[R_d u_R' - u \right]\Big|_{R=R_d^+}, \quad & u|_{R=R_g} = 0, \end{aligned} \tag{B.3}$$

where

$$T_0 \equiv T_0(R) = \begin{cases} T_{g0}(R_g) - T_{d0}(R) & \text{when } R \leq R_d \\ T_{g0}(R_g) - T_{g0}(R) & \text{when } R_d < R \leq R_g, \end{cases}.$$

Conditions (B.3) need to be amended by the boundary condition at $R = 0$. Since $T - T_{g0}$ is finite at $R = 0$, it follows that $u|_{R=0} = 0$.

We look for the solution to (B.1) in the form

$$u = \sum_{n=1}^{\infty} \Theta_n(t) v_n(R), \tag{B.4}$$

where the functions $v_n(R)$ form the full set of nontrivial solutions of the eigenvalue problem

S. Sazhin, *Droplets and Sprays*, DOI: 10.1007/978-1-4471-6386-2,

$$\frac{\mathrm{d}^2 v}{\mathrm{d}R^2} + a^2\lambda^2 v = 0 \tag{B.5}$$

subject to boundary conditions

$$\left.\begin{aligned} &v|_{R=0} = v|_{R=R_g} = 0 \\ &v|_{R=R_d^-} = v|_{R=R_d^+} \\ &k_l\left[R_d v_R' - v\right]\Big|_{R=R_d^-} = k_g\left[R_d v_R' - v\right]\Big|_{R=R_d^+} \end{aligned}\right\}, \tag{B.6}$$

where

$$a = \frac{1}{\sqrt{\kappa}} = \begin{cases} \sqrt{\frac{c_l\rho_l}{k_l}} \equiv a_d & \text{when } R \le R_d \\ \sqrt{\frac{c_{pg}\rho_g}{k_g}} \equiv a_g & \text{when } R_d < R \le R_g. \end{cases} \tag{B.7}$$

Note that λ has dimension $1/\sqrt{\text{time}}$. We look for the solution of (B.5) in the form

$$v(R) = \begin{cases} A\sin(\lambda a_d R) & \text{when } R \le R_d \\ B\sin(\lambda a_g(R - R_g)) & \text{when } R_d < R \le R_g. \end{cases} \tag{B.8}$$

The function (B.8) satisfies boundary conditions (B.6) at $R = 0$. On substituting function (B.8) into boundary conditions (B.6) at $R = R_d$, we obtain

$$A\sin(\lambda a_d R_d) = B\sin(\lambda a_g(R_d - R_g)), \tag{B.9}$$

$$\begin{aligned} &Ak_l\left[R_d\lambda a_d\cos(\lambda a_d R_d) - \sin(\lambda a_d R_d)\right] \\ &\quad = Bk_g\left[R_d\lambda a_g\cos(\lambda a_g(R_d - R_g)) - \sin(\lambda a_g(R_d - R_g))\right]. \end{aligned} \tag{B.10}$$

Condition (B.9) is satisfied when

$$\left.\begin{aligned} A &= \left[\sin(\lambda a_d R_d)\right]^{-1} \\ B &= \left[\sin(\lambda a_g(R_d - R_g))\right]^{-1} \end{aligned}\right\}. \tag{B.11}$$

On substituting (B.11) into (B.10), we obtain

$$k_l\left[R_d\lambda a_d\cot(\lambda a_d R_d) - 1\right] = k_g\left[R_d\lambda a_g\cot(\lambda a_g(R_d - R_g)) - 1\right]. \tag{B.12}$$

Using the definitions of a_d and a_g, Eq. (B.12) can be simplified to

$$\sqrt{k_l c_l \rho_l}\cot(\lambda a_d R_d) - \sqrt{k_g c_{pg}\rho_g}\cot(\lambda a_g(R_d - R_g)) = \frac{k_l - k_g}{R_d\lambda}. \tag{B.13}$$

Equation (B.13) allows us to find a countable set of positive eigenvalues λ_n which can be arranged in ascending order $0 < \lambda_1 < \lambda_2 < \cdots$. Note that the negative solutions $-\lambda_n$ also satisfy (B.13), since both sides of this equation are odd functions of λ. However, $\lambda = 0$ does not satisfy this equation. On substituting these values of λ_n into expression (B.8) and recalling expressions (B.11), we obtain the expressions for the eigenfunctions v_n in the form

$$v_n(R) = \begin{cases} \frac{\sin(\lambda_n a_d R)}{\sin(\lambda_n a_d R_d)} & \text{when } R \le R_d \\ \frac{\sin(\lambda_n a_g (R-R_g))}{\sin(\lambda_n a_g (R_d-R_g))} & \text{when } R_d < R \le R_g \end{cases}. \tag{B.14}$$

It can be shown (see Appendix C) that the functions $v_n(R)$ are orthogonal with weight

$$b = \begin{cases} k_l a_d^2 = c_l \rho_l & \text{when } R \le R_d \\ k_g a_g^2 = c_{pg} \rho_g & \text{when } R_d < R \le R_g \end{cases}.$$

This means that $\int_0^{R_g} v_n(R) v_m(R) b \mathrm{d}R = \delta_{nm} ||v_n||^2$, where

$$\delta_{nm} = \begin{cases} 1 & \text{when } n = m \\ 0 & \text{when } n \ne m \end{cases}.$$

The proof of completeness of this set of functions is much more complicated (it is based on the methods of functional analysis and properties of Banach spaces [3]). Implicitly, the fact that this set is complete could be supported by the agreement between the predictions of formula (3.95) without thermal radiation, in the limit $R_g = \infty$ and $T_{g0} = \text{const}$, and those of Cooper [1].

The norm of v_n with weight b is calculated as

$$\begin{aligned} ||v_n||^2 &= \int_0^{R_g} v_n^2 b \mathrm{d}R = \int_0^{R_b} \left[\frac{\sin(\lambda_n a_d R)}{\sin(\lambda_n a_d R_d)} \right]^2 c_l \rho_l \mathrm{d}R \\ &\quad + \int_{R_b}^{R_g} \left[\frac{\sin(\lambda_n a_g (R - R_g))}{\sin(\lambda_n a_g (R_d - R_g))} \right]^2 c_{pg} \rho_g \mathrm{d}R \\ &= \frac{c_l \rho_l}{2 \sin^2(\lambda_n a_d R_d)} \left[R_d - \frac{\sin(2\lambda_n a_d R_d)}{2\lambda_n a_d} \right] \\ &\quad + \frac{c_{pg} \rho_g}{2 \sin^2(\lambda_n a_g (R_d - R_g))} \left[R_g - R_d + \frac{\sin(2\lambda_n a_d (R_d - R_g))}{2\lambda_n a_g} \right] \\ &= \frac{c_l \rho_l R_d}{2 \sin^2(\lambda_n a_d R_d)} + \frac{c_{pg} \rho_g (R_g - R_d)}{2 \sin^2(\lambda_n a_g (R_d - R_g))} - \frac{k_l - k_g}{2 R_d \lambda_n^2}. \end{aligned} \tag{B.15}$$

In deriving (B.15), we took into account (B.13). Since all functions v_n satisfy boundary conditions (B.6), the function u defined by expression (B.4) satisfies boundary conditions (B.3). Let us expand $RP(t, R)$ in a series over v_n:

$$RP(t, R) = \sum_{n=1}^{\infty} p_n(t) v_n(R), \tag{B.16}$$

where

$$p_n(t) = \frac{1}{||v_n||^2} \int_0^{R_g} RP(t, R) v_n(R) b \mathrm{d}R.$$

Since $P(t, R) = 0$ at $R > R_d$, the latter formula can be simplified to

$$p_n(t) = \frac{c_l \rho_l}{||v_n||^2} \int_0^{R_d} RP(t, R) v_n(R) \mathrm{d}R.$$

On substituting (B.4) and (B.16) into (B.1)), we obtain

$$\sum_{n=1}^{\infty} \Theta_n^{'}(t) v_n(R) = -\sum_{n=1}^{\infty} \Theta_n(t) \lambda_n^2 v_n(R) + \sum_{n=1}^{\infty} p_n(t) v_n(R). \tag{B.17}$$

In deriving (B.17), we took into account that the functions $v_n(R)$ satisfy (B.5) for $\lambda = \lambda_n$. Equation (B.17) is satisfied if and only if

$$\Theta_n^{'}(t) = -\lambda_n^2 \Theta_n(t) + p_n(t). \tag{B.18}$$

The initial condition for $\Theta_n(t)$ can be obtained after substituting expression (B.4) into the initial condition (B.2) for u:

$$\sum_{n=1}^{\infty} \Theta_n(0) v_n(R) = -T_0 R. \tag{B.19}$$

From the orthogonality of v_n with weight b, we obtain from (B.19) that

$$\Theta_n(0) = \frac{1}{||v_n||^2} \int_0^{R_g} (-T_0 R) v_n(R) b \mathrm{d}R.$$

$$= -\frac{c_l \rho_l}{||v_n||^2 \sin(\lambda_n a_d R_d)} \int_0^{R_d} T_0(R) R \sin(\lambda_n a_d R) \mathrm{d}R$$

$$- \frac{c_{pg}\rho_g}{||v_n||^2 \sin(\lambda_n a_g (R_d - R_g))} \int_{R_d}^{R_g} T_0(R) R \sin(\lambda_n a_g (R - R_g)) \mathrm{d}R.$$

If $T_{g0}(R) = T_{g0}(R_g) = \text{const}$ and $T_{d0}(R) = \text{const}$, then

$$T_0 \equiv T_0(R) = \begin{cases} T_{g0}(R_g) - T_{b0}(R_d) & \text{when } R \le R_d \\ 0 & \text{when } R_d < R \le R_g, \end{cases}$$

and the expression for $\Theta_n(0)$ can be further simplified to

$$\Theta_n(0) = -\frac{c_l \rho_l T_0}{||v_n||^2 \sin(\lambda_n a_d R_d)} \int_0^{R_d} R \sin(\lambda_n a_d R) \mathrm{d}R$$

$$= \frac{T_0 \sqrt{k_l c_l \rho_l}}{\lambda_n ||v_n||^2} \left[R_d \cot(\lambda_n a_d R_d) - \frac{1}{\lambda_n a_d} \right]. \tag{B.20}$$

The solution to (B.18) subject to the initial condition (B.20) can be written as

$$\Theta_n(t) = \exp\left(-\lambda_n^2 t\right) \Theta_n(0) + \int_0^t \exp\left(-\lambda_n^2 (t - \tau)\right) p_n(\tau) \mathrm{d}\tau. \tag{B.21}$$

Equation (3.95) follows from the definition of u and Eqs. (B.4) and (B.21). Equation (3.8) follows from (3.95) in the limit when $P = 0$.

References

1. Cooper, F. (1977). Heat transfer from a sphere to an infinite medium. *International Journal Heat and Mass Transfer*, *20*, 991–993.
2. Sazhin, S. S., Krutitskii, P. A., Martynov, S. B., Mason, D., Heikal, M. R., & Sazhina, E. M. (2007). Transient heating of a semitransparent spherical body. *International Journal of Thermal Science*, *46*(5), 444–457.
3. Vladimirov, V. S. (1971). *Equations of mathematical physics*. New York: Marcel Dekker.

Appendix C
Proof of Orthogonality of $v_n(R)$ with the Weight b

Using expressions (B.14) for $v_n(R)$, we can write, for $n \neq m$ [1],

$$
\begin{aligned}
I_{nm} \equiv \int_0^{R_g} v_n(R) v_m(R) b \mathrm{d}R &= \frac{k_l a_d^2}{\sin(\lambda_m a_d R_d)\sin(\lambda_m a_d R_d)} \int_0^{R_d} \sin(\lambda_n a_d R)\sin(\lambda_m a_d R)\mathrm{d}R \\
&+ \frac{k_g a_g^2}{\sin(\lambda_n a_g (R_d - R_g))\sin(\lambda_m a_g (R_d - R_g))} \int_{R_d}^{R_g} \sin(\lambda_n a_g (R - R_g))\sin(\lambda_m a_g (R - R_g))\mathrm{d}R \\
&= \frac{k_l a_d^2}{2\sin(\lambda_n a_d R_d)\sin(\lambda_m a_d R_d)} \left[\frac{\sin((\lambda_n - \lambda_m) a_d R_d)}{(\lambda_n - \lambda_m) a_d} - \frac{\sin((\lambda_n + \lambda_m) a_d R_d)}{(\lambda_n + \lambda_m) a_d} \right] \\
&- \frac{k_g a_g^2}{2\sin(\lambda_n a_g (R_d - R_g))\sin(\lambda_m a_g (R_d - R_g))} \\
&\times \left[\frac{\sin((\lambda_n - \lambda_m) a_g (R_d - R_g))}{(\lambda_n - \lambda_m) a_g} - \frac{\sin((\lambda_n + \lambda_m) a_g (R_d - R_g))}{(\lambda_n + \lambda_m) a_g} \right] \\
&= \frac{1}{2(\lambda_n - \lambda_m)} \left[\frac{k_l a_d \sin((\lambda_n - \lambda_m) a_d R_d)}{\sin(\lambda_n a_d R_d)\sin(\lambda_m a_d R_d)} - \frac{k_g a_g \sin((\lambda_n - \lambda_m) a_g (R_d - R_g))}{\sin(\lambda_n a_g (R_d - R_g))\sin(\lambda_m a_g (R_d - R_g))} \right] \\
&+ \frac{1}{2(\lambda_n + \lambda_m)} \left[- \frac{k_l a_d \sin((\lambda_n + \lambda_m) a_d R_d)}{\sin(\lambda_n a_d R_d)\sin(\lambda_m a_d R_d)} + \frac{k_g a_g \sin((\lambda_n + \lambda_m) a_g (R_d - R_g))}{\sin(\lambda_n a_g (R_d - R_g))\sin(\lambda_m a_g (R_d - R_g))} \right] \\
&= \frac{\left[k_l a_d \left(\cot(\lambda_m a_d R_d) - \cot(\lambda_n a_d R_d)\right) - k_g a_g \left(\cot(\lambda_m a_d (R_d - R_g)) - \cot(\lambda_n a_d (R_d - R_g))\right)\right]}{2(\lambda_n - \lambda_m)} \\
&+ \frac{\left[-k_l a_d \left(\cot(\lambda_m a_d R_d) + \cot(\lambda_n a_d R_d)\right) + k_g a_g \left(\cot(\lambda_m a_d (R_d - R_g)) + \cot(\lambda_n a_d (R_d - R_g))\right)\right]}{2(\lambda_n + \lambda_m)}.
\end{aligned}
$$

Using Eq. (B.13), we can write

$$
\begin{aligned}
I_{nm} &= \frac{1}{2(\lambda_n - \lambda_m)} \left(\frac{k_l - k_g}{R_d \lambda_m} - \frac{k_l - k_g}{R_d \lambda_n} \right) - \frac{1}{2(\lambda_n + \lambda_m)} \left(\frac{k_l - k_g}{R_d \lambda_m} + \frac{k_l - k_g}{R_d \lambda_n} \right) \\
&= \frac{k_l - k_g}{2 R_d \lambda_n \lambda_m} - \frac{k_l - k_g}{2 R_d \lambda_n \lambda_m} = 0.
\end{aligned}
$$

S. Sazhin, *Droplets and Sprays*, DOI: 10.1007/978-1-4471-6386-2,

Reference

1. Sazhin, S. S., Krutitskii, P. A., Martynov, S. B., Mason, D., Heikal, M. R., & Sazhina, E. M. (2007). Transient heating of a semitransparent spherical body. *International Journal of Thermal Science*, *46*(5), 444–457.

Appendix D
Derivation of Formula (4.86)

In this appendix, the details of the derivation of formula (4.86) are presented, following [4].

Preliminary Analysis

Following Kartashov [1], we introduce a new variable

$$r = R/R_d(t)$$

and a new function

$$F(t,r) = u(t,R).$$

This new variable allows us to reduce the problem with a moving boundary to one with a stationary boundary. Note that

$$0 \leq r \leq 1 \quad \text{when} \quad 0 \leq R \leq R_d(t).$$

Since

$$u'_t = F'_t + F'_r r'_t = F'_t - r\frac{R'_d(t)}{R_d(t)}F'_r; \quad u'_R = F'_r r'_R = \frac{F'_r}{R_d(t)}; \quad u''_{RR} = \frac{F''_{rr}}{R_d^2(t)},$$

we can rewrite (4.79) as

$$R_d^2(t)F'_t = \kappa F''_{rr} + rR'_d(t)R_d(t)F'_r. \tag{D.1}$$

Equation (D.1) is identical to the one studied in [2], where the distribution of temperature in the melting region was considered (plane problem).

Equation (D.1) is to be solved at $t \in [0, t_e]$ (or $t \in [t_0, t_0 + \Delta t]$) and $0 \leq r \leq 1$. Initial and boundary conditions for this equation can be represented as

S. Sazhin, *Droplets and Sprays*, DOI: 10.1007/978-1-4471-6386-2,

$$F|_{t=0} = R_{d0} r T_0(r R_{d0}), \quad 0 \le r \le 1,$$

$$F|_{r=0} = 0, \quad \left(F_r' + \tilde{H}(t)F\right)\Big|_{r=1} = \tilde{\mu}(t), \quad 0 \le t \le t_e \quad (\text{or } t \in [t_0, t_0 + \Delta t]),$$

where $\tilde{H}(t) = H(t)R_d(t)$, $\tilde{\mu}(t) = M(t)R_d^2(t)$.

Following Kartashov [1], we introduce the new unknown function $W(t, r)$ via the relation

$$F(t, r) = \frac{1}{\sqrt{R_d(t)}} \exp\left[-\frac{R_d'(t)R_d(t)}{4\kappa} r^2\right] W(t, r). \tag{D.2}$$

From (D.2), we obtain the following expressions for the derivatives:

$$F_t' = \left\{ -\left[\frac{1}{2} R_d^{-3/2}(t) R_d'(t) + R_d^{-1/2}(t) \left(\frac{(R_d'(t))^2 + R_d(t) R_d''(t)}{4\kappa} r^2 \right)\right] W(t, r) + R_d^{-1/2}(t) W_t'(t, r) \right\} \exp\left[-\frac{R_d'(t)R_d(t)}{4\kappa} r^2\right],$$

$$F_r' = \left\{ -\frac{2r R_d'(t) R_d(t)}{4\sqrt{R_d(t)}\kappa} W(t, r) + \frac{1}{\sqrt{R_d(t)}} W_r'(t, r) \right\} \exp\left[-\frac{R_d'(t)R_d(t)}{4\kappa} r^2\right],$$

$$F_{rr}'' = \left\{ \frac{2}{\sqrt{R_d(t)}} \left[-\frac{R_d(t)R_d'(t)}{4\kappa} + 2r^2 \left(\frac{R_d(t)R_d'(t)}{4\kappa}\right)^2 \right] W(t, r) - \frac{4r}{\sqrt{R_d(t)}} \frac{R_d(t)R_d'(t)}{4\kappa} W_r'(t, r) + \frac{1}{\sqrt{R_d(t)}} W_{rr}''(t, r) \right\} \exp\left[-\frac{R_d'(t)R_d(t)}{4\kappa} r^2\right].$$

Recalling (4.83), we can see that $\mathrm{d}^2 R_d/\mathrm{d}t^2 = 0$. Keeping in mind this condition, substitution of the above-mentioned expressions for the derivatives into (D.1) gives

$$R_d^2(t) W_t'(t, r) = \kappa W_{rr}''(t, r), \tag{D.3}$$

where $t \in [0, t_e]$ (or $t \in [t_0, t_0 + \Delta t]$), $0 \le r \le 1$.

In the case of nonzero $\mathrm{d}^2 R_d/\mathrm{d}t^2$ and $P(R)$, Eq. (D.3) would need to be replaced by the following equation (cf. 8.149 in [1]):

$$R_d^2(t) W_t'(t, r) = \kappa W_{rr}''(t, r) + \left[\frac{r^2}{4\kappa}\right] R_d^3 \frac{\mathrm{d}^2 R_d}{\mathrm{d}t^2} W(t, r) + \frac{R_d^2 R}{q_K(r, t)} P(R), \tag{D.4}$$

where

$$q_K(r,t) = \frac{1}{\sqrt{R_d(t)}} \exp\left[-\frac{R_d'(t) R_d(t)}{4\kappa} r^2 \right].$$

Equation (D.4) reduces to (D.3) in the limit when $\mathrm{d}^2 R_d/\mathrm{d}t^2 = 0$ and $P(R) = 0$. Equation (D.3) is to be solved subject to initial and boundary conditions

$$W(t,r)|_{t=0} = W_0(r) \equiv R_{d0}^{3/2} r T_0(r R_{d0}) \exp\left[\frac{R_d'(0) R_{d0}}{4\kappa} r^2 \right], \tag{D.5}$$

$$0 \le r \le 1,$$

$$W(t,r)|_{r=0} = 0, \tag{D.6}$$

$$t \in [0, t_e], \quad (\text{or } t \in [t_0, t_0 + \Delta t])$$

$$\left[W_r'(t,r) + H_0(t) W(t,r) \right]\Big|_{r=1} = \mu_0(t) \equiv \tilde{\mu}(t) \sqrt{R_d(t)} \exp\left[\frac{R_d'(t) R_d(t)}{4\kappa} \right], \tag{D.7}$$

$$t \in [0, t_e] \quad (\text{or } t \in [t_0, t_0 + \Delta t]),$$

where

$$H_0(t) = \tilde{H}(t) - \frac{R_d'(t) R_d(t)}{2\kappa} = \frac{h(t)}{k_l} R_d(t) - 1 - \frac{R_d'(t) R_d(t)}{2\kappa}.$$

Analytical Solution

Let us now simplify the problem further by assuming that $H_0(t) \equiv h_0 = \text{const} > -1$. Since $h = k_g/R_d(t)$ for stationary droplets, the term

$$\frac{h(t)}{k_l} R_d(t)$$

can be simplified to k_g/k_l, and this ratio does not depend on t. Except at the final stage of droplet evaporation, in Diesel engine-like conditions it is typically much larger than

$$\frac{R_d'(t) R_d(t)}{2\kappa}.$$

Hence the time-dependence of $H_0(t)$ during the time step in this case can be ignored. This assumption will be relaxed later.

Our next goal is to find such a change of variables that the inhomogeneous boundary condition (D.7) is replaced by a homogeneous one. This is achieved by the introduction of the new function $V(t, r)$ via the relation

$$W(t, r) = V(t, r) + \frac{\mu_0(t)}{1 + h_0} r. \tag{D.8}$$

Equation (D.8) allows us to rearrange (D.3) to

$$R_d^2(t) V_t'(t, r) = \kappa V_{rr}''(t, r) - \frac{\mu_0'(t)}{1 + h_0} R_d^2(t) r, \tag{D.9}$$

$$t \in [0, t_e] \ \ (\text{or } t \in [t_0, t_0 + \Delta t]), \quad 0 \le r \le 1.$$

The initial and boundary conditions for (D.9) can be represented as

$$V(t, r)|_{t=0} = W_0(r) - \frac{\mu_0(0)}{1 + h_0} r, \quad 0 \le r \le 1,$$

$$V(t, r)|_{r=0} = 0, \quad \left[V_r'(t, r) + h_0 V(t, r) \right] \Big|_{r=1} = 0,$$

$$t \in [0, t_e] \ \ (\text{or } t \in [t_0, t_0 + \Delta t]).$$

In the absence of thermal radiation, (D.9) is identical to Equation (12) in [3]. Following the analysis presented in that paper, we look for the solution of (D.9) in the form

$$V(t, r) = \sum_{n=1}^{\infty} \Theta_n(t) v_n(r), \tag{D.10}$$

where the functions $v_n(r)$ form the full set of nontrivial solutions of the equation

$$\frac{\mathrm{d}^2 v}{\mathrm{d}r^2} + \lambda^2 v = 0, \quad 0 \le r \le 1, \tag{D.11}$$

subject to boundary conditions

$$v|_{r=0} = \left(\frac{\mathrm{d}v}{\mathrm{d}r} + h_0 v \right) \Big|_{r=1} = 0.$$

The general solution to (D.11), namely

$$v(r) = A \cos \lambda r + B \sin \lambda r, \tag{D.12}$$

satisfies the boundary conditions when $A = 0$ and

$$\lambda \cos \lambda + h_0 \sin \lambda = 0. \tag{D.13}$$

The solution to (D.13) gives a set of positive eigenvalues λ_n numbered in ascending order ($n = 1, 2, ...$). If $h_0 = 0$, then $\lambda_n = \pi(n - \frac{1}{2})$. Assuming that $B = 1$, the expressions for the eigenfunctions $v_n(r)$ can be written as

$$v_n(r) = \sin \lambda_n r \qquad (n = 1, 2, ...). \tag{D.14}$$

The solution $\lambda = 0$ is excluded, since it leads to the trivial solution $v_n(r) = 0$.

The value of B is implicitly accounted for by the coefficients $\Theta_n(t)$ in the series (D.10). The functions $v_n(r)$ form a full set of eigenfunctions which are orthogonal for $r \in [0, 1]$. The orthogonality of functions $v_n(r)$ follows from the relation

$$\int_0^1 v_n(r) v_m(r) \mathrm{d}r = \delta_{nm} ||v_n||^2, \tag{D.15}$$

where

$$\delta_{nm} = \begin{cases} 0 & n \neq m \\ 1 & n = m \end{cases},$$

$$||v_n||^2 = \frac{1}{2}\left(1 - \frac{\sin 2\lambda_n}{2\lambda_n}\right) = \frac{1}{2}\left(1 + \frac{h_0}{h_0^2 + \lambda_n^2}\right). \tag{D.16}$$

It is easy to see that the expression for $V(t, r)$ in the form (D.10) with $v_n(r)$ defined by (D.14) satisfies the boundary conditions for (D.9).

The orthogonality of the functions $v_n(r)$ allows us to expand known functions in (D.9) and the initial conditions in the series

$$f(r) \equiv -r/(1 + h_0) = \sum_{n=1}^{\infty} f_n v_n(r), \tag{D.17}$$

$$W_0(r) = \sum_{n=1}^{\infty} q_n v_n(r), \tag{D.18}$$

where

$$f_n = \frac{1}{||v_n||^2} \int_0^1 f(r) v_n(r) \mathrm{d}r = -\frac{\sin \lambda_n}{||v_n||^2 \lambda_n^2},$$

$$q_n = \frac{1}{||v_n||^2} \int_0^1 W_0(r) v_n(r) \mathrm{d}r.$$

Although both series (D.17) and (D.18) converge, the speed of convergence of series (D.17) can be slow. This can create a problem with applications of the new approach to practical engineering problems.

If $T_0(R)$ is a continuously twice differentiable function (this was assumed at the very beginning of the analysis), then this property is retained by the function $W_0(r)$. In this case, one can show that [3]

$$|q_n| < \frac{\text{const}}{\lambda_n^2}. \tag{D.19}$$

Using (D.10) and (D.17), (D.9) can be rewritten as

$$\sum_{n=1}^{\infty}\left(R_d^2(t)\frac{\mathrm{d}\Theta_n(t)}{\mathrm{d}t}+\Theta_n(t)\kappa\lambda_n^2\right)v_n(r)=\sum_{n=1}^{\infty}\left(f_nR_d^2(t)\frac{\mathrm{d}\mu_0(t)}{\mathrm{d}t}\right)v_n(r). \tag{D.20}$$

Both sides of (D.20) are Fourier series with respect to the functions $v_n(r)$. Two Fourier series are equal if and only if their coefficients are equal. This implies that

$$R_d^2(t)\frac{\mathrm{d}\Theta_n(t)}{\mathrm{d}t}+\Theta_n(t)\kappa\lambda_n^2=f_nR_d^2(t)\frac{\mathrm{d}\mu_0(t)}{\mathrm{d}t}. \tag{D.21}$$

Equation (D.21) is to be solved subject to the initial condition

$$\Theta_n(0)=q_n+\mu_0(0)f_n. \tag{D.22}$$

To simplify the notation, henceforward it is assumed that $t_0=0$.

The general solution to the homogeneous equation

$$R_d^2(t)\frac{\mathrm{d}\Theta_n(t)}{\mathrm{d}t}+\Theta_n(t)\kappa\lambda_n^2=0 \tag{D.23}$$

can be represented as

$$\ln\left(\Theta_n(t)/\Theta_n(0)\right)=-\kappa\lambda_n^2\int_0^t\frac{\mathrm{d}t}{R_d^2(t)}. \tag{D.24}$$

The initial condition (D.22) was used. Assuming that $R_d(t)$ is a linear function of t given by (4.86), solution (D.24) can be represented in a more explicit form:

$$\Theta_n(t)=\Theta_n(0)\exp\left[\frac{\kappa\lambda_n^2}{\alpha R_{d0}^2}\left(\frac{1}{1+\alpha t}-1\right)\right]. \tag{D.25}$$

One can see that the function

$$\Theta_{n\,(\mathrm{part})}(t) = f_n \int_0^t \frac{\mathrm{d}\mu_0(\tau)}{\mathrm{d}\tau} \exp\left[\frac{\kappa\lambda_n^2}{\alpha R_{d0}^2}\left(\frac{1}{1+\alpha t} - \frac{1}{1+\alpha\tau}\right)\right]\mathrm{d}\tau \tag{D.26}$$

satisfies (D.21). Hence, this function can be considered a particular solution to (D.21). Integration by parts in (D.26) allows us to represent $\Theta_{n\,(\mathrm{part})}(t)$ as

$$\Theta_{n\,(\mathrm{part})}(t) = f_n \left\{\mu_0(t) - \mu_0(0)\exp\left[-\frac{\kappa\lambda_n^2 t}{R_{d0}R_d(t)}\right]\right.$$
$$\left. - \exp\left[\frac{\kappa\lambda_n^2}{\alpha R_{d0}R_d(t)}\right]\int_0^t \frac{\mu_0(\tau)\kappa\lambda_n^2}{R_d^2(\tau)}\exp\left[-\frac{\kappa\lambda_n^2}{\alpha R_{d0}R_d(\tau)}\right]\mathrm{d}\tau\right\}. \tag{D.27}$$

Using (D.25) and (D.26), the solution to (D.21) can be represented as

$$\Theta_n(t) = \Theta_n(0)\exp\left[\frac{\kappa\lambda_n^2}{\alpha R_{d0}^2}\left(\frac{1}{1+\alpha t} - 1\right)\right]$$
$$+ f_n \int_0^t \frac{\mathrm{d}\mu_0(\tau)}{\mathrm{d}\tau}\exp\left[\frac{\kappa\lambda_n^2}{\alpha R_{d0}^2}\left(\frac{1}{1+\alpha t} - \frac{1}{1+\alpha\tau}\right)\right]\mathrm{d}\tau. \tag{D.28}$$

Using (D.27) and (D.22), we can write an alternative formula for $\Theta_n(t)$:

$$\Theta_n(t) = q_n \exp\left[-\frac{\kappa\lambda_n^2 t}{R_{d0}R_d(t)}\right] + f_n\mu_0(t)$$
$$- f_n\kappa\lambda_n^2 \int_0^t \frac{\mu_0(\tau)}{R_d^2(\tau)}\exp\left[\frac{\kappa\lambda_n^2}{\alpha R_{d0}}\left(\frac{1}{R_d(t)} - \frac{1}{R_d(\tau)}\right)\right]\mathrm{d}\tau. \tag{D.29}$$

Note that $\Theta_n(t)$ in the form (D.28) satisfies (D.21), while $\Theta_n(t)$ in the form (D.29) does not satisfy it. This is related to the fact that (D.21) was derived under the assumption that the series (D.10), after being substituted into (D.9), can be differentiated term by term (the derivative of the series is equal to the series of the derivatives). This assumption is valid when $\Theta_n(t)$ takes the form (D.28), but it is not valid when $\Theta_n(t)$ is of the form (D.29), since

$$\mu_0(t)\frac{\mathrm{d}^2}{\mathrm{d}r^2}\left(\sum_{n=1}^{\infty} f_n v_n\right) \neq \mu_0(t)\sum_{n=1}^{\infty} f_n \frac{\mathrm{d}^2 v_n}{\mathrm{d}r^2}$$

(the latter series diverges). Note that the series (D.10) satisfies (D.9) regardless of whether $\Theta_n(t)$ is taken in the form (D.28) or in the form (D.29).

In the limit $\alpha t \ll 1$, Eq. (D.28) reduces to (A11) of [3] if the effects of thermal radiation are ignored. The latter equation was derived assuming a stationary droplet boundary during the time step of integration.

On substituting (D.28) or (D.29) into (D.10), we obtain the required solution for $V(t, r)$. Since exponential terms are less than or equal to 1, from the definition of f_n and condition (D.19), we can see that the series (A.13) converges absolutely and uniformly, since $\lambda_n^{-2} < n^{-2}$ for $n > 1$ (see [3]).

The final equation for temperature distribution inside the droplet can be represented as

$$T(R) = \frac{1}{R\sqrt{R_d(t)}} \exp\left[-\frac{\alpha R_{d0} R^2}{4\kappa R_d(t)}\right] \left[\sum_{n=1}^{\infty} \Theta_n(t) \sin\left(\lambda_n \frac{R}{R_d(t)}\right) + \frac{\mu_0(t)}{1+h_0} \frac{R}{R_d(t)}\right], \tag{D.30}$$

where the Θ_n are given by (D.28) or (D.29).

Note that strictly speaking, (D.30) is an implicit function of droplet temperature, since α depends on droplet surface temperature T_s. Hence, an iteration process would be required. However, as follows from calculations performed in [4], except at the very final stage of droplet evaporation, for sufficiently small time steps, the value of T_s can be taken equal to that obtained at the end of the previous time step. This allows us to consider (D.30) an explicit formula for $T(R)$.

Equation (D.30) is identical to formula (4.86).

Analysis of the Case $h_0 \neq$ const

Let us now relax our assumption that $H_0(t) \equiv h_0 = \text{const} > -1$ and assume that

$$H_0(t) = h_0 + h_1(t), \tag{D.31}$$

where $h_0 = \text{const} > -1$. Note that many of the following equations are greatly simplified when $h_0 = 0$. In view of (D.31), we can rewrite the boundary condition at $r = 1$ for (D.3) in the form

$$\left[W_r'(t, r) + h_0 W(t, r)\right]\Big|_{r=1} = \mu_0(t) - h_1(t) W(t, 1) \equiv \hat{\mu}_0(t). \tag{D.32}$$

Assuming that $\hat{\mu}_0(t)$ is known, we can formally use the previously obtained solutions (D.8) and (D.10) to represent the solution to problem (D.3)–(D.7) in the form

$$W(t,r) = \frac{\hat{\mu}_0(t)}{1+h_0} r + V(t,r) = \sum_{n=1}^{\infty} \sin(\lambda_n r) q_n \exp\left[-\frac{\kappa \lambda_n^2 t}{R_{d0} R_d(t)}\right]$$
$$- \sum_{n=1}^{\infty} \sin(\lambda_n r) f_n \kappa \lambda_n^2 \int_0^t \frac{\hat{\mu}_0(\tau)}{R_d^2(\tau)} \exp\left[\frac{\kappa \lambda_n^2}{\alpha R_{d0}} \left(\frac{1}{R_d(t)} - \frac{1}{R_d(\tau)}\right)\right] \mathrm{d}\tau. \tag{D.33}$$

In contrast to the previous case of $H_0(t) = \text{const}$, Equation (D.33) does not give us an explicit solution for $W(t, r)$, since $\hat{\mu}_0(t)$ depends on $W(t, 1)$.

Equation (D.33) can be represented in a more compact form:

$$W(t,r) = \mathscr{V}(t,r) - \int_0^t \hat{\mu}_0(\tau) G(t,\tau,r) \mathrm{d}\tau, \tag{D.34}$$

where

$$\mathscr{V}(t,r) = \sum_{n=1}^{\infty} \sin(\lambda_n r) q_n \exp\left[-\frac{\kappa \lambda_n^2 t}{R_{d0} R_d(t)}\right]$$

$$G(t,\tau,r) = -\sum_{n=1}^{\infty} \sin(\lambda_n r) \frac{\kappa \sin(\lambda_n)}{R_d^2(\tau) ||v_n||^2} \exp\left[\frac{\kappa \lambda_n^2}{\alpha R_{d0}} \left(\frac{1}{R_d(t)} - \frac{1}{R_d(\tau)}\right)\right].$$

Explicit expressions for f_n are used in these formulas. Both functions $\mathscr{V}(t, r)$ and $G(t, \tau, r)$ are assumed to be known.

Keeping in mind (D.32), we can rewrite (D.34) as

$$W(t,r) = \mathscr{V}(t,r) - \int_0^t \left[\mu_0(\tau) - h_1(\tau) W(\tau, 1)\right] G(t,\tau,r) \mathrm{d}\tau. \tag{D.35}$$

This is an integral representation for a solution to problem (D.3)–(D.7) for time-dependent $H_0(t)$ given by (D.31). For $r = 1$, the integral representation (D.35) reduces to the Volterra integral equation of the second kind for the function $W(t, 1)$:

$$W(t,1) = \mathscr{V}(t,1) - \int_0^t \left[\mu_0(\tau) - h_1(\tau) W(\tau, 1)\right] G(t,\tau,1) \mathrm{d}\tau. \tag{D.36}$$

One can show that (see Eq. 28 of [3])

$$\sin^2 \lambda_n = \frac{\lambda_n^2}{\lambda_n^2 + h_0^2}. \tag{D.37}$$

From (D.16) and (D.37), we obtain

$$G(t, \tau, 1) = -\frac{2\kappa}{R_d^2(\tau)} \sum_{n=1}^{\infty} \frac{\lambda_n^2}{h_0^2 + h_0 + \lambda_n^2} \exp\left[\frac{\kappa\lambda_n^2}{\alpha R_{d0}} \left(\frac{1}{R_d(t)} - \frac{1}{R_d(\tau)}\right)\right]. \quad \text{(D.38)}$$

Let us now introduce a new function $G_1(t, \tau, r)$ defined as

$$G_1(t, \tau, r) = R_d^2(\tau) G(t, \tau, r), \quad \text{(D.39)}$$

where $0 \leq \tau \leq t$.

The convergence of the series in $G_1(t, \tau, r)$ for $(t - \tau) \in [\delta, -1/\alpha)$, where δ is an arbitrary small positive number, is proven in Appendix E. Moreover, in the same appendix, the validity of the estimate

$$|G_1(t, \tau, r)| \leq \tilde{c}/\sqrt{t - \tau}, \qquad t - \tau \in (0, t_0] \quad \text{(D.40)}$$

is shown, where $\tilde{c}$ is a constant, and t_0 is an arbitrary fixed time in the range $(0, t_e)$ (or $(0, \Delta t)$).

All results in Appendix E are applicable to the first series in (D.33) if we assume that $\tau = 0$.

Equation (D.36) has a unique solution, although this solution cannot be found in an explicit form. The numerical solution can be found as described in Appendix F. Once the solution to this equation has been found, we can substitute it into the integral representation (D.35) and obtain the required solution to the initial and boundary value problem (D.3)–(D.7). The required distribution of T is found to be

$$T(t, R) = \frac{1}{R\sqrt{R_d(t)}} \exp\left[-\frac{R_d'(t) R^2}{4\kappa R_d(t)}\right] W(t, R/R_d(t)). \quad \text{(D.41)}$$

When $h_1(t) = 0$ and $\alpha t \ll 1$, this solution reduces to that given by equation (16) of [3]. Note that in the case of $h_0 = 0$, we have $\lambda_n = \pi(n - (1/2))$ and $||v_n||^2 = 1/2$ in all equations.

References

1. Kartashov, E. M. (2001). *Analytical methods in the heat transfer theory in solids.* Moscow: Vysshaya Shkola (in Russian).
2. Savović, S., & Caldwell, J. (2003). Finite difference solution of one-dimensional Stefan problem with periodic boundary conditions. *International Journal of Heat and Mass Transfer*, 46, 2911–2916.

3. Sazhin, S. S., Krutitskii, P. A., Abdelghaffar, W. A., Mikhalovsky, S. V., Meikle, S. T., & Heikal, M. R. (2004). Transient heating of diesel fuel droplets. *International Journal Heat and Mass Transfer*, *47*, 3327–3340.
4. Sazhin, S. S., Krutitskii, P. A., Gusev, I. G., & Heikal, M. R. (2010). Transient heating of an evaporating droplet. *International Journal Heat and Mass Transfer*, *53*, 2826–2836.

Appendix E
Convergence of the Series in $G_1(t, \tau, r)$

In this appendix, the convergence of the series in $G_1(t, \tau, r)$ from the previous appendix is investigated, following [1].

Let us assume that

$$0 \leq \tau \leq t < t_e = -1/\alpha$$

and introduce the new function

$$f(t, \tau) \equiv -\frac{1}{\alpha R_{d0}} \left(\frac{1}{R_d(t)} - \frac{1}{R_d(\tau)} \right) = \frac{t - \tau}{R_d(t) R_d(\tau)}. \tag{E.1}$$

In the case of a time step, t_e needs to be replaced by Δt. As was done earlier, to simplify the notation it is assumed that t_0 (the start of the time step) is equal to zero. This comment and assumption apply to both Appendices E and F. Note that

$$f(t, \tau) \geq \frac{t - \tau}{R_{d0}^2}, \tag{E.2}$$

since $\alpha < 0$ and $R_d(t) \leq R_{d0}$.

It follows from (D.16) and the estimate $\lambda_n > n$ for $n > 1$ that $||v_n||^2 > 1/4$ for $n > 1$. Therefore,

$$||v_n||^2 \geq c_0, \quad n \geq 1, \tag{E.3}$$

where $c_0 = \min \left\{ || \; v_1 \; ||^2, 1/4 \right\}$ is a positive constant.

Condition (E.2) allows us to make the following estimate:

$$\exp \left[-\kappa \lambda_n^2 f(t, \tau) \right] \leq \exp \left[-\kappa n^2 \frac{t - \tau}{R_{d0}^2} \right], \quad n > 1, \tag{E.4}$$

where we took into account that $\lambda_n > n$ for $n > 1$. Using (E.4), one can conclude that the series in $G_1(t, \tau, r)$ converges absolutely and uniformly to a continuous function for $(t - \tau, r) \in [\delta, -1/\alpha) \times [0, 1]$ for any small $\delta > 0$ since:

$$\exp\left[-\kappa n^2 \frac{t-\tau}{R_{d0}^2}\right] \le \exp\left[-\kappa n^2 \frac{\delta}{R_{d0}^2}\right], \quad |\sin\lambda_n r| \le 1. \tag{E.5}$$

Indeed, each term with $n > 1$ in the series in $G_1(t, \tau, r)$ for $(t - \tau, r) \in [\delta, -1/\alpha) \times [0, 1]$ can be majorised by the corresponding term of the convergent number series

$$\kappa c_0^{-1} \exp\left(-\kappa n^2 \frac{\delta}{R_{d0}^2}\right).$$

Now we estimate $G_1(t, \tau, r)$ for small $t - \tau > 0$. Inequalities (E.3) and (E.4) allow us to write

$$\begin{aligned} |G_1(t, \tau, r)| &\le c_0^{-1}\kappa \left\{1 + \sum_{n=2}^{\infty} \exp\left[-\kappa n^2 f(t, \tau)\right]\right\} \\ &\le c_0^{-1}\kappa \left\{1 + \sum_{n=2}^{\infty} \exp\left[-\kappa n^2 (t-\tau)/R_{d0}^2\right]\right\} \equiv \tilde{G}(t-\tau). \end{aligned} \tag{E.6}$$

The sum $\sum_{n=2}^{\infty} \exp\left[-\kappa n^2 (t-\tau)/R_{d0}^2\right]$ can be considered a sum of areas of polygons of unit width placed under the curve $\exp\left[-\kappa y^2 (t-\tau)/R_{d0}^2\right]$. This sum is less than the area under this curve. Hence,

$$\begin{aligned} \sum_{n=2}^{\infty} \exp\left[-\kappa n^2 (t-\tau)/R_{d0}^2\right] &< \int_1^{\infty} \exp\left[-\kappa y^2 (t-\tau)/R_{d0}^2\right] \mathrm{d}y \\ &< \int_0^{\infty} \exp\left[-\kappa y^2 (t-\tau)/R_{d0}^2\right] \mathrm{d}y = \frac{R_{d0}}{\sqrt{\kappa(t-\tau)}} \int_0^{\infty} \exp\left[-z^2\right] \mathrm{d}z \\ &= \frac{R_{d0}\sqrt{\pi}}{2\sqrt{\kappa(t-\tau)}}. \end{aligned} \tag{E.7}$$

On substituting (E.7) into (E.6) we obtain:

$$\begin{aligned} |G_1(t, \tau, r)| &\le \tilde{G}(t-\tau) < c_0\kappa \left[1 + \frac{R_{d0}\sqrt{\pi}}{2\sqrt{\kappa(t-\tau)}}\right] < \tilde{c}/\sqrt{t-\tau}, \\ & t-\tau \in (0, t_{00}], \end{aligned} \tag{E.8}$$

for any small fixed $t_{00} \in (0, -1/\alpha)$. The new constant $\tilde{c}$ depends on t_{00}. Inequality (E.8) holds uniformly for $r \in [0, 1]$.

Reference

1. Sazhin, S. S., Krutitskii, P. A., Gusev, I. G., & Heikal, M. R. (2010). Transient heating of an evaporating droplet. *International Journal Heat and Mass Transfer, 53*, 2826–2836.

Appendix F
Numerical Solution of Equation (D.36)

In this appendix, the scheme of the numerical solution of (D.36) is described, following [1].

Let $\psi(t) \equiv W(t, 1)$ and rewrite (D.36) as

$$\psi(t) = \mathscr{V}(t, 1) - \int_0^t [\mu_0(\tau) - h_1(\tau)\psi(\tau)]\, G(t, \tau, 1)\mathrm{d}\tau. \tag{F.1}$$

We look for the solution of (F.1) for $t \in [0, \hat{t}]$, where $\hat{t}$ is a constant, $\hat{t} < t_e$. Let $\Delta t = \hat{t}/N$ and $t_n = n\Delta t$, where N is the total number of time steps, $n = 0, 1, \ldots, N$ is the number of the current time step. Note that $t_0 = 0$ and $t_N = \hat{t}$. Discretisation of (F.1) gives

$$\psi(t_n) = \mathscr{V}(t_n, 1) - \sum_{j=1}^{n} \int_{t_{j-1}}^{t_j} [\mu_0(\tau) - h_1(\tau)\psi(\tau)]\, G(t_n, \tau, 1)\mathrm{d}\tau, \tag{F.2}$$

where $n = 1, \ldots, N$. Note that $\psi(t_0) = \psi(0) = \mathscr{V}(0, 1) = W_0(1)$ is a known constant.

The first $(n - 1)$ integrals in this sum can be approximated as

$$\int_{t_{j-1}}^{t_j} [\mu_0(\tau) - h_1(\tau)\psi(\tau)]\, G(t_n, \tau, 1)\mathrm{d}\tau$$
$$\approx \left\{\mu_0(\tau_j) - h_1(\tau_j)\left[\psi(t_j) + \psi(t_{j-1})\right]/2\right\} G(t_n, \tau_j, 1)\Delta t, \tag{F.3}$$

where $j = 1, 2, ..., n - 1$, $\tau_j = t_j - \frac{1}{2}\Delta t$. Approximation (F.3) is valid, since all functions in the integrand are continuous, and we look for the solution in the class of continuous functions.

S. Sazhin, *Droplets and Sprays*, DOI: 10.1007/978-1-4471-6386-2,

In approximation (F.3), the known functions are taken at $\tau = \tau_j$ (middle of the range $[t_{j-1}, t_j]$), while the unknown functions are taken as the averages of the values at the endpoints t_{j-1} and t_j.

The last term in the sum in (F.2) requires special investigation, since the kernel $G(t_n, \tau, 1)$ in the integrand becomes singular when $\tau \to t_n - 0$ (see estimate (E.8)). All other functions in this integrand, including the unknown function $\psi(t)$, are assumed continuous. Hence, we can write

$$\int_{t_{n-1}}^{t_n} [\mu_0(\tau) - h_1(\tau)\psi(\tau)]\, G(t_n, \tau, 1)\mathrm{d}\tau$$
$$\approx \left\{\mu_0(\tau_n) - h_1(\tau_n)\frac{\psi(t_n) + \psi(t_{n-1})}{2}\right\} \int_{t_{n-1}}^{t_n} G(t_n, \tau, 1)\mathrm{d}\tau. \qquad \text{(F.4)}$$

In view of series (D.38), we can write

$$\int_{t_{n-1}}^{t_n} G(t_n, \tau, 1)\mathrm{d}\tau$$
$$= -2\kappa \sum_{m=1}^{\infty} \frac{\lambda_m^2}{h_0^2 + h_0 + \lambda_m^2} \int_{t_{n-1}}^{t_n} \frac{1}{R_d^2(\tau)} \exp\left[\frac{\kappa\lambda_m^2}{\alpha R_{d0}}\left(\frac{1}{R_d(t_n)} - \frac{1}{R_d(\tau)}\right)\right] \mathrm{d}\tau$$
$$= -2\kappa \sum_{m=1}^{\infty} \frac{\lambda_m^2}{h_0^2 + h_0 + \lambda_m^2} \frac{1}{\kappa\lambda_m^2} \exp\left[\frac{\kappa\lambda_m^2}{\alpha R_{d0}}\left(\frac{1}{R_d(t_n)} - \frac{1}{R_d(\tau)}\right)\right]\Bigg|_{\tau=t_{n-1}}^{\tau=t_n}$$
$$= -2\kappa \sum_{m=1}^{\infty} \frac{\lambda_m^2}{h_0^2 + h_0 + \lambda_m^2} \frac{1}{\kappa\lambda_m^2} \left\{1 - \exp\left[\frac{\kappa\lambda_m^2}{\alpha R_{d0}}\left(\frac{1}{R_d(t_n)} - \frac{1}{R_d(t_{n-1})}\right)\right]\right\}$$
$$= -2 \sum_{m=1}^{\infty} \frac{1}{h_0^2 + h_0 + \lambda_m^2} \left\{1 - \exp\left[\frac{-\kappa\lambda_m^2\Delta t}{R_d(t_n)R_d(t_{n-1})}\right]\right\}$$
$$= -\frac{1}{1 + h_0} + 2 \sum_{m=1}^{\infty} \frac{1}{h_0^2 + h_0 + \lambda_m^2} \exp\left[\frac{-\kappa\lambda_m^2\Delta t}{R_d(t_n)R_d(t_{n-1})}\right] \equiv g_n. \qquad \text{(F.5)}$$

If $h_0 = 0$ then $\lambda_m = \pi(m - (1/2))$ in series (F.5). The combination of formulas (F.3)–(F.5) allows us to present the discretised form of (F.1) (Equation (F.2)) as follows:

$$\psi(t_n) = \mathscr{V}(t_n, 1) - \{\mu_0(\tau_n) - h_1(\tau_n)\,[\psi(t_n) + \psi(t_{n-1})]/2\}\, g_n \\ - \sum_{j=1}^{n-1} \{\mu_0(\tau_j) - h_1(\tau_j)\,[\psi(t_j) + \psi(t_{j-1})]/2\}\, G(t_n, \tau_j, 1)\Delta t, \tag{F.6}$$

where $n = 1, 2, \ldots, N$, and g_n is given by series (F.5).

Equation (F.6) can be rearranged into a form particularly convenient for numerical analysis:

$$\psi(t_n) = \frac{1}{1 - 0.5h_1(\tau_n)g_n} \Bigg\{ \mathscr{V}(t_n, 1) - \left[\mu_0(\tau_n) - \frac{h_1(\tau_n)\psi(t_{n-1})}{2}\right] g_n \\ - \sum_{j=1}^{n-1} \{\mu_0(\tau_j) - h_1(\tau_j)\,[\psi(t_j) + \psi(t_{j-1})]/2\}\, G(t_n, \tau_j, 1)\Delta t \Bigg\}. \tag{F.7}$$

For $n = 1$, the sum in formula (F.7) is equal to zero, and $\psi(t_0)$ is a known constant (see above). This allows us to calculate $\psi(t_1)$ explicitly using formula (F.7). Once $\psi(t_1)$ has been calculated, we can use formula (F.7) for calculation of $\psi(t_2)$, etc. At the nth step, formula (F.7) is used for calculation of $\psi(t_n)$ using the values of $\psi(t_0)$, $\psi(t_1)$, ..., $\psi(t_{n-1})$ calculated at the previous steps. At this step, all terms in the sum $\sum_{j=1}^{n-1}$ are already known.

Once we have obtained the numerical solution to the integral equation (D.36), we are in a position to calculate numerically the function $W(t, r)$ using its integral representation (D.34), where

$$\hat{\mu}_0(t) = \mu_0(t) - h_1(t)\psi(t).$$

Using the same discretisation by t and τ as above, we can present the discretised form of this representation as

$$W(\hat{t}, r) = \mathscr{V}(\hat{t}, r) - \sum_{j=1}^{N} \int_{t_{j-1}}^{t_j} \hat{\mu}_0(\tau) G(\hat{t}, \tau, r) \mathrm{d}\tau \\ = \mathscr{V}(\hat{t}, r) - \sum_{j=1}^{N-1} \frac{\hat{\mu}_0(t_{j-1}) + \hat{\mu}_0(t_j)}{2} G(\hat{t}, \tau_j, r)\Delta t - \frac{\hat{\mu}_0(t_{N-1}) + \hat{\mu}_0(t_N)}{2} \\ \times \int_{t_{N-1}}^{t_N} G(t_N, \tau, r) \mathrm{d}\tau. \tag{F.8}$$

Note that $t_N = \hat{t}$. If $N = 1$, then the sum in (F.8) is equal to zero. The last integral in (F.8) is improper and needs to be calculated separately. Keeping in mind the definition of $G(t, \tau, r)$ and almost repeating the derivation of (F.5), we can write

$$\int\limits_{t_{N-1}}^{t_N} G(t_N, \tau, r)\mathrm{d}\tau = -2\sum_{m=1}^{\infty} \frac{h_0^2 + \lambda_m^2}{h_0^2 + h_0 + \lambda_m^2} \frac{\sin\lambda_m \sin\lambda_m r}{\lambda_m^2} \times \left\{1 - \exp\left[\frac{-\kappa\lambda_m^2 \Delta t}{R_d(t_N)R_d(t_{N-1})}\right]\right\}$$

$$= -\frac{r}{1+h_0} + 2\sum_{m=1}^{\infty} \frac{h_0^2 + \lambda_m^2}{h_0^2 + h_0 + \lambda_m^2} \frac{\sin\lambda_m \sin\lambda_m r}{\lambda_m^2} \times \exp\left[\frac{-\kappa\lambda_m^2 \Delta t}{R_d(t_N)R_d(t_{N-1})}\right].$$

On substituting the latter equation into (F.8), and recalling the definition of $\hat{\mu}_0(t_j)$, we obtain the required value of $W(\hat{t}, r)$.

Reference

1. Sazhin, S. S., Krutitskii, P. A., Gusev, I. G., & Heikal, M. R. (2010). Transient heating of an evaporating droplet. *International Journal Heat and Mass Transfer*, *53*, 2826–2836.

Appendix G
Numerical Calculation of the Improper Integrals

The integrals in (4.104) and (4.116) have the same type of integrable singularity as the integral in (4.98). The following analysis will focus on the latter equation, which will enable us to simplify the notation, following [1]. Let us rewrite this equation as

$$v(R,\hat{t}) = \sum_{j=1}^{N} \int_{t_{j-1}}^{t_j} v(\tau) G(\hat{t},\tau,R) \mathrm{d}\tau, \tag{G.1}$$

where $\hat{t} = t_N$, $t_n = n\Delta t$, $n = 0, 1, 2, \ldots, N$, $\Delta t = \hat{t}/N$. In all integrals, we can replace $v(\tau)$ with the average values over the corresponding time interval $(v(t_{j-1}) + v(t_j))/2$. Moreover, in all integrals except the last one, we can replace $G(\hat{t},\tau,R)$ with $G(\hat{t},\tau_j,R)$, where $\tau_j = (t_{j-1}+t_j)/2$. As a result, (G.1) can be represented in a more explicit form:

$$v(R,\hat{t}) = \sum_{j=1}^{N-1} \frac{v(t_{j-1}) + v(t_j)}{2} G(\hat{t},\tau_j,R)\Delta t + \frac{v(t_{N-1}) + v(t_N)}{2} \int_{t_{N-1}}^{t_N} G(\hat{t},\tau,R) \mathrm{d}\tau. \tag{G.2}$$

Firstly, we assume that an R chosen a priori is not equal to $R_d(\hat{t})$. In this case, $G(\hat{t},\tau,R)$, as defined by (4.99), approaches 0 as $\tau \to \hat{t} - 0$. Hence the singularity in the integrand is not present, and the last time step can be treated like all the previous time steps. This allows us to simplify (G.2) to

$$v(R,\hat{t}) = \sum_{j=1}^{N} \frac{v(t_{j-1}) + v(t_j)}{2} G(\hat{t},\tau_j,R)\Delta t. \tag{G.3}$$

When $R = R_d(\hat{t})$, the first exponent in (4.99) tends to 1 as $\tau \to \hat{t} - 0$. This leads to a singularity $(\hat{t}-\tau)^{-1/2}$ in the integrand in (G.2). As a result, the integral in this equation can be represented as

S. Sazhin, *Droplets and Sprays*, DOI: 10.1007/978-1-4471-6386-2,

$$\int_{t_{N-1}}^{t_N} G(\hat{t}, \tau_N, R)\mathrm{d}\tau = \frac{\sqrt{\kappa}}{2\sqrt{\pi}} \left\{ \exp\left[-\frac{(R - R_d(\tau_N))^2}{4\kappa(\hat{t} - \tau_N)} \right] \int_{t_{N-1}}^{t_N} \frac{\mathrm{d}\tau}{\sqrt{\hat{t} - \tau}} \right.$$
$$\left. - \exp\left[-\frac{(R + R_d(\tau_N))^2}{4\kappa(\hat{t} - \tau_N)} \right] \frac{\Delta t}{\sqrt{\hat{t} - \tau_N}} \right\}$$
$$= \frac{\sqrt{\kappa}}{\sqrt{2\pi}} \left\{ \sqrt{2} \exp\left[-\frac{(R - R_d(\tau_N))^2}{2\kappa \Delta t} \right] \right.$$
$$\left. - \exp\left[-\frac{(R + R_d(\tau_N))^2}{2\kappa \Delta t} \right] \right\} \sqrt{\Delta t}. \quad \text{(G.4)}$$

The latter equation allows us to simplify the equation for $v(R, \hat{t})$ for $R = R_d(\hat{t})$ to

$$v(R, \hat{t}) = \sum_{j=1}^{N-1} \frac{v(t_{j-1}) + v(t_j)}{2} G(\hat{t}, \tau_j, R)\Delta t + \frac{(v(t_{N-1}) + v(t_N))\sqrt{\kappa}}{2\sqrt{2\pi}}$$
$$\times \left\{ \sqrt{2} \exp\left[-\frac{(R - R_d(\tau_N))^2}{2\kappa \Delta t} \right] - \exp\left[-\frac{(R + R_d(\tau_N))^2}{2\kappa \Delta t} \right] \right\} \sqrt{\Delta t}. \quad \text{(G.5)}$$

Reference

1. Sazhin, S. S., Krutitskii, P. A., Gusev, I. G., & Heikal, M. (2011). Transient heating of an evaporating droplet with presumed time evolution of its radius. *International Journal Heat and Mass Transfer*, *54*, 1278–1288.

Appendix H
Derivation of Formula (5.18)

In what follows, the details of the solution of (5.1) ($Y_{li}(t, R)$) for $t \geq 0$ and $0 \leq R < R_d$ are presented, following [1], assuming that $R_d = \text{const}$. In light of the physical background of the problem, we will look for a solution which is continuously twice differentiable in the whole domain.

Let us rewrite the boundary condition (5.9) in the form

$$\left.\left(\frac{\partial Y_{li}}{\partial R} - \frac{\alpha_m}{D_l} Y_{li}\right)\right|_{R=R_d} = -\frac{\alpha_m \varepsilon_i(t)}{D_l}, \tag{H.1}$$

where

$$\alpha_m = \frac{D_v \rho_{\text{total}} \ln(1 + B_M)}{\rho_l R_d}. \tag{H.2}$$

The initial condition is $Y_{li}(t = 0) = Y_{li0}(R)$.

We look for the solution to (5.1) in the form

$$Y_{li}(t, R) = y(t, R) + \varepsilon(t), \tag{H.3}$$

where the subscript i at ε is omitted to simplify the notation.

On substituting (H.3) into (5.1) and the boundary condition (H.1), we can rewrite this equation and the corresponding boundary and initial conditions in the form

$$\frac{\partial y}{\partial t} = D_l \left(\frac{\partial^2 y}{\partial R^2} + \frac{2}{R}\frac{\partial y}{\partial R}\right) - \frac{\mathrm{d}\varepsilon(t)}{\mathrm{d}t}, \tag{H.4}$$

$$\left.\left(\frac{\partial y}{\partial R} - \frac{\alpha_m}{D_l} y\right)\right|_{R=R_d} = 0, \tag{H.5}$$

$$y|_{t=0} = Y_{li0}(R) - \varepsilon(0) \equiv Y_{li0}(R) - \varepsilon_0. \tag{H.6}$$

S. Sazhin, *Droplets and Sprays*, DOI: 10.1007/978-1-4471-6386-2,

Introduction of the new variable

$$u(t, R) = y(t, R)R$$

as in Appendix C allows us to rewrite (H.4) and the corresponding boundary and initial conditions in the form

$$\frac{\partial u}{\partial t} = D_l \frac{\partial^2 u}{\partial R^2} - R\frac{\mathrm{d}\varepsilon(t)}{\mathrm{d}t}, \tag{H.7}$$

$$u|_{R=0} = \left(\frac{\partial u}{\partial R} + \frac{h_0}{R_d}u\right)\bigg|_{R=R_d} = 0, \tag{H.8}$$

$$u|_{t=0} = R\left(Y_{li0}(R) - \varepsilon_0\right), \tag{H.9}$$

where

$$h_0 = -\left(1 + \frac{\alpha_m R_d}{D_l}\right).$$

Note that the change of variable from y to u leads to the need for a second boundary condition at $R = 0$. Our assumption that the solution is continuously twice differentiable implies that y is finite everywhere in the domain $0 \leq R < R_d$, whence the boundary condition (H.8) at $R = 0$. The solution to problem (H.7)–(H.9) for $h_0 > -1$ was discussed in Appendix C. Here the focus is on the case $h_0 < -1$, which is directly relevant to the problem of diffusion of species inside droplets.

We look for the solution to (H.7) in the form

$$u = \sum_{n=0}^{\infty} \Theta_n(t) v_n(R), \tag{H.10}$$

where $v_n(R)$ is the full set of nontrivial solutions to the equation

$$\frac{\partial^2 v}{\partial R^2} + pv = 0, \tag{H.11}$$

subject to the boundary conditions

$$v|_{R=0} = \left(\frac{\partial v}{\partial R} + \frac{h_0}{R_d}v\right)\bigg|_{R=R_d} = 0. \tag{H.12}$$

Equation (H.11) with boundary conditions (H.12) is the well-known Sturm–Liouville problem. Our first task is to find eigenvalues p for this problem. The cases $p = 0$, $p < 0$, and $p > 0$ are considered separately.

The Sturm–Liouville Problem for $p = 0$

For $p = 0$, the general solution to (H.11) can be written as $v = A + BR$. The condition $v|_{R=0} = 0$ implies that $A = 0$. The boundary condition at $R = R_d$ leads to the equation $B(1 + h_0) = 0$. Since $h_0 \in (-\infty, -1)$, the latter equation is satisfied only when $B = 0$. This leads to the trivial solution $v = 0$, which is disregarded in our analysis. Hence, (H.11) has no nontrivial solutions for $p = 0$.

The Sturm–Liouville Problem for $p < 0$

Assuming that $p = -\lambda^2 < 0$, we write the general solution to (H.11) as

$$v(R) = A\cosh\left(\lambda\frac{R}{R_d}\right) + B\sinh\left(\lambda\frac{R}{R_d}\right), \tag{H.13}$$

where A and B are arbitrary constants.

The boundary condition at $R = 0$ (see (H.12)) implies that $A = 0$. The boundary condition at $R = R_d$ leads to the following equation:

$$\frac{B}{R_d}\left(\lambda\cosh\lambda + h_0\sinh\lambda\right) = 0. \tag{H.14}$$

The number B in this equation is not equal to zero, since we do not consider the trivial solution $v = 0$. Hence, (H.14) can be rewritten as

$$\tanh\lambda = -\frac{\lambda}{h_0}. \tag{H.15}$$

It is easy to show that for $h_0 \in (-\infty, -1)$, Eq. (H.15) has three solutions $\lambda = 0$ and $\pm\lambda_0$, where $\lambda_0 \in (0, +\infty)$, and it has no solutions for $h_0 > -1$. The solution $\lambda = 0$ leads to the trivial solution $v = 0$, which is disregarded in our analysis. The solutions $\lambda = \pm\lambda_0$ lead to solutions (H.13) (eigenfunctions), which differ only by the sign of B. Since the value of the coefficient B is determined by the normalisation condition only (see below), the solution $\lambda = -\lambda_0$ can be disregarded. Hence, we can conclude that the solution to (H.15) gives only one eigenvalue, $\lambda = \lambda_0 > 0$, and the corresponding eigenfunction

$$v_0(R) = \sinh\left(\lambda_0\frac{R}{R_d}\right), \tag{H.16}$$

where the normalisation leading to $B = 1$ has been chosen.

A direct calculation of the integrals, taking into account condition (H.15), leads to the following expression for the norm of v_0:

$$||v_0||^2 = \int_0^{R_d} v_0^2(R)\mathrm{d}R = -\frac{R_d}{2}\left[1 + \frac{h_0}{h_0^2 - \lambda_0^2}\right]. \tag{H.17}$$

The Sturm–Liouville Problem for $p > 0$

Assuming that $p = \lambda^2 > 0$, we write the general solution to (H.11) as

$$v(R) = A\cos\left(\lambda\frac{R}{R_d}\right) + B\sin\left(\lambda\frac{R}{R_d}\right), \tag{H.18}$$

where A and B are arbitrary constants.

The boundary condition at $R = 0$ (see (H.12)) implies that $A = 0$. The boundary condition at $R = R_d$ leads to the following equation:

$$\frac{B}{R_d}\left(\lambda\cos\lambda + h_0\sin\lambda\right) = 0, \tag{H.19}$$

where we note that B in this equation is not equal to zero, since we do not consider the trivial solution $v = 0$. Hence, (H.19) can be rewritten as

$$\tan\lambda = -\frac{\lambda}{h_0}. \tag{H.20}$$

As in the case $p < 0$, we disregard the solutions to this equation corresponding to zero and negative λ. A countable set of positive solutions to this equation (positive eigenvalues) λ_n are arranged in ascending order:

$$0 < \lambda_1 < \lambda_2 < \lambda_3 < \cdots .$$

The corresponding eigenfunctions are represented as

$$v_n(R) = \sin\left(\lambda_n\frac{R}{R_d}\right), \tag{H.21}$$

where the normalisation leading to $B = 1$ is chosen as in the case $p < 0$.

A direct calculation of the integrals, taking into account condition (H.20), leads to the following expression for the norm of v_n for $n \geq 1$:

$$||v_n||^2 = \int_0^{R_d} v_n^2(R)\mathrm{d}R = \frac{R_d}{2}\left[1 + \frac{h_0}{h_0^2 + \lambda_0^2}\right]. \tag{H.22}$$

The norm (H.22) differs from the norm chosen in Appendix C by the factor R_d. This does not affect the final solution.

Orthogonality of the Eigenfunctions

The orthogonality of functions v_n ($n \geq 1$) was proven in Appendix C. To prove that functions v_0 and v_n ($n \geq 1$) are orthogonal, we need to calculate the following integral:

$$I = \int_{0}^{R_d} \sinh\left(\lambda_0 \frac{R}{R_d}\right) \sin\left(\lambda_n \frac{R}{R_d}\right) \mathrm{d}R, \tag{H.23}$$

where $n \geq 1$.

Using integration by parts twice in calculating the integral on the right-hand side of (H.23), we rearrange this equation to

$$I = -\frac{R_d}{\lambda_n}\left[\sinh\lambda_0 \cos\lambda_n - \frac{\lambda_0}{\lambda_n}\cosh\lambda_0 \sin\lambda_n + \frac{\lambda_0^2}{\lambda_n R_d} I\right], \tag{H.24}$$

where I on the right-hand side of this equation is the same as in (H.23).

Equation (H.24) can be rearranged to yield

$$I = -\frac{\frac{R_d}{\lambda_n}\left[\frac{\sinh\lambda_0}{\cosh\lambda_0} - \frac{\lambda_0}{\lambda_n}\frac{\sin\lambda_n}{\cos\lambda_n}\right]\cosh\lambda_0 \cos\lambda_n}{1 + \left(\frac{\lambda_0}{\lambda_n}\right)^2}. \tag{H.25}$$

Keeping in mind Eqs. (H.15) and (H.20), we can see that I defined by (H.25) is equal to zero. This implies that the functions v_n are orthogonal for $n \geq 0$, and we can write

$$\int_{0}^{R_d} v_n(R) v_m(R) \mathrm{d}R = \delta_{nm} ||v_n||^2, \tag{H.26}$$

where $n \geq 0$ and $m \geq 0$, $||v_n||^2$ is defined by expression (H.17) when $n = 0$ and expression (H.22) when $n \geq 1$.

Expansion of R in a Fourier Series with Respect to Functions v_n

Before we are able to use (H.10) for the solution of (H.7), we need to find the representation of R as a Fourier series with respect to functions v_n:

$$R = \sum_{n=0}^{\infty} Q_n v_n(R), \tag{H.27}$$

where

$$Q_n = \frac{1}{||v_n||^2} \int_{0}^{R_d} R v_n(R) \mathrm{d}R. \tag{H.28}$$

Direct calculation of the integrals on the right-hand side of expression (H.28) leads to the following explicit expressions for Q_n:

$$Q_n = \begin{cases} -\frac{1}{||v_0||^2}\left(\frac{R_d}{\lambda_0}\right)^2 (1+h_0)\sinh\lambda_0 & \text{when} \quad n=0 \\ \frac{1}{||v_n||^2}\left(\frac{R_d}{\lambda_n}\right)^2 (1+h_0)\sin\lambda_n & \text{when} \quad n\geq 1 \end{cases} \tag{H.29}$$

Calculation of Coefficients $\Theta_n(t)$ in Expression (H.10)

On substituting expressions (H.10) and (H.27) into (H.7), we rewrite the latter equation in the form

$$\sum_{n=0}^{\infty}\Theta_n'(t)v_n(R) = D_l\sum_{n=0}^{\infty}\Theta_n(t)v_n''(R) - \varepsilon'(t)\sum_{n=0}^{\infty}Q_n v_n(R), \tag{H.30}$$

where

$$\Theta_n' = \frac{d\Theta_n}{dt}; \quad v_n''(R) = \frac{d^2v_n}{dR^2}; \quad \varepsilon'(t) = \frac{d\varepsilon}{dt} \equiv \varepsilon'.$$

Since the expansion in the series with respect to v_n (Fourier series) is unique, (H.30) is satisfied only when it is satisfied for each term in this expansion. From

$$v_0'' = \left(\frac{\lambda_0}{R_d}\right)^2 v_0 \quad \text{and} \quad v_n'' = -\left(\frac{\lambda_n}{R_d}\right)^2 v_n \quad (n\geq 1),$$

we can see that this implies that

$$\Theta_0'(t) = D_l\left(\frac{\lambda_0}{R_d}\right)^2\Theta_0(t) - \varepsilon' Q_0, \tag{H.31}$$

$$\Theta_n'(t) = -D_l\left(\frac{\lambda_n}{R_d}\right)^2\Theta_n(t) - \varepsilon' Q_n, \tag{H.32}$$

when $n\geq 1$.

Equations (H.31) and (H.32) need to be solved subject to the initial conditions for $\Theta_n(t)$ $(n\geq 0)$. To find these initial conditions, we substitute expression (H.10) into the initial condition (H.9) and expand $RY_{li0}(R)$ into a Fourier series with respect to v_n. Since the expansion with respect to v_n is unique, this leads us to the following equation for $\Theta_n(0)$:

$$\Theta_n(0) = q_{in} - \varepsilon(0)Q_n, \tag{H.33}$$

where Q_n is defined by (H.28) or (H.29),

$$q_{in} = \frac{1}{||v_n||^2}\int_0^{R_d} RY_{li0}(R)v_n(R)dR, \tag{H.34}$$

$n\geq 0$.

The solutions to (H.31) and (H.32) subject to the initial condition (H.33) is represented in the form

$$\Theta_0(t) = \exp\left[D_l\left(\frac{\lambda_0}{R_d}\right)^2 t\right]\left[q_{i0} - \varepsilon(0)Q_0\right] - Q_0\int\limits_0^t \frac{d\varepsilon(\tau)}{d\tau}\exp\left[D_l\left(\frac{\lambda_0}{R_d}\right)^2 (t-\tau)\right]d\tau, \qquad (H.35)$$

$$\Theta_n(t) = \exp\left[-D_l\left(\frac{\lambda_n}{R_d}\right)^2 t\right]\left[q_{in} - \varepsilon(0)Q_n\right] - Q_n\int\limits_0^t \frac{d\varepsilon(\tau)}{d\tau}\exp\left[-D_l\left(\frac{\lambda_n}{R_d}\right)^2 (t-\tau)\right]d\tau, \qquad (H.36)$$

where $n \geq 1$.

The Final Solution

On substituting expressions (H.16), (H.21), (H.35), and (H.36) into expression (H.10), we can represent the final solution to (5.18) satisfying boundary condition (5.9) and the corresponding initial condition in the form

$$Y_{li} = \varepsilon_i + \frac{1}{R}\Bigg\{\Bigg[\exp\left[D_l\left(\frac{\lambda_0}{R_d}\right)^2 t\right]\left[q_{i0} - \varepsilon_i(0)Q_0\right] - Q_0\int\limits_0^t \frac{d\varepsilon_i(\tau)}{d\tau}\exp\left[D_l\left(\frac{\lambda_0}{R_d}\right)^2 (t-\tau)\right]d\tau\Bigg]\sinh\left(\lambda_0\frac{R}{R_d}\right) + \sum_{n=1}^{\infty}\Bigg[\exp\left[-D_l\left(\frac{\lambda_n}{R_d}\right)^2 t\right]\left[q_{in} - \varepsilon_i(0)Q_n\right] - Q_n\int\limits_0^t \frac{d\varepsilon_i(\tau)}{d\tau}\exp\left[-D_l\left(\frac{\lambda_n}{R_d}\right)^2 (t-\tau)\right]d\tau\Bigg]\sin\left(\lambda_n\frac{R}{R_d}\right)\Bigg\}, \qquad (H.37)$$

where Q_n, q_{in}, λ_0, and λ_n ($n \geq 1$) are defined by (H.29), (H.34), (H.15), and (H.20) respectively; the subscript $_i$ at ε has been restored.

Note that expression (H.37) contains a term which increases exponentially with time. This, however, does not lead to an unphysical solution to (5.18), since this equation is valid only for $0 \le Y_{li} \le 1$. Once Y_{li} reaches one of its limiting values, it remains equal to that value.

If we apply Solution (H.37) to individual short timesteps, we can ignore the time dependence of $\frac{\mathrm{d}\varepsilon_i(\tau)}{\mathrm{d}\tau}$ during this timestep and assume that

$$\frac{\mathrm{d}\varepsilon_i(t)}{\mathrm{d}t} = \left.\frac{\mathrm{d}\varepsilon_i(t)}{\mathrm{d}t}\right|_{t=0} \equiv \varepsilon_i'.$$

This allows us to simplify expression (H.37) to

$$\begin{aligned} Y_{li} = \varepsilon_i + \frac{1}{R}\Bigg\{ &\left[\exp\left[D_l\left(\frac{\lambda_0}{R_d}\right)^2 t\right]\left[q_{i0} - Q_0\left(\varepsilon_i(0) + \frac{R_d^2}{D_l\lambda_0^2}\varepsilon_i'\right)\right]\right. \\ &\left. + Q_0\frac{R_d^2}{D_l\lambda_0^2}\varepsilon_i'\right]\sinh\left(\lambda_0\frac{R}{R_d}\right) + \sum_{n=1}^{\infty}\left[\exp\left[-D_l\left(\frac{\lambda_n}{R_d}\right)^2 t\right]\right. \\ &\times\left[q_{in} - Q_n\left(\varepsilon_i(0) - \frac{R_d^2}{D_l\lambda_n^2}\varepsilon_i'\right)\right] \\ &\left. - Q_n\frac{R_d^2}{D_l\lambda_n^2}\varepsilon_i'\right]\sin\left(\lambda_n\frac{R}{R_d}\right)\Bigg\}. \end{aligned} \tag{H.38}$$

If we ignore the time-dependence of ε_i, then expression (H.38) is simplified to

$$\begin{aligned} Y_{li} = \varepsilon_i + \frac{1}{R}\Bigg\{ &\exp\left[D_l\left(\frac{\lambda_0}{R_d}\right)^2 t\right]\left[q_{i0} - \varepsilon_i(0)Q_0\right]\sinh\left(\lambda_0\frac{R}{R_d}\right) \\ &+ \sum_{n=1}^{\infty}\exp\left[-D_l\left(\frac{\lambda_n}{R_d}\right)^2 t\right]\left[q_{in} - \varepsilon(0)Q_n\right]\sin\left(\lambda_n\frac{R}{R_d}\right)\Bigg\}. \end{aligned} \tag{H.39}$$

The assumption that the time dependence of ε_i can be ignored during the time step can be used in the analysis of most practically important cases. This allows us to base our calculations on expression (H.39).

Expression (H.39) is identical to formula (5.18)

Reference

1. Sazhin, S. S., Elwardany, A., Krutitskii, P. A., Deprédurand, V., Castanet, G., Lemoine, F., et al. (2011). Multi-component droplet heating and evaporation: Numerical simulation versus experimental data. *International Journal of Thermal Science*, *50*, 1164–1180.

Appendix I
Derivation of Formula (5.24)

In what follows, formula (5.24) is derived, following [1].

As in Appendix D, the following parameters and functions are introduced:

$$r = R/R_d(t), \quad (0 \le r \le 1), \quad F(t,r) = RY_{li}(t, rR_{d0}),$$

$$F(t,r) = \frac{1}{\sqrt{R_d(t)}} \exp\left[-\frac{R_d'(t)R_d(t)}{4D_l} r^2\right] W(t,r). \tag{I.1}$$

These allow us to rewrite (5.1) and the corresponding initial condition and boundary condition (5.9) as

$$R_d^2(t) W_t'(t,r) = D_l W_{rr}''(t,r), \tag{I.2}$$

where $t \ge 0$,

$$W(t,r)|_{t=0} = W_0(r) \equiv R_{d0}^{3/2} rY_{li0}(rR_{d0}) \exp\left[\frac{R_d'(0)R_{d0}}{4D_l} r^2\right], \tag{I.3}$$

$$W(t,r)|_{r=0} = 0, \tag{I.4}$$

$$\left[W_r'(t,r) + H_0(t)W(t,r)\right]\Big|_{r=1} = \mu_0(t) \equiv -\frac{\alpha_m \varepsilon_i (R_d(t))^{5/2}}{D_l} \exp\left[\frac{R_d'(t)R_d(t)}{4D_l}\right], \tag{I.5}$$

where

$$H_0(t) = -\frac{\alpha_m}{D_l} R_d(t) - 1 - \frac{R_d'(t)R_d(t)}{2D_l}.$$

S. Sazhin, *Droplets and Sprays*, DOI: 10.1007/978-1-4471-6386-2,

Condition (I.4) is an additional boundary condition, which follows from the requirement that $Y_{li}(t, R)$ be a continuously twice differentiable function. In deriving (I.2), we took into account that $\mathrm{d}^2R_d/\mathrm{d}t^2 = 0$.

Further simplification of (I.2) and the corresponding initial and boundary conditions is possible when we apply this equation to a short time step. In this case, we can ignore the time-dependence of $H_0(t)$ and assume that $H_0(t) \equiv h_0 = \text{const}$. Using Eq. (4.83), we can write

$$h_0 = -\frac{\alpha_m}{D_l}R_d(t) - \frac{\alpha R_{d0}}{2D_l}R_d(t) - 1 \approx -\frac{\alpha_m}{D_l}R_{d0} - \frac{\alpha R_{d0}}{2D_l}R_{d0} - 1 < -1. \qquad \text{(I.6)}$$

It is essential for us to retain both α_m and α in the expression for h_0 if we intend to compare our results with those of conventional analysis when $\alpha = 0$ but $\alpha_m \neq 0$.

Our next goal is to find a change of variables whereby the inhomogeneous boundary condition (I.5) is replaced by the corresponding homogeneous one. This is achieved by the introduction of the new function $V(t, r)$ via the relation

$$W(t, r) = V(t, r) + \frac{\mu_0(t)}{1 + h_0}r. \qquad \text{(I.7)}$$

Formula (I.7) allows us to rearrange (I.2) to obtain

$$R_d^2(t)V_t'(t, r) = D_l V_{rr}''(t, r) - \frac{\mu_0'(t)}{1 + h_0}R_d^2(t)r. \qquad \text{(I.8)}$$

The initial and boundary conditions for (I.8) are represented as

$$V(t, r)|_{t=0} = W_0(r) - \frac{\mu_0(0)}{1 + h_0}r,$$

$$V(t, r)|_{r=0} = 0, \quad \left[V_r'(t, r) + h_0 V(t, r)\right]\Big|_{r=1} = 0.$$

As in Appendix D, we look for the solution of (I.8) in the form

$$V(t, r) = \sum_{n=0}^{\infty} \Theta_n(t) v_n(r), \qquad \text{(I.9)}$$

where the functions $v_n(r)$ form a full set of nontrivial solutions to the equation

$$\frac{\mathrm{d}^2 v}{\mathrm{d}r^2} + pv = 0, \quad 0 \leq r \leq 1, \qquad \text{(I.10)}$$

subject to boundary conditions

$$v|_{r=0} = \left(\frac{\mathrm{d}v}{\mathrm{d}r} + h_0 v \right) \bigg|_{r=1} = 0. \tag{I.11}$$

For $p = 0$, Eq. (I.10) has no nontrivial solutions satisfying the boundary conditions (I.11). For $p \equiv -\lambda^2 < 0$, this equation has the solution

$$v_0(r) = \sinh(\lambda_0 r), \tag{I.12}$$

where λ_0 is the solution to the equation

$$\tanh \lambda = -\frac{\lambda}{h_0}. \tag{I.13}$$

The latter equation has three solutions (positive, negative, and zero), where we keep in mind that $h_0 < -1$. We are interested in the positive solution to this equation only (cf. Appendix H).

Note that this solution does not exist in the case of the heat conduction equation, when h_0 is greater than -1.

For $p \equiv \lambda^2 > 0$, Eq. (I.10) has the solutions

$$v_n(r) = \sin(\lambda_n r) \tag{I.14}$$

for $n \geq 1$, where λ_n are the solutions to the equation

$$\tan \lambda = -\frac{\lambda}{h_0}. \tag{I.15}$$

As in the case of $p < 0$, we disregard the solutions to this equation corresponding to zero and negative λ. A countable set of positive solutions to this equation (positive eigenvalues) λ_n is arranged in ascending order:

$$0 < \lambda_1 < \lambda_2 < \lambda_3 < \cdots .$$

It can be shown that the functions $v_n(r)$, $n \geq 0$, are orthogonal for $0 \leq r \leq 1$ (cf. Appendix H).

The completeness of the set of functions $v_n(r)$ for $n \geq 0$ has been tested. Namely, we considered various functions not belonging to this set and found that the Fourier expansions of those functions on the set of $\{v_n(r)\}_{n=0}^{\infty}$ coincide with the functions themselves. If the set of functions is not complete, then the Fourier expansion of an arbitrary function constructed based on this set does not coincide with that function.

The norms of the functions $v_n(r)$ for $n \geq 0$ are given by the following expression:

$$||v_n||^2 = \int_0^1 v_n^2(r)\mathrm{d}r = \frac{(-1)^{\delta_{n.0}}}{2}\left[1 + \frac{h_0}{h_0^2 + (-1)^{\delta_{n.0}}\lambda_n^2}\right], \tag{I.16}$$

where $\delta_{n,0}$ is Kronecker's delta symbol.

Note that the norm in (I.16) differs from those in (H.17) and (H.22) by the factor R_d, but it coincides with the one chosen in Appendix C. This does not affect the final results, but note that expression (5.24) is used for the norms appearing in expressions (H.17) and (H.22).

Recalling that the functions $v_n(r)$ for $n \geq 0$ are orthogonal and assuming that the set of these functions is complete, we can write

$$f(r) \equiv -r/(1 + h_0) = \sum_{n=0}^{\infty} f_n v_n(r), \tag{I.17}$$

$$W_0(r) = \sum_{n=0}^{\infty} q_n v_n(r), \tag{I.18}$$

where

$$f_n = \frac{1}{||v_n||^2}\int_0^1 f(r)v_n(r)\mathrm{d}r = \begin{cases} \frac{1}{||v_0||^2\lambda_0^2}\sinh\lambda_0 & \text{when} \quad n = 0 \\ -\frac{1}{||v_n||^2\lambda_n^2}\sin\lambda_n & \text{when} \quad n \geq 1 \end{cases},$$

$$q_n = \frac{1}{||v_n||^2}\int_0^1 W_0(r)v_n(r)\mathrm{d}r.$$

Note that the expressions for f_n and q_n differ from those for f_{Yn} and q_{Yn} in formula (5.24) by the factor R_d, since the latter formula was derived using formulas (H.17) and (H.22).

Using (I.9) and (I.17), Eq. (I.8) can be rewritten as

$$\sum_{n=0}^{\infty}\left(R_d^2(t)\frac{\mathrm{d}\Theta_n(t)}{\mathrm{d}t} + (-1)^{\delta_{n,0}}\Theta_n(t)D_l\lambda_n^2\right)v_n(r) = \sum_{n=0}^{\infty}\left(f_n R_d^2(t)\frac{\mathrm{d}\mu_0(t)}{\mathrm{d}t}\right)v_n(r). \tag{I.19}$$

Both sides of (I.19) are Fourier series with respect to the functions $v_n(r)$. Two Fourier series are equal if and only if their coefficients are equal. This implies that

$$R_d^2(t)\frac{\mathrm{d}\Theta_n(t)}{\mathrm{d}t} + (-1)^{\delta_{n,0}}\Theta_n(t)D_l\lambda_n^2 = f_n R_d^2(t)\frac{\mathrm{d}\mu_0(t)}{\mathrm{d}t}. \tag{I.20}$$

Equation (I.20) is to be solved subject to the initial condition

$$\Theta_n(0) = q_n + \mu_0(0)f_n. \tag{I.21}$$

The general solution to the homogeneous equation

$$R_d^2(t)\frac{\mathrm{d}\Theta_n(t)}{\mathrm{d}t} + (-1)^{\delta_{n,0}}\Theta_n(t)D_l\lambda_n^2 = 0 \tag{I.22}$$

can be represented as

$$\ln\left(\Theta_n(t)/\Theta_n(0)\right) = -(-1)^{\delta_{n,0}}D_l\lambda_n^2\int_0^t \frac{\mathrm{d}t}{R_d^2(t)}. \tag{I.23}$$

Assuming that $R_d(t)$ is a linear function of t given by (4.83), solution (I.23) can be represented in a more explicit form:

$$\Theta_n(t) = \Theta_n(0)\exp\left[\frac{(-1)^{\delta_{n,0}}D_l\lambda_n^2}{\alpha R_{d0}^2}\left(\frac{1}{1+\alpha t} - 1\right)\right]. \tag{I.24}$$

One can see that the function

$$\Theta_{n\,(\text{part})}(t) = f_n\int_0^t \frac{\mathrm{d}\mu_0(\tau)}{\mathrm{d}\tau}\exp\left[(-1)^{\delta_{n,0}}\frac{D_l\lambda_n^2}{\alpha R_{d0}^2}\left(\frac{1}{1+\alpha t} - \frac{1}{1+\alpha\tau}\right)\right]\mathrm{d}\tau \tag{I.25}$$

satisfies (I.20). Hence, this function can be considered a particular solution of (I.20). Integration by parts in expression (I.25) allows us to represent $\Theta_{n\,(\text{part})}(t)$ as

$$\begin{aligned}\Theta_{n\,(\text{part})}(t) = f_n\Bigg\{&\mu_0(t) - \mu_0(0)\exp\left[-(-1)^{\delta_{n,0}}\frac{D_l\lambda_n^2 t}{R_{d0}R_d(t)}\right]\\ &- \exp\left[(-1)^{\delta_{n,0}}\frac{D_l\lambda_n^2}{\alpha R_{d0}R_d(t)}\right]\int_0^t (-1)^{\delta_{n,0}}\frac{\mu_0(\tau)D_l\lambda_n^2}{R_d^2(\tau)}\\ &\times\exp\left[-(-1)^{\delta_{n,0}}\frac{D_l\lambda_n^2}{\alpha R_{d0}R_d(\tau)}\right]\mathrm{d}\tau\Bigg\}.\end{aligned} \tag{I.26}$$

Using (I.24) and (I.26), the solution to (I.20) can be presented as

$$\Theta_n(t) = \Theta_n(0)\exp\left[(-1)^{\delta_{n,0}}\frac{D_l\lambda_n^2}{\alpha R_{d0}^2}\left(\frac{1}{1+\alpha t}-1\right)\right]$$
$$+ f_n\int_0^t \frac{\mathrm{d}\mu_0(\tau)}{\mathrm{d}\tau}\exp\left[(-1)^{\delta_{n,0}}\frac{D_l\lambda_n^2}{\alpha R_{d0}^2}\left(\frac{1}{1+\alpha t}-\frac{1}{1+\alpha\tau}\right)\right]\mathrm{d}\tau. \quad \text{(I.27)}$$

Recalling (I.26) and (I.21), we can write an alternative formula for $\Theta_n(t)$:

$$\Theta_n(t) = q_n\exp\left[-(-1)^{\delta_{n,0}}\frac{D_l\lambda_n^2 t}{R_{d0}R_d(t)}\right]+f_n\mu_0(t)$$
$$- f_n(-1)^{\delta_{n,0}}D_l\lambda_n^2\int_0^t \frac{\mu_0(\tau)}{R_d^2(\tau)}\exp\left[(-1)^{\delta_{n,0}}\frac{D_l\lambda_n^2}{\alpha R_{d0}}\left(\frac{1}{R_d(t)}-\frac{1}{R_d(\tau)}\right)\right]\mathrm{d}\tau. \quad \text{(I.28)}$$

Since solution (I.28) is applied to a very short time step, changes of $\mu_0(\tau)$ in the integrand before the exponential term can be ignored. This allows us to simplify solution (I.28) to (see Appendix J)

$$\Theta_n(t) = \left[q_n+f_n\mu_0(0)\right]\exp\left[-(-1)^{\delta_{n,0}}\frac{D_l\lambda_n^2 t}{R_{d0}R_d(t)}\right]+f_n\mu_0(t)-f_n\mu_0(0). \quad \text{(I.29)}$$

Note that $\Theta_n(t)$ in the form (I.27) satisfies (I.20), while $\Theta_n(t)$ in the form (I.29) does not. This is related to the fact that (I.20) was derived under the assumption that the series (I.9), after being substituted into (I.8), can be differentiated term by term (derivative of the series is equal to the series of derivatives). This assumption is valid when $\Theta_n(t)$ is taken in the form (I.27), but it is not valid when $\Theta_n(t)$ is taken in the form (I.29), since

$$\mu_0(t)\frac{\mathrm{d}^2}{\mathrm{d}r^2}\left(\sum_{n=0}^{\infty} f_n v_n\right) \neq \mu_0(t)\sum_{n=0}^{\infty} f_n\frac{\mathrm{d}^2 v_n}{\mathrm{d}r^2}$$

(the series on the right-hand side of this formula diverges). Note that the series (I.9) satisfies (I.8) regardless of whether $\Theta_n(t)$ is taken in the form (I.27) or the form (I.29).

Recalling expressions (I.17) and (I.29), we see that (I.9) can be rewritten as

$$V(t,r) = \sum_{n=0}^{\infty}\breve{\Theta}_n(t)v_n(r) - \frac{\mu_0(t)}{1+h_0}\frac{R}{R_d(t)} + \frac{\mu_0(0)}{1+h_0}\frac{R}{R_d(t)}, \quad \text{(I.30)}$$

where

$$\breve{\Theta}_n(t) = \left[q_n+f_n\mu_0(0)\right]\exp\left[-(-1)^{\delta_{n,0}}\frac{D_l\lambda_n^2 t}{R_{d0}R_d(t)}\right]. \quad \text{(I.31)}$$

The final equation for the mass fraction inside the droplet can be represented as

$$Y_{li}(R) = \frac{1}{R\sqrt{R_d(t)}} \exp\left[-\frac{\alpha R_{d0} R^2}{4D_l R_d(t)}\right] \left[\sum_{n=1}^{\infty} \breve{\Theta}_n(t) \sin\left(\lambda_n \frac{R}{R_d(t)}\right) + \breve{\Theta}_0(t) \sinh\left(\lambda_0 \frac{R}{R_d(t)}\right) + \frac{\mu_0(0)}{1+h_0} \frac{R}{R_d(t)}\right], \qquad \text{(I.32)}$$

where $\breve{\Theta}_n$ are given by expressions (I.31).

On substituting expression (I.31) into expression (I.32), we can rearrange the latter expression for the short time step to

$$Y_{li}(R) = \frac{\alpha_m \varepsilon_i \exp\left[\frac{\alpha R_{d0}}{4D_l}\left(\frac{R_{d0} R_d(t) - R^2}{R_d(t)}\right)\right]}{\alpha_m + \frac{\alpha R_{d0}}{2}} \frac{R_{d0}^{5/2}}{R_d^{5/2}(t)} + \frac{1}{R\sqrt{R_d(t)}} \exp\left[-\frac{\alpha R_{d0} R^2}{4D_l R_d(t)}\right]$$
$$\times \left[\sum_{n=1}^{\infty} [q_n + f_n \mu_0(0)] \exp\left[-\frac{D_l \lambda_n^2 t}{R_{d0} R_d(t)}\right] \sin\left(\lambda_n \frac{R}{R_d(t)}\right) + [q_0 + f_0 \mu_0(0)] \exp\left[\frac{D_l \lambda_0^2 t}{R_{d0} R_d(t)}\right] \sinh\left(\lambda_0 \frac{R}{R_d(t)}\right)\right]. \qquad \text{(I.33)}$$

Solution (I.33) is identical to formula (5.24).

Note that expression (I.33) was derived under the assumption that $H_0(t) \equiv h_0 =$ const. It is essential for us to retain both α and α_m in expression (I.33) to be able to compare the prediction of this expression with that of the conventional approach to modelling species diffusion when $\alpha = 0$ but $\alpha_m \neq 0$ during the time step. In the latter case, expression (I.33) is further simplified to

$$Y_{li}(R) = \varepsilon_i + \frac{1}{R\sqrt{R_d(t)}} \left[\sum_{n=1}^{\infty} [q_n + f_n \mu_0(0)] \exp\left[-\frac{D_l \lambda_n^2 t}{R_{d0} R_d(t)}\right] \sin\left(\lambda_n \frac{R}{R_d(t)}\right) + [q_0 + f_0 \mu_0(0)] \exp\left[\frac{D_l \lambda_0^2 t}{R_{d0} R_d(t)}\right] \times \sinh\left(\lambda_0 \frac{R}{R_d(t)}\right)\right]. \qquad \text{(I.34)}$$

This expression is identical to (5.18) (see Appendix H). Note that the norm of v_n ($||v_n||^2$) in Appendix H is dimensional.

Let us now relax our assumption that $H_0(t) \equiv h_0 =$ const and assume that

$$H_0(t) = h_0 + h_1(t), \qquad \text{(I.35)}$$

where $h_0 = \text{const} < -1$. In view of (I.35), we can rewrite the boundary condition at $r = 1$ for (I.2) in the form

$$\left[W_r'(t, r) + h_0 W(t, r)\right]\Big|_{r=1} = \mu_0(t) - h_1(t)W(t, 1) \equiv \hat{\mu}_0(t). \tag{I.36}$$

Assuming that $\hat{\mu}_0(t)$ is known, we can formally use the previously obtained solutions (I.7) and (I.9) to represent the solution to problem (I.2)–(I.5) in the form

$$\begin{aligned} W(t, r) = \frac{\hat{\mu}_0(t)}{1 + h_0} r + V(t, r) &= \sum_{n=0}^{\infty} v_n(r) q_n \exp\left[-(-1)^{\delta_{n,0}} \frac{D_l \lambda_n^2 t}{R_{d0} R_d(t)}\right] \\ &- \sum_{n=0}^{\infty} v_n(r)(-1)^{\delta_{n,0}} f_n D_l \lambda_n^2 \times \int_0^t \frac{\hat{\mu}_0(\tau)}{R_d^2(\tau)} \\ &\times \exp\left[(-1)^{\delta_{n,0}} \frac{D_l \lambda_n^2}{\alpha R_{d0}} \left(\frac{1}{R_d(t)} - \frac{1}{R_d(\tau)}\right)\right] \mathrm{d}\tau, \end{aligned} \tag{I.37}$$

where expression (I.28) for $\Theta_n(t)$ was used.

In contrast to the previous case of $H_0(t) = \text{const}$, expression (I.37) does not give us an explicit solution for $W(t, r)$, since $\hat{\mu}_0(t)$ depends on $W(t, 1)$.

Expression (I.37) can be represented in a more compact form:

$$W(t, r) = \mathscr{V}(t, r) - \int_0^t \hat{\mu}_0(\tau) G(t, \tau, r) \mathrm{d}\tau, \tag{I.38}$$

where

$$\mathscr{V}(t, r) = \sum_{n=0}^{\infty} v_n(r) q_n \exp\left[-(-1)^{\delta_{n,0}} \frac{D_l \lambda_n^2 t}{R_{d0} R_d(t)}\right],$$

$$\begin{aligned} G(t, \tau, r) = &- \sum_{n=0}^{\infty} v_n(r) \\ &\times \frac{D_l v_n(1)}{R_d^2(\tau) ||v_n||^2} \exp\left[(-1)^{\delta_{n,0}} \frac{D_l \lambda_n^2}{\alpha R_{d0}} \left(\frac{1}{R_d(t)} - \frac{1}{R_d(\tau)}\right)\right]. \end{aligned}$$

Explicit expressions for f_n were used in these formulas. Both functions $\mathscr{V}(t, r)$ and $G(t, \tau, r)$ are assumed to be known.

Using (I.36), we can rewrite expression (I.38) as

$$W(t, r) = \mathcal{V}(t, r) - \int_0^t [\mu_0(\tau) - h_1(\tau)W(\tau, 1)]\, G(t, \tau, r)\mathrm{d}\tau. \tag{I.39}$$

This is an integral representation for a solution to problem (I.2)–(I.5) for time-dependent $H_0(t)$ given by (I.35). For $r = 1$, the integral representation (I.39) reduces to a Volterra integral equation of the second kind for the function $W(t, 1)$:

$$W(t, 1) = \mathcal{V}(t, 1) - \int_0^t [\mu_0(\tau) - h_1(\tau)W(\tau, 1)]\, G(t, \tau, 1)\mathrm{d}\tau. \tag{I.40}$$

One can show that

$$\frac{v_n^2(r=1)}{||v_n||^2} = \frac{2(-1)^{\delta_{n,0}}\lambda_n^2}{(-1)^{\delta_{n,0}}\lambda_n^2 + h_0^2 + h_0}. \tag{I.41}$$

Keeping in mind Equation (I.41), we obtain

$$G(t, \tau, 1) = -\frac{2D_l}{R_d^2(\tau)} \sum_{n=0}^{\infty} \frac{(-1)^{\delta_{n,0}}\lambda_n^2}{h_0^2 + h_0 + (-1)^{\delta_{n,0}}\lambda_n^2} \times \exp\left[(-1)^{\delta_{n,0}} \frac{D_l\lambda_n^2}{\alpha R_{d0}} \left(\frac{1}{R_d(t)} - \frac{1}{R_d(\tau)}\right)\right]. \tag{I.42}$$

Equation (I.40) has a unique solution, although this solution cannot be found in an explicit form. The numerical solution is similar to the one described in Appendix F. Once the solution to this equation has been found, we can substitute it into the integral representation (I.39) and obtain the required solution to the initial and boundary value Problem (I.2)–(I.5). The required distribution of Y_{li} was found to be

$$Y_{li}(t, R) = \frac{1}{R\sqrt{R_d(t)}} \exp\left[-\frac{R_d'(t)R^2}{4D_l R_d(t)}\right] W(t, R/R_d(t)). \tag{I.43}$$

Reference

1. Gusev, I. G., Krutitskii, P. A., Sazhin, S. S., & Elwardany, A. (2012). A study of the species diffusion equation in the presence of the moving boundary. *International Journal Heat and Mass Transfer*, *55*, 2014–2021.

Appendix J
Derivation of Formula (I.29)

Since solution (I.28) is applied to a very short time step, changes of $\mu_0(\tau)$ in the integrand before the exponential term can be ignored. This allows us to simplify solution (I.28) to [1]

$$\Theta_n(t) = q_n \exp\left[-\frac{(-1)^{\delta_{n,0}} D_l t \lambda_n^2}{R_{d0} R_d(t)}\right] + f_n \mu_0(t) - f_n \mu_0(0) \Bigg[(-1)^{\delta_{n,0}} D_l \lambda_n^2 \times \int_0^t \frac{1}{R_d(\tau)^2} \exp\left[(-1)^{\delta_{n,0}} \frac{D_l \lambda_n^2}{\alpha R_{d0}} \left(\frac{1}{R_d(t)} - \frac{1}{R_d(\tau)}\right)\right] \mathrm{d}\tau\Bigg]. \quad \text{(J.1)}$$

$\mu_0(\tau)$ in the integrand of solution (I.28) is taken at the beginning of the time step.

Recalling that

$$\mathrm{d}(R_d(\tau)^{-1}) = -\frac{R_d'}{R_d^2(\tau)} \mathrm{d}\tau,$$

we can rearrange the last term in expression (J.1) to [1]

$$\begin{aligned}
(-1)^{\delta_{n,0}} D_l \lambda_n^2 \int_0^t & \frac{1}{R_d(\tau)^2} \exp\left[(-1)^{\delta_{n,0}} \frac{D_l \lambda_n^2}{\alpha R_{d0}} \left(\frac{1}{R_d(t)} - \frac{1}{R_d(\tau)}\right)\right] \mathrm{d}\tau \\
&= \frac{(-1)^{\delta_{n,0}} D_l \lambda_n^2}{R_d'} \exp\left[(-1)^{\delta_{n,0}} \frac{D_l \lambda_n^2}{\alpha R_{d0}} \left(\frac{1}{R_d(t)}\right)\right] \\
&\quad \times \int_0^t \frac{R_d'}{R_d(\tau)^2} \exp\left[(-1)^{\delta_{n,0}} \frac{D_l \lambda_n^2}{\alpha R_{d0}} \left(-\frac{1}{R_d(\tau)}\right)\right] \mathrm{d}\tau \\
&= \frac{(-1)^{\delta_{n,0}} D_l \lambda_n^2}{R_d'} \exp\left[(-1)^{\delta_{n,0}} \frac{D_l \lambda_n^2}{\alpha R_{d0}} \left(\frac{1}{R_d(t)}\right)\right]
\end{aligned}$$

S. Sazhin, *Droplets and Sprays*, DOI: 10.1007/978-1-4471-6386-2,

$$\times \frac{\alpha R_{d0}}{(-1)^{\delta_{n,0}} D_l \lambda_n^2} \left(\exp\left[(-1)^{\delta_{n,0}} \frac{D_l \lambda_n^2}{\alpha R_{d0}} \left(-\frac{1}{R_d(t)} \right) \right] \right.$$
$$\left. - \exp\left[(-1)^{\delta_{n,0}} \frac{D_l \lambda_n^2}{\alpha R_{d0}} \left(-\frac{1}{R_d(0)} \right) \right] \right)$$
$$= \left(1 - \exp\left[(-1)^{\delta_{n,0}} \frac{D_l \lambda_n^2}{\alpha R_{d0}} \left(\frac{1}{R_d(t)} - \frac{1}{R_d(0)} \right) \right] \right)$$
$$= \left(1 - \exp\left[-(-1)^{\delta_{n,0}} \frac{D_l \lambda_n^2 t}{R_{d0} R_d(t)} \right] \right). \qquad (J.2)$$

In deriving (J.2), we took into account (4.83).

On substituting equation (J.2) into expression (J.1), we obtain formula (I.29).

Reference

1. Gusev, I. G., Krutitskii, P. A., Sazhin, S. S., & Elwardany, A. (2012). A study of the species diffusion equation in the presence of the moving boundary. *International Journal Heat and Mass Transfer*, *55*, 2014–2021.

Appendix K
Approximations for Alkane Fuel Properties

In what follows the approximations for the temperature dependencies of density, viscosity, heat capacity and thermal conductivity for liquid alkanes (C_nH_{2n+2}) with $5 \leq n \leq 25$, as inferred from published approximations and data, are described, following [1, 4].

Liquid Density

Following [6], the temperature-dependence of the density of liquid alkanes for $5 \leq n \leq 25$ is approximated as

$$\rho_l(T) = 1000 A_\rho B_\rho^{-\left(1-\frac{T}{T_c}\right)^{C_\rho}}, \tag{K.1}$$

where the numerical values of A_ρ, B_ρ, and C_ρ for individual values of n are given in [6]. These were approximated by the following expressions [1]:

$$\begin{cases} A_\rho = 0.00006196104 \times n + 0.234362 \\ B_\rho = 0.00004715697 \times n^2 - 0.00237693 \times n + 0.2768741 \\ C_\rho = 0.000597039 \times n + 0.2816916 \end{cases} \tag{K.2}$$

The following approximations for the critical and boiling temperatures of alkanes were suggested in [4], using data provided in [3]:

$$T_c(n) = a_c + b_c n + c_c n^2 + d_c n^3, \tag{K.3}$$

$$T_b(n) = a_b + b_b n + c_b n^2 + d_b n^3, \tag{K.4}$$

where the coefficients are presented in Table K.1.

The values of T_c and T_b inferred from [3] and their approximations (K.3) and (K.4) for $5 \leq n \leq 25$ are shown in Fig. K.1 [4]. As one can see from this figure, both T_c and T_b are monotonically increasing functions of n.

The range of applicability of approximation (K.1) depends on the values of n. For $n = 5$, this range was determined to be 143.42–469.65 K; for $n = 10$, this range

S. Sazhin, *Droplets and Sprays*, DOI: 10.1007/978-1-4471-6386-2,

Table K.1 Values of coefficients in expressions (K.3) and K.4)

Coefficient	a_c	b_c	c_c	d_c
Value	242.3059898052	55.9186659144	−2.1883720897	0.0353374481
Coefficient	a_b	b_b	c_b	d_b
Value	118.3723701848	44.9138126355	−1.4047483216	0.0201382787

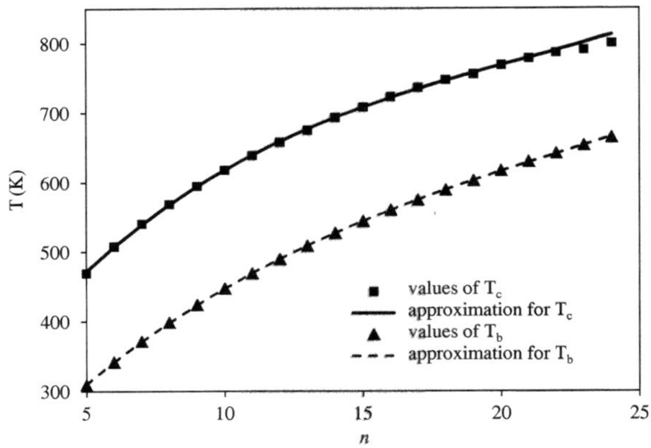

Fig. K.1 Plots of T_c and T_b, and their approximations (K.3) and (K.4), versus n. Reprinted from Ref. [4]. Copyright Elsevier (2011)

was determined to be 243.49–618.45 K; for $n = 25$, this range was determined to be 315.15–850.13 K [6] (the upper limits are critical temperatures of the components). Recalling that the contribution of alkanes with n close to 25 is relatively small, we will assume that formulas (K.1) and (K.2) are valid in the whole range from room temperature to close to the critical temperature.

Plots of ρ_l versus n for $T = 300$ K and $T = 450$ K, as inferred from approximation (K.1) with coefficients A_ρ, B_ρ, and C_ρ given by [6] (filled squares and filled triangles), and approximated by formulas (K.2) (solid and dashed curves) are shown in Fig. K.2. As follows from this figure, the agreement between the values of liquid density predicted by approximation (K.1), with the values of the coefficients given in [6] and approximated by formulas (K.2), looks almost ideal.

Liquid Viscosity

Following [2], the temperature-dependence of the dynamic viscosity of liquid alkanes for $4 \le n \le 44$ is approximated as

$$\mu_l(n, T) = 10^{-3} \left[10^{[100(0.01\,T)^{b(n)}]} - 0.8 \right], \qquad \text{(K.5)}$$

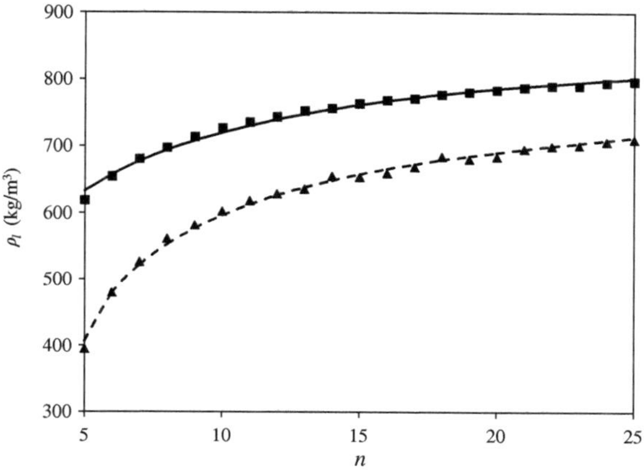

Fig. K.2 Plots of ρ_l versus n for $T = 300$ K and $T = 450$ K, as inferred from approximation (K.1) with coefficients A_ρ, B_ρ, and C_ρ given by [6] (*filled squares* for $T = 300$ K and *filled triangles* for $T = 450$ K), and approximated by formulas (K.2) (*solid curve* for $T = 300$ K and *dashed curve* for $T = 450$ K). Reprinted from Ref. [1]. Copyright Elsevier (2012)

where

$$b(n) = -5.745 + 0.616\,\ln(n) - 40.468\,n^{-1.5}. \tag{K.6}$$

The temperature range of the applicability of approximations (K.5) and (K.6) was not explicitly specified in [2], but the author of that paper demonstrated good agreement between the predictions of these approximations and experimental data in the range of temperatures from 10 to 100 °C.

Plots of μ_l versus n for $T = 300$ K and $T = 450$ K, as inferred from approximation (K.5) (solid and dashed curves), and the corresponding values of μ_l in the range $5 \leq n \leq 12$, inferred from [8] (filled squares ($T = 300$ K) and filled triangles ($T = 450$ K)), are shown in Fig. K.3. As follows from this figure, the agreement between the values of liquid viscosity predicted by approximation (K.5) and the results presented on the NIST website [8] looks almost ideal. Note that the values of dynamic viscosity affect droplet heating and evaporation only via the corrections to the values of thermal conductivity and diffusivity in the Effective Thermal Conductivity and Effective Diffusivity (ETC/ED) models. In most practically important cases, the influence of viscosity on the final results is expected to be very weak.

Heat Capacity

Following [5], the temperature-dependence of the heat capacity of liquid alkanes for $2 \leq n \leq 26$ is approximated as

$$c_l(n, T) = 1000\left[\frac{43.9 + 13.99(n-1) + 0.0543(n-1)T}{M(n)}\right], \tag{K.7}$$

where $M(n) = 14n + 2$ is the molar mass of alkanes.

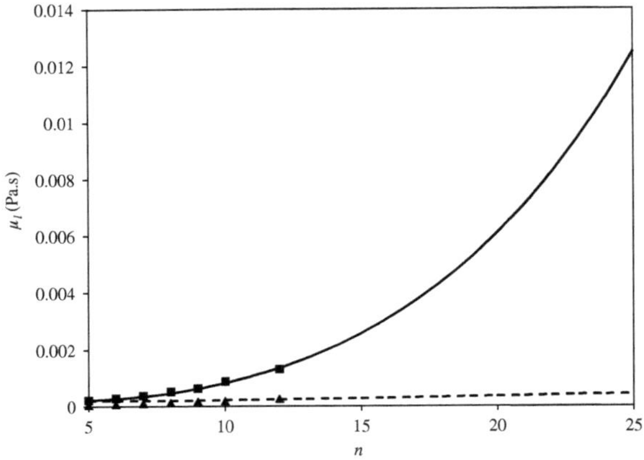

Fig. K.3 Plots of μ_l versus n for $T = 300$ K and $T = 450$ K, as inferred from approximation (K.5) (*solid* ($T = 300$ K) and *dashed* ($T = 450$ K) *curves*), and the corresponding values of μ_l in the range $5 \leq n \leq 12$, inferred from [8] (*filled squares* ($T = 300$ K) and *filled triangles* ($T = 450$ K)). Reprinted from Ref. [1]. Copyright Elsevier (2012)

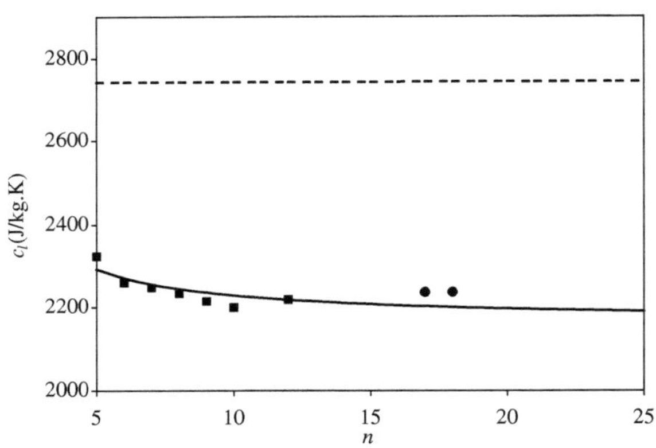

Fig. K.4 Plots of c_l versus n for $T = 300$ K and $T = 450$ K, as inferred from approximation (K.7) (*solid* ($T = 300$ K) and *dashed* ($T = 450$ K) *curves*), and the corresponding experimental values of c_l for $T = 300$ K in the range $5 \leq n \leq 18$, inferred from [8] (*filled squares*) and [5] (*filled circles* •). Reprinted from Ref. [1]. Copyright Elsevier (2012)

The temperature range of applicability of (K.7) was not clearly identified in [5] for all n, except to say that this approximation is not valid at temperatures close to the temperature of fusion. For $n = 16$ and $n = 17$, these ranges were identified as 340–400 K and 335–400 K respectively. In the case of $n = 16$ and $n = 25$, the temperatures of fusion are equal to 295.1 K and 329.25 K respectively. However, since the contribution of the alkanes with $n > 16$ is very small, it is assumed that approximation (K.7) is valid in the whole temperature range from room temperature onwards.

Plots of c_l versus n for $T = 300$ K and $T = 450$ K, as inferred from approximation (K.7) (solid and dashed curves), and the corresponding experimental values of c_l for $T = 300$ K in the range $5 \le n \le 18$, inferred from [8] (filled squares) and [5] (filled circles •), are shown in Fig. K.4. As follows from this figure, the agreement between the values of the liquid heat capacity predicted by approximation (K.7) and the experimental results for $T = 300$ K looks almost ideal.

Thermal Conductivity

Following [7], the temperature-dependence of thermal conductivity of liquid alkanes for $5 \le n \le 20$ is approximated as

$$k_l(n, T) = 10^{\left[A_k + B_k\left(1 - \frac{T}{T_c}\right)^{2/7}\right]}, \quad \text{(K.8)}$$

where T_c are critical temperatures as in approximation (K.1), the numerical values of A_k and B_k for individual values of n are given in [7]. These values were approximated by the following expressions:

$$\begin{cases} A_k = 0.002911 \times n^2 - 0.071339 \times n - 1.319595 \\ B_k = -0.002498 \times n^2 + 0.058720 \times n + 0.710698 \end{cases} \quad \text{(K.9)}$$

Although approximations (K.8) and (K.9) were derived for $5 \le n \le 20$, they were used in the whole range $5 \le n \le 25$. Possible errors imposed by these approximations in the range $21 \le n \le 25$ are expected to have a very small effect on the final results, since the mass fractions of alkanes in this range of n are very small in Diesel fuel and negligible in gasoline fuel.

The range of applicability of approximation (K.8) depends on the values of n. For $n = 5$, this range was determined to be 143–446 K; for $n = 10$, this range was determined to be 243–588 K; for $n = 20$, this range was determined to be 310–729 K [7]. Since the contribution of alkanes with $n \ge 20$ is relatively small, it was assumed that approximations (K.8) and (K.9) are valid in the whole range from room temperature until close to the critical temperature, as in the case of approximations (K.1) and (K.2).

Plots of k_l versus n for $T = 300$ K and $T = 450$ K, as inferred from approximation (K.8) with coefficients A_k and B_k given by [7] [filled squares ($T = 300$ K) and filled triangles ($T = 450$ K)] and approximated by formulas (K.9) (solid and dashed curves) are shown in Fig. K.5. In the same figure, the values of k_l inferred from [8] (squares ($T = 300$ K) and triangles $\triangle$ ($T = 450$ K)) are shown. As follows from this figure, the agreement between the values of thermal conductivity predicted by approximation (K.8) with the values of the coefficients given in [7] and approximated by formulas (K.9) looks almost ideal. Both these values agree well with the data reported in [8].

Note that during calculations, a small number of lighter components inside droplets could have temperatures exceeding their critical temperatures. In this case, the values of saturation pressure, latent heat of evaporation, density, viscosity, heat capacity, and thermal conductivity were assumed equal to those at $T = T_c$. This

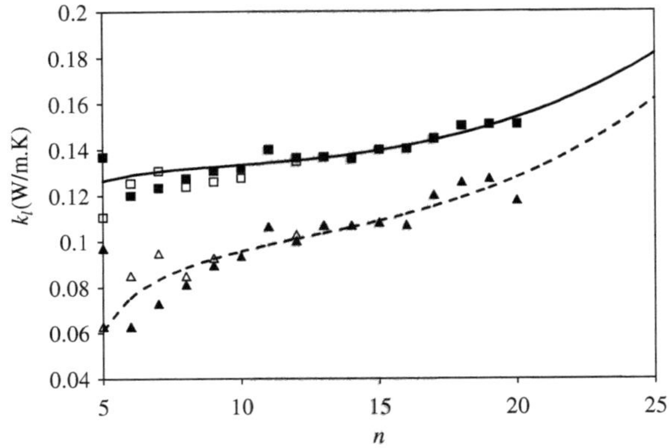

Fig. K.5 Plots of k_l versus n for $T = 300$ K and $T = 450$ K, as inferred from approximation (K.8) with coefficients A_k and B_k given by [7] (*filled squares* ($T = 300$ K) and *filled triangles* ($T = 450$ K)) and approximated by formulas (K.9) (*solid* ($T = 300$ K) and *dashed* ($T = 450$ K) *curves*); the values of k_l inferred from [8] (*squares* ($T = 300$ K) and *triangles* △ ($T = 450$ K)). Reprinted from Ref. [1]. Copyright Elsevier (2012)

assumption allows us to avoid the analysis of heat and mass transfer in supercritical conditions without imposing significant errors in our analysis due to the fact that the number of components affected by this assumption is very small.

References

1. Elwardany, A. E., & Sazhin, S. S. (2012). A quasi-discrete model for droplet heating and evaporation: application to Diesel and gasoline fuels. *Fuel*, *97*, 685–694.
2. Mehrotra, A. K. (1994). Correlation and prediction of the viscosity of pure hydrocarbons. *The Canadian Journal of Chemical Engineering*, *72*, 554–557.
3. Poling, B. E., Prausnitz, J. M., & O'Connell, J. (2000). *The properties of gases and liquids*. New York: McGraw-Hill.
4. Sazhin, S. S., Elwardany, A., Sazhina, E. M., & Heikal, M. R. (2011). A quasi-discrete model for heating and evaporation of complex multicomponent hydrocarbon fuel droplets. *International Journal Heat and Mass Transfer*, *54*, 4325–4332.
5. van Miltenburg, J. C. (2000). Fitting the heat capacity of liquid n-alkanes: new measurements of n-heptadecane and n-octadecane. *Thermochimica Acta*, *343*, 57–62.
6. Yaws, C. L. (Ed.). (2008). *Thermophysical properties of chemicals and hydrocarbons*. Beaumont. USA: William Andrew Inc.

7. Yaws, C. L. (1995). *Handbook of thermal conductivity. Organic compounds*, C_5 to C_7 (Vol. 2) and *organic compounds*, C_8 to C_{28} (Vol. 3). Houston, London, Paris, Zurich, Tokyo: Gulf Publishing Company.
8. http://webbook.nist.gov/chemistry/.

Appendix L
Tikhonov's Theorem

Tikhonov's theorem [2, 3] justifies the system order reduction using traditional tools in the theory of singularly perturbed systems. This theorem is formulated, following [1], for a simplified case of an autonomous system with one small parameter:

$$\frac{dx}{d\tau} = f(x, y) \tag{L.1}$$

$$\varepsilon \frac{dy}{d\tau} = g(x, y), \tag{L.2}$$

where $x(0) = x_0, y(0) = y_0, f : D \to \mathscr{R}^m$, and $g : D \to \mathscr{R}^n$ are continuous vector functions, D is an open set in $\mathscr{R}^{n+m}$, $(x_0, y_0) \in D$, and ε is a small positive parameter.

Definitions The vector differential equation

$$\frac{dy}{d\bar{\tau}} = g(x, y)\,, \tag{L.3}$$

where $\bar{\tau} = \tau/\varepsilon$ and x is considered a parameter, is called a *boundary layer equation*. The m-dimensional surface $\mathscr{S}$ given by

$$g(x, y) = 0 \tag{L.4}$$

is called a *slow surface*. It describes all steady states of the boundary layer Eq. (L.3). This surface is given as a graph of a function, and thus there exists a continuous mapping $\phi : X \to \mathscr{R}^n$ from some compact set X in $\mathscr{R}^m$ such that $(x, \phi(x)) \in D$ for all $x \in X$ and $\mathscr{S} = \{(x, y) : y = \phi(x),\ x \in X\}$.

Following the Lyapunov stability theory, referring to systems depending on parameters, the steady state $y = \phi(x)$ of (L.3) is called:

S. Sazhin, *Droplets and Sprays*, DOI: 10.1007/978-1-4471-6386-2,

1. Stable if for any $\mu > 0$ there exists such η that any solution $y(\bar{\tau})$ of (L.3) with $\|y(0) - \phi(x)\| < \eta$ may be extended on $\bar{\tau} > 0$ and it satisfies $\|y(\bar{\tau}) - \phi(x)\| < \mu$.

The steady state is asymptotically stable if it is stable and $\lim_{\bar{\tau}\to\infty} y(\bar{\tau}) = \phi(x)$ for all solutions satisfying the condition $\|y(0) - \phi(x)\| < \eta$.

2. Attractive if a domain of attractivity exists, i.e., if some neighbourhood $\mathscr{V}$ of $\mathscr{S}$ such that any solution $y(\bar{\tau})$ of (L.3) with $y(0) \in \mathscr{V}$ may be extended for all $\bar{\tau} > 0$ and $\lim_{\bar{\tau}\to\infty} y(\tau) = \phi(x)$.

The domain of attractivity of the steady state $y = \phi(x)$ is uniform in X if there exists $a > 0$ such that for all $x \in X$, the ball $\mathscr{B} = \{x \in \mathscr{R}^m : \|y - \phi(x)\| \leq a\}$ with centre in $\phi(x)$ with radius a is the domain of attractivity for $\phi(x)$. The stable and attractive steady state is asymptotically stable.

The following assumptions are made:

(H1) For all $x \in X$, the boundary layer equation (L.3) has a unique solution for a given initial value.

(H2) For all $x \in X$, $y = \phi(x)$ is an *isolated root* of (L.4), i.e., $g(x, \phi(x)) = 0$, and there exists a positive number $\delta > 0$ such that the conditions $x \in X$, $\|y - \phi(x)\| < \delta$, and $y \neq \phi(x)$ imply $g(x, y) \neq 0$. This does not mean that (L.4) has no roots other than $\phi(x)$.

(H3) For any $x \in X$, the point $y = \phi(x)$ is an asymptotically stable steady state of (L.3), and the domain of attractivity $y = \phi(x)$ is uniform with respect to X.

(H4) The system

$$\frac{dx}{d\tau} = f(x, \phi(x)) \tag{L.5}$$

with a given initial condition has a unique solution.

(H5) For any interior point x_0 in X, the point y_0 belongs to the domain of attractivity of the steady state $y = \phi(x)$.

Remarks Let us consider the following initial value problem for the boundary layer equation:

$$\frac{dy}{d\bar{\tau}} = g(x_0, y) \quad y(0) = y_0, \tag{L.6}$$

where x_0 is an arbitrary interior point in X. Let $\bar{y}(\bar{\tau})$ be the solution to this problem, and the function $\bar{y}(\bar{\tau})$ is defined for all $\bar{\tau} \geq 0$ and $\lim_{\bar{\tau}\to\infty} \bar{y}(\bar{\tau}) = \phi(x_0)$.

The reduction of (L.1) leads to the initial value problem

$$\frac{dx}{d\tau} = f(x, \phi(x)), \quad x(0) = x_0. \tag{L.7}$$

Let $\bar{x}(\tau)$ be the solution to the reduced problem for $\tau \in I = [0, T)$, where $T \in (0, \infty]$ is the maximal interval where this solution is defined.

The Tikhonov theorem is formulated as follows:

If assumptions (H1)–(H5) *are valid, then the solution* $(x(\tau, \varepsilon), y(\tau, \varepsilon))$ *to the initial value problem* (L.1)–(L.2) *exists in* $[0, T]$, *and the following conditions hold:*

$$\lim_{\varepsilon \to 0} x(\tau, \varepsilon) = \bar{x}(\tau), \quad 0 \le \tau \le T; \tag{L.8}$$

$$\lim_{\varepsilon \to 0} y(\tau, \varepsilon) = \phi(\bar{x}(\tau)), \quad 0 < \tau \le T. \tag{L.9}$$

The convergence in (L.8) *and* (L.9) *is uniform in the interval* $0 \le \tau \le T$ *for* $x(\tau, \varepsilon)$ *and in any interval* $0 < \tau_1 \le \tau \le T$ for $y(\tau, \varepsilon)$.

References

1. Sazhin, S. S., Shchepakina, E. A., & Sobolev, V. A. (2010). Order reduction of a non-Lipschitzian model of monodisperse spray ignition. *Mathematical and Computer Modelling*, *52*, 529–537.
2. Tikhonov, A. N. (1952). Systems of differential equations containing small parameters in the derivatives. *Matematicheskii Zbornik*, *31*, 575–586 (in Russian).
3. Vasil'eva, A. B., Butuzov, V. F., & Kalachev, L. V. (1995). *The boundary function method for singular perturbation problems*, SIAM Studies in Applied Mathematics *14*.

MIX
Papier aus verantwortungsvollen Quellen
Paper from responsible sources
FSC® C105338

If you have any concerns about our products,
you can contact us on
ProductSafety@springernature.com

In case Publisher is established outside the EU,
the EU authorized representative is:
Springer Nature Customer Service Center GmbH
Europaplatz 3, 69115 Heidelberg, Germany

Printed by Libri Plureos GmbH
in Hamburg, Germany